DYNAMIC DESCRIPTION TECHNOLOGY OF FRACTURED VUGGY CARBONATE GAS RESERVOIRS

DYNAMIC DESCRIPTION TECHNOLOGY OF FRACTURED VUGGY CARBONATE GAS RESERVOIRS

TONGWEN JIANG

HEDONG SUN

XINGLIANG DENG

ELSEVIER

G|P P|

Gulf Professional Publishing
An imprint of Elsevier

Gulf Professional Publishing is an imprint of Elsevier
50 Hampshire Street, 5th Floor, Cambridge, MA 02139, United States
The Boulevard, Langford Lane, Kidlington, Oxford, OX5 1GB, United Kingdom

Library of Congress Cataloging-in-Publication Data
A catalog record for this book is available from the Library of Congress

British Library Cataloguing-in-Publication Data
A catalogue record for this book is available from the British Library

ISBN: 978-0-12-818324-3

For information on all Gulf Professional publications visit our website at https://www.elsevier.com/books-and-journals

Publisher: Brian Romer
Senior Acquisition Editor: Katie Hammon
Senior Editorial Project Manager: Andrae Akeh
Production Project Manager: Kamesh Ramajogi
Cover Designer: Victoria Pearson

Typeset by SPi Global, India

Contents

About the Author

Tongwen Jiang, Ph.D., is a AAPG member, born in 1968. He is now a Professor of Engineering, having earned his Ph.D. degree in Oil and Gas Field Development Engineering from the Southwest Petroleum Institute in 1996. Previously, he worked for Tarim Oilfield Company since 1996 and has gained extensive experience in reservoir engineering and management, spanning over 20 years. He has published more than 30 papers on oil and gas reservoir studies, including condensate gas reservoirs and fractured reservoirs. He is the author of four books published by Petroleum Industry Press.

E-mail: jiangtw-tlm@petrochina.com.cn

Tel: +86-996-2171090

Address: P.O. Box 78, Korla, Xingjiang 841000, PR China

Hedong Sun, Ph.D. is a SPE member and a Professor of Reservoir Engineering earning his Ph.D. from Xi'an Jiaotong University in 2004. Since 2004, he has been a Research Engineer at the Research Institute of Petroleum Exploration and Development of Petrochina. Hedong has more than 20 years of reservoir engineering experience, with a focus on well test analysis and production decline analysis. He has published more than 60 papers in peer-reviewed journals and SPE conferences. He is the author of two books published by Elsevier, including "Advanced Production Decline Analysis and Application," and "Well Test Analysis for Multilayered Reservoirs with Formation Crossflow."

E-mail: sunhed@petrochina.com.cn
Tel: +86-10-83596743
Address: P.O. Box 44, Wanzhuang, Langfang, Hebei Province 065007, PR China

Xingliang Deng, he is a Senior Engineer and Director of the Department of Oil and Gas Reservoir Evaluation of the Research Institute of Petroleum Exploration & Development, Tarim Oilfield. Since he graduated from the China University of Petroleum in 1992, he has worked in the Tarim Oilfield, and specialized in reservoir descriptions, well position deployment, and development programs for fractured-vuggy carbonate reservoirs, including the Tazhong1, Lungu, Halahatang, and Yingmaili reservoirs. He holds a BS degree in Petroleum and Natural Gas Geological Exploration Technology from China University of Petroleum, with a Ph.D. degree in Petroleum Geology from Nanjing University. He has published more than 30 papers in peer-reviewed journals, and presented at SPE conferences. He is an author of one book published by the Petroleum Industry Press.

E-mail: dxl-tlm@petrochina.com.cn
Tel: +86-996-2172487
Address: Research Institute of Petroleum Exploration & Development of Tarim Oilfield, Korla, Xinjiang 841000, PR China

Acknowledgment

We would like to thank the editorial and production staff of Elsevier for their work and professionalism, most notably Katie Hammon, Andrae Akeh, Gabriela D. Capille, Kamesh Ramajogi, and Victoria Pearson.

Thanks also go to Weiping Ouyang, Wenchao Liu, Baohua Chang, Shiyin Li, Peng Wang, and Wen Cao for their assistance and help during the proofreading of the book.

We also acknowledge financial support from China National Major Science and Technology Project "Large Oil/Gas Field and Coalbed Methane Development"—"Key recovery enhancement techniques for carbonate hydrocarbon reservoirs in Tarim Basin" (No. 2016ZX05053).

Introduction

Since the beginning of the 21st century, China's natural gas industry has been in a stage of rapid development, with the gas production growing rapidly from 302×10^8 m^3 in 2001 to 1480×10^8 m^3 in 2017, mainly due to the exploration and development of fractured-vuggy carbonate gas reservoirs, especially those in the Tazhong I gas field and the Sinian in the Gaoshiti-Moxi block of the Anyue gas field, Northwestern China. However, these reservoirs are characterized by strong heterogeneity, diverse flow mechanisms, and complex fluid properties. All these make the static description of reservoirs an extremely challenging task, thereby impeding reserves estimation, development design, and production performance forecasting. To develop these reservoirs more rationally, it is critical to combine both the static and dynamic data to improve the accuracy of reservoir characterization. Through years of effort, a series of dynamic description techniques have been developed for reservoir performance evaluation, performance-based reserves estimation, and deliverability evaluation for these highly heterogeneous carbonate gas reservoirs.

The dynamic description of gas reservoirs refers to the process during which the gas reservoirs are characterized comprehensively and accurately to obtain the well and reservoir parameters. The process is based on well test and production performance data using gas reservoir engineering methods, including modern well test analysis methods and advanced production decline analysis methods. These critical techniques can improve the accuracy of complex gas reservoir characterization for the purposes of production performance forecasting and subsequent rational development.

In this book, the typical characteristics of fractured-vuggy carbonate gas reservoirs are illustrated. On this basis, the dynamic description technique based on lifecycle well test analysis and advanced production decline analysis is systematically introduced, and its application in reservoir performance evaluation, performance-based reserves estimation, performance forecasting, and other aspects are expounded upon. The book is divided into five chapters.

Chapter 1—Typical characteristics of fractured vuggy carbonate gas reservoirs illustrates the geological setting and characteristics of fractured-vuggy carbonate gas reservoirs and the typical characteristics of reservoir development.

Chapter 2—Introduction to the dynamic description technique and dynamic monitoring of gas reservoirs illustrates the connotation, status, role, and the methodology of the dynamic description technique of gas reservoirs, and cautions in performance monitoring.

Chapter 3—Well test analysis methods of fractured-vuggy carbonate gas reservoir presents the three-dimensional numerical well test analysis method and lifecycle well test analysis method to address such problems as the diversity of well test curves, the ambiguity of interpretation results, and the complexity of analysis models, and expounds upon the application of well test analysis methods in gas reservoir description.

Chapter 4—Reserves estimation methods of fractured-vuggy carbonate gas reservoirs presents the reserves estimation methods of fractured-vuggy gas reservoirs, and illustrates the single-well performance-based reserves, recoverable reserves, and producing reserves estimation methods for fractured-vuggy carbonate gas reservoirs.

Chapter 5—Performance forecasting method of fractured vuggy carbonate gas reservoir provides examples to explain the principles and processes of well and unit performance forecasting methods, including the gas reservoir engineering method based on the material balance theory and PVT test data, the numerical well test method based on well testing data and geological cognition, and the advanced production decline analysis method based on production performance data.

This book is a summary of the authors' research. It reflects the promotion and elevation of gas reservoir engineering theory and field practice. Many achievements in this discipline have been applied to Tazhong I and other similar gas fields in preliminary assessment and development design, and have facilitated the scientific, predictable, and beneficial development of these gas fields. We hope this publication can contribute to the development of complex gas reservoirs in China.

In reference to any incorrect statements in this book that are due to limitations in the authors' knowledge and experience, your comments and feedback are warmly welcomed.

January 1, 2018

CHAPTER 1

Typical characteristics of fractured vuggy carbonate gas reservoirs

Contents

1.1 Typical characteristics of gas reservoir geology

1.1.1 Geological setting

1.1.1.1 Disappearance of primary pores in ancient marine carbonate reservoirs—A long process from deposition to diagenesis

Globally, large-scale oil and gas exploration and development of marine carbonate rocks are concentrated in the Upper Paleozoic and Cenozoic marine carbonate strata in the Middle East, North America, Australia, and other regions. Even the Paleozoic reservoirs, with primary pores in dominance, have roughly consistent geological characteristics with the porous sandstone reservoirs. In the Ordos Basin, the burial depth of Ordovician reservoirs, as the key target for gas exploration of marine carbonate rocks, mainly ranges from 2500 to 4000 m, and exceeds 4500 m locally in the Tianhuan Depression in the west. In the Sichuan Basin, the Sinian-Cambrian reservoirs, as the key target in the Anyue gas field, are 4500–6000 m deep, and are mostly dolomite reservoirs containing secondary pores controlled by the ancient bioherm-beach complex. In the Tarim Basin, marine carbonate rocks are mainly distributed in the Lower Paleozoic strata; especially in the ancient Ordovician and Cambrian strata, the target reservoirs are usually deeper than

Dynamic Description Technology of Fractured Vuggy Carbonate Gas Reservoirs
https://doi.org/10.1016/B978-0-12-818324-3.00001-2

6000 m (up to 7700 m in the Halahatang region). In the Tarim Basin, the Early Paleozoic carbonate reservoirs, which were formed on a small craton and experienced extremely complex evolution, have low matrix porosity and exhausted primary pores as a result of the long-term strong diagenesis; the karst reservoirs and dolomite reservoirs controlled by multiphase unconformities and fault systems act as the main effective reservoirs, and the fractured vuggy and cave reservoirs reflect strong heterogeneity.

1.1.1.2 Development of various fractures created due to multiphase tectonic movements

Ancient marine carbonate rocks generally experienced multiphase tectonic movements, giving rise to multiple faults and fractures (or even complex fracture networks locally). For example, the Tazhong Uplift in the Tarim Basin, an inherited paleo-uplift developed from the Cambrian-Ordovician giant folded anticline, originated at the end of the Early Ordovician, and basically formed before the Devonian, and it was dominated by tectonic migration and transformation after the Early Hercynian. In the Early Caledonian, when the intercratonic carbonate platform grew, there were only small normal faults in local areas, and the Tazhong region was connected with the Tabei region to form a large stable platform. In the Middle-Late Caledonian, when the paleo-uplift was formed, with the regional tectonic stress field changing from S-N extension to S-N extrusion, the EW-trending basement thrust fault composed of multiorder faults was formed, which determined the overall tectonic framework. In the Early Hercynian, when the paleo-uplift was reworked, the Tazhong Uplift was triggered by the intense compression from the southwest to further rise and form with multiple NE-trending strike-slip faults emerging in various types and sizes. Controlled by the horizontal position and vertical displacement of the strike-slip faults, fractures were created in greatly variable dimensions. The faults and their associated fractures in this period are one of the main factors for reworking karst reservoirs, and provide major pathways for oil and gas charging. In the Late Hercynian-Yanshanian, the monolithic uplifting stage of the paleo-uplift, especially in the late stage of Early Permian, intermediate-basic volcanic eruption, and magmatic intrusion occurred in the west of the Tazhong region, leading to the doming of the Carboniferous strata and the thin top and thick flank of the Lower Permian strata at structural highs, and giving rise to the drape anticline and synsedimentary anticline. Fractures originated from the volcanic eruption, and magmatic intrusion were developed in the carbonates. In the Himalayan, when the paleo-uplift was buried rapidly, tensional faults were developed locally under an extensional tectonic setting, with fractures in moderate sizes.

1.1.1.3 Development of diverse karsts formed due to multiphase tectonic movements

The marine carbonate reservoirs in China are old, and have a complex geological history. The regional tectonic setting is definitely the essential factor for determining the reservoir type and evolution in the basin. The tectonic extent and velocity also play a certain role in controlling the karst development. The rapid tectonic ups and downs and sea-level changes offer little chance for the development of karst reservoirs. However, the multiphase tectonic movements and intermittent sea-level changes are both favorable for the long-term karstification, leading to the diversity of karstification, including penecontemporaneous dissolution, interlayer karstification, buried-hill karstification, and hydrothermal karstification.

In the platform-basin region of the Tarim Basin, all paleo-karsts in the Early Caledonian (Lower Ordovician Penglaiba Formation) and Middle Caledonian (top of the Yingshan Formation, top of the Yijianfang Formation, top of the Tumuxiuke Formation, and top of the Lianglitage Formation) were developed on an intercratonic paleo-uplift, but displayed different features due to distinct tectonic intensities in these periods. On the northern slope of the Tazhong region, the Yingshan Formation karsts were developed in the Early-Middle Caledonian, above which the Yijianfang Formation and Tumuxiuke Formation are missing, and were only exposed for a relatively short period of time (only 5–10 million years), reflecting the features of interlayer karstification. In the Tazhong I slope-break belt, the Upper Ordovician reef-beach complex represents the contemporaneous or penecontemporaneous karsts that were formed by the upward building and exposure of the Late Ordovician Lianglitage Formation reef-beach complex at the platform margin, relative fall of sea level, and leaching of meteoric fresh water in the early stage of Middle Caledonian. In the Lunnan region, northern Tarim Basin, from the Late Caledonian to the Early Hercynian, the Ordovician paleo-uplift elevated continuously under the action of compressional stress for hundreds of million years; as a result, the strata overlying the Ordovician Yingshan Formation were completely denudated, but replaced by the Carboniferous strata that contain buried-hill karsts. The early buried karsts evolved into hydrothermal karsts when hydrothermal fluids with acid gases (e.g., CO_2, H_2S, and SO_2) penetrated carbonate rocks along faults and fractures to create new secondary pores during the late tectonic movements. Such hydrothermal karstification may appear in any tectonic period after the sedimentation and diagenesis of carbonate rocks, and karst reservoirs distribute along faults through strata.

1.1.1.4 Distribution of fractured vuggy reservoirs induced by faults, fractures, and karsts

Faults and fractures determine the permeability, connectivity, and direction of ancient carbonate rocks, and further control the surface/subsurface run-offs and their directions. The early faults and fractures contributed to the development of karst caves significantly, resulting in zonal and even linear distribution of fractured vuggy reservoirs along the fracture zone (Fig. 1.1), and large karst caves are always developed along the fracture zone, or at the inflection and intersection of faults. The development depth and scale of faults control that of karsts, and the multiphase tectonic movements determine the multiphase karsts. The exploration practices and research results of ancient carbonate rocks in China indicate the control of fractures

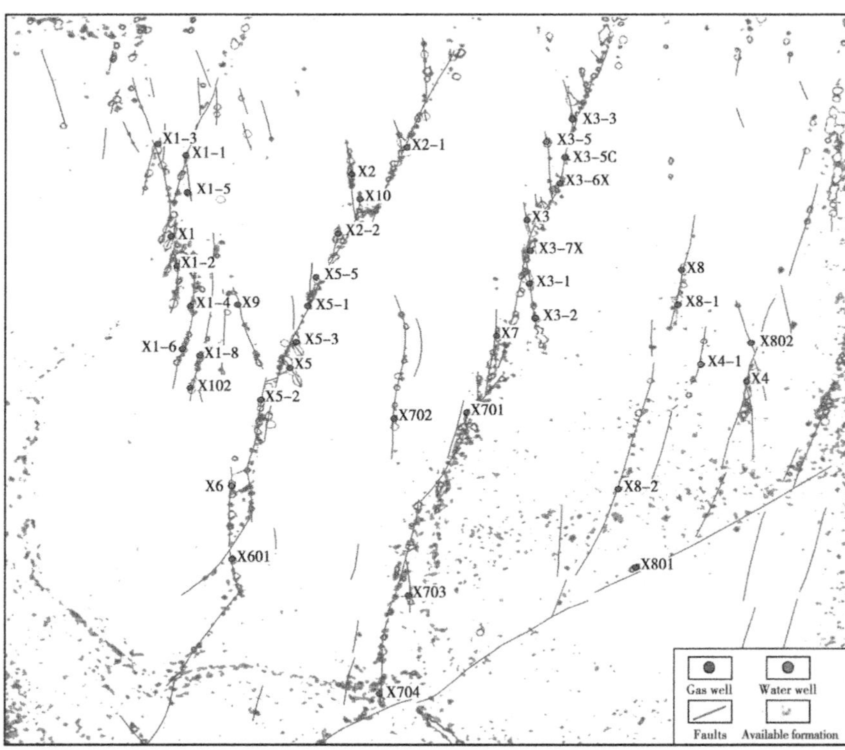

Fig. 1.1 Superimposition of faults for Ordovician reservoirs in the X block, Halahatang.

on the distribution of prolific oil/gas wells and blocks. With fractures, carbonate rocks, whether deposited in a high-energy facies belt or not, could form the vuggy, fractured vuggy, and large cave reservoirs. The accumulation space composed of fractures, vugs, and caves, and the strong heterogeneity, are two major characteristics of the karst reservoirs. There is a close relationship between fractures and dissolved vugs. On one hand, preexisting fractures controlled the formation of karst reservoirs. On the other hand, the late karst collapse induced numerous secondary fractures.

1.1.2 Geological characteristics

1.1.2.1 Types of accumulation space and the complex and diverse reasons for their formations

1.1.2.1.1 Caves

Caves are mainly formed by karstification with diverse patterns, including deep hydrothermal karstification, and buried-hill karstification. The deep hydrothermal karstification, which is extensive in carbonate rocks in the Tarim Basin, results in rich karst caves in a large dimension, forming effective accumulation spaces. The buried-hill karstification, which is popular in the Tarim Basin, refers to the entering of organic acids and carbon dioxide together with oil and gas into reservoirs during the process of hydrocarbon accumulation.

1.1.2.1.2 Vugs

Vugs, in this book, refer to small dissolved vugs appearing in honeycombs, beads, and other forms, unfilled or fully filled with calcite and mud, or occasionally with fluorite and celestite of hydrothermal origin. They are formed by the freshwater karstification during deposition, mainly creating intragranular dissolved pores and moldic pores, which are rarely preserved with severe cementation and filling.

1.1.2.1.3 Fractures

By origins, fractures are classified into structural fractures, dissolved fractures, and diagenetic fractures.

The structural fractures are considered a product of tectonic stress, and the most dominant in the basin, existing as shear fractures, followed by extensional fractures. According to the occurrence, they are divided into high-angle oblique fractures, vertical fractures, and horizontal fractures, with vertical fractures and microfissures in dominance. Most of the fractures formed earlier are filled or partly filled with calcite, mud, or bitumen.

In local areas, multiphase fractures in different occurrences intersect each other to form fracture networks and break the rocks, thus greatly improving the porosity and permeability.

The dissolved fractures are formed from the early fractures that are enlarged by surface water and groundwater dissolution. For these extensive fractures, the fracture surfaces usually expand irregularly due to dissolution, and have granular transparent white calcite crystals, or crystal clusters in intact crystallines on the walls.

The diagenetic fractures, or pressolutional fractures, mainly appearing as suture lines, are formed by compaction and pressure dissolution as a result of sedimentary loading, which is related to the formation pressure, temperature, and mud content in limestone. Most of the suture lines are parallel to the bed plane in a sawtooth shape, generally a few millimeters wide, and have been filled with or dissolved by calcite, mud, or bitumen. According to the fluorescence thin section observation, some suture lines have strong fluorescence, indicating effective accumulated spaces.

1.1.2.2 Propagation of caves, vugs, and fractures

1.1.2.2.1 Caves

By diameter, carbonate caves are roughly divided into centimeter-size and meter-size caves. The centimeter-size caves, with a diameter of 2–1000 mm, can be analyzed by way of core observation. Fig. 1.2 shows the pictures of carbonate cores collected from the Tazhong region, where the caves are visible. Fig. 1.2A illustrates the sparite grainstone in dominance, with unfilled-partly filled karst caves between 10 and 40 mm in diameter. Fig. 1.2B illustrates the sparite calcarenite in dominance, with karst caves between 20 and 60 mm in diameter, and filled with grayish green mud later.

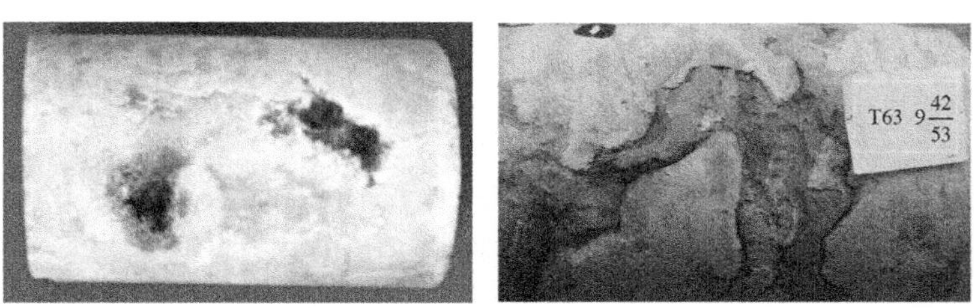

(A) (B)

Fig. 1.2 Cores from the Tazhong region. (A) Unfilled-partly filled karst caves, Well D518. (B) Caves filled with mud, Well T63.

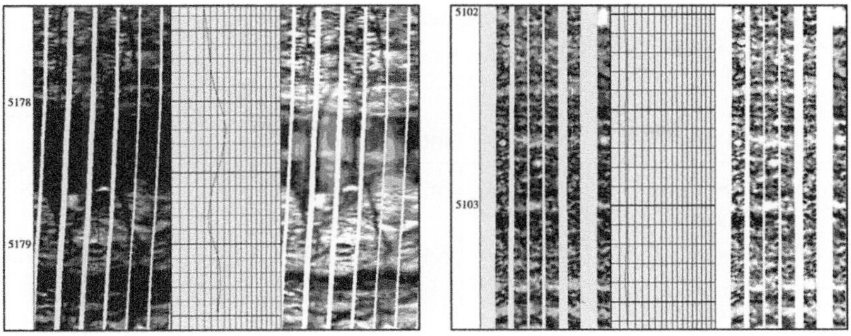

(A) (B)

Fig. 1.3 Imaging logging responses of karst caves in the Tazhong region. (A) Caves partly filled with mud and breccia, Well D462. (B) Layer-parallel karst caves, Well D434.

The centimeter-size caves can also be investigated by way of imaging logging. Caves with different filling patterns have distinct geological characteristics, and thus reflect significantly variable imaging logging responses. In Fig. 1.3A, the caves filled with mud reveal dark spots with high resistivity, and correspond to high GR values. In Fig. 1.3B, the karst caves are clear, reflecting honeycomb spots on the image.

The meter-size caves are generally larger than 1000 mm in diameter, and some of them can extend even a thousand meters. These caves are often characterized by using the drilling and logging data, outcrops, and seismic data. Table 1.1 shows the statistics of drilling anomalies in the carbonate reservoirs in the Tazhong region. Drilling break and fluid leakage occurred in the intervals where meter-size caves were encountered, with the diameter of 0.3–5 m.

The meter-size caves can be observed directly on field outcrops. Fig. 1.4 shows the outcrop in the Tarim Basin. Large meter-size caves with the diameter of 1.5–2.5 m are visible and accessible. These caves are less connected laterally and mainly consist of carbonate rocks, with strong heterogeneity.

Some caves extend as far as over 1000 m, and they are usually described by using seismic data. The large meter-size caves generally reflect bead-like responses on the seismic profile. In the Tarim Basin, the carbonate reservoirs distribute widely in the plane, and some meter-size caves are connected up to a thousand meters. Fig. 1.5 shows the kilometer-size large karst caves in the D15 cluster (extending 5 km) in the Tazhong region.

Table 1.1 Statistics of drilling break and fluid leakage in Lower Ordovician Yingshan Formation, the Tazhong region

No.	Drilling period	Well	Interval (m)	Drilling break (m^3)	Fluid leakage (m^3)
1	Mar. 28–Nov. 6, 2006	T84	5631.16–5666.29		148.3
2	Feb.27–Sep. 8, 2006	T721	5463.14–5505		4488.83
3	Jan. 8–May 28, 2007	D3	6529		4.6
4	Mar. 6–July 25, 2007	D17	6665–6700		95.84
5	Dec. 12, 2006–May 17, 2008	D1	6375.46–6389.37		125.2
6	Feb. 21–Sep. 5, 2008	D8	6127–6145.58	4.3	3776.3
7	Dec. 29, 2008–May 25, 2009	D10	6299.8–6309.7	0.33	11.2
8	Nov. 13, 2008–April 3, 2009	D11	6459.65–6475.61	4.69	1196.53
9	May 14–Sep. 07, 2008	T726	5531.04–5534.09		235.9
10	Oct. 15, 2008–Jan. 30, 2009	D6	6172.73		118.46
11	Oct. 15, 2008–Jan. 30, 2009	D21	5869		12.4
12	June 20, 2009–Sep. 21, 2009	D22	5736.66		121
13	April 30–Sep. 10, 2009	D12	6187.75–6189.9	2.25	2771.2
14	Sep. 27, 2008–Mar. 3, 2009	D9	6212–6214		21.1
15	Jan. 30–May 30, 2010	D44	5858.5–5858.9	0.4	287.22
16	June 24–Nov. 25, 2009	D23	5945–5946		3.44
17	June 3–Nov. 25, 2009	D111	6110.85		27.75
18	Mar. 10–June 10, 2010	T23C	5543.61		24.28
19	Feb. 3–Mar. 26, 2010	T201C	5779		17.1

Fig. 1.4 Outcrop with meter-size caves, the Tarim Basin.

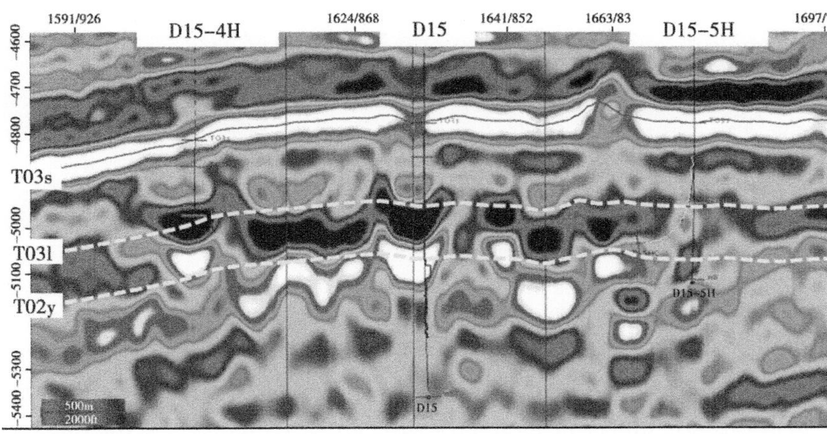

Fig. 1.5 Seismic profile of kilometer-size carves in the D15 cluster, the Tazhong region.

1.1.2.2.2 Vugs

Vugs in the reservoirs are generally in millimeter size, with the diameter of 0.01–2 mm, and they can only be observed by casting thin sections. Fig. 1.6 shows the typical carbonate vugs in the Tazhong region. Microscopically, abundant vugs are observed, and they are dominantly intragranular dissolved pores and moldic pores, as well as intergranular dissolved pores. Most of these vugs comprise effective accumulation spaces.

1.1.2.2.3 Fractures

By extension length, the carbonate fractures are graded into millimeter-size, centimeter-size, and meter-size. The millimeter-size fractures, with the extension length of 0.1–10 mm, can be observed by casting thin sections. Fig. 1.7 shows the typical carbonate fractures in the Tazhong region. Microscopically, the fractures are mainly composed of structural fractures and

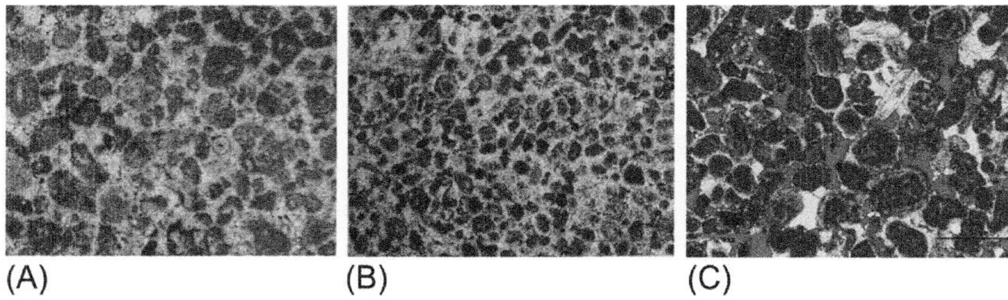

Fig. 1.6 Microscopic vugs in Yingshan Formation carbonate rocks, the Tazhong region. (A) Intragranular dissolved pores, Well D9. (B) Moldic pores, Well D9. (C) Intergranular dissolved pores, Well D203.

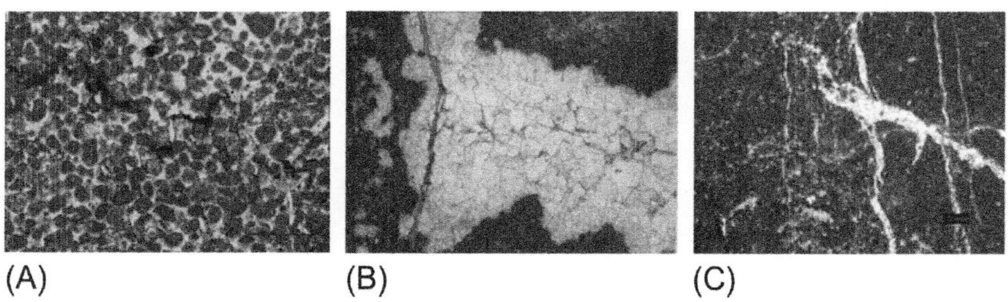

Fig. 1.7 Microscopic fractures in carbonate rocks, the Tazhong region. (A) Dissolved fractures, Well D203. (B) Unfilled fractures, Well D7. (C) Structural dissolved fractures, Well D104.

suture lines, and unfilled structural fractures are observed. These fractures act as flow pathways and accumulation spaces in carbonate reservoirs.

The centimeter-size fractures, with the extension length of 10–1000 mm, are usually characterized by way of core observation and imaging logging. Fig. 1.8 shows the cores with centimeter-size fractures. In Fig. 1.8A, the micritic calcarenite core contains dissolved vugs and intersected multiphase fractures. In Fig. 1.8B, the sparite calcarenite core contains an oblique fracture, about 100 mm long and 20–40 mm wide, which is fully filled with dark gray mud. The centimeter-size fractures appear as sinusoidal dark bands on logging images. Fig. 1.9 shows the logging images of carbonate rocks in the Tazhong region, revealing clearly the existence of centimeter-size fractures.

The meter-size fractures, with the extension length of over 1000 mm, are generally observed and described by using outcrops and seismic data. Fig. 1.10 shows the outcrop with meter-size fractures in the Tarim Basin. It can be seen that there are multiple large fractures and faults extending a few meters to dozens of meters, most of which cut across the strata, and multiphase fractures and faults are incised locally.

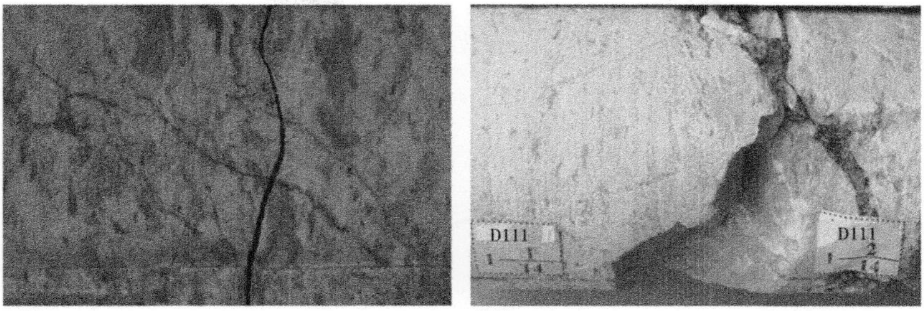

(A) (B)

Fig. 1.8 Cores with centimeter-size fractures, the Tazhong region. (A) Dissolved vugs and multiphase fracture, Well T12. (B) Fractures filled with mud, Well D111.

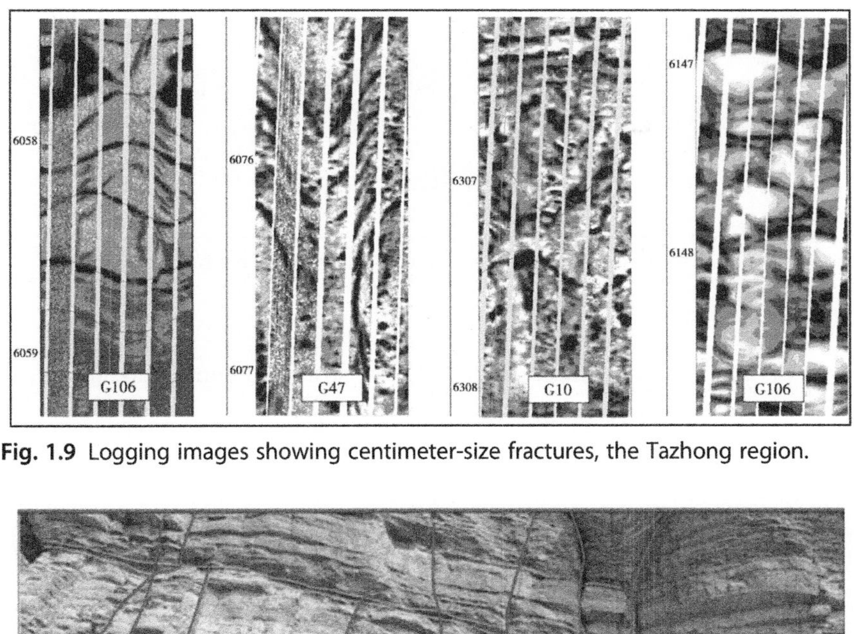

Fig. 1.9 Logging images showing centimeter-size fractures, the Tazhong region.

Fig. 1.10 Outcrop with centimeter-size fractures, the Tazhong region.

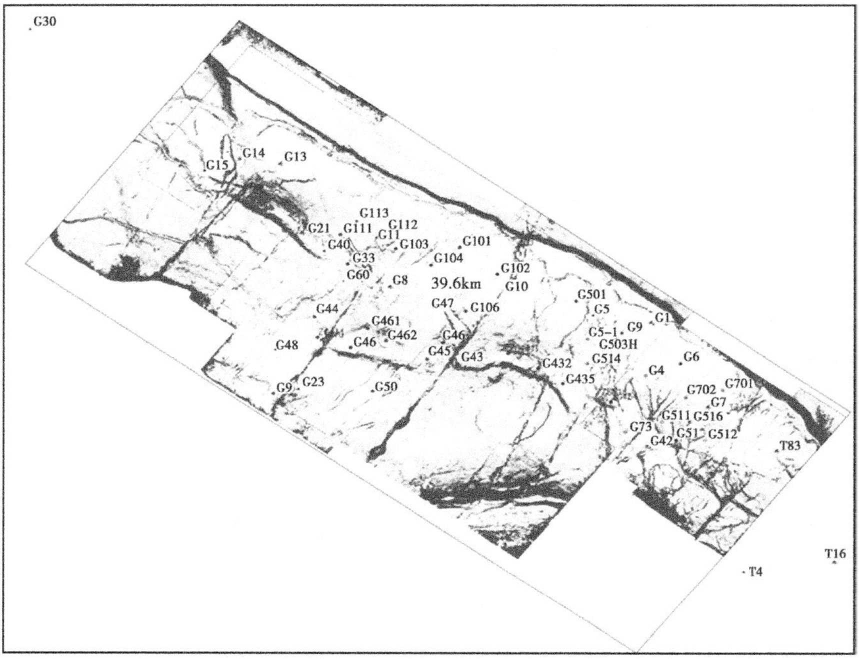

Fig. 1.11 Coherence map of local limestone top in the eastern Tazhong region.

Faults and fractures can be predicted by seismic coherence and curvature techniques. In the Tazhong region, for example, the coherence analysis reveals that large fracture zones extend several kilometers to dozens of kilometers throughout the eastern region (Fig. 1.11); some basin-forming faults are found across the center-western region (Fig. 1.12). These faults act as the important hydrocarbon charging and migration pathways and accumulation spaces in carbonate reservoirs.

1.1.2.3 Types of reservoirs by accumulation space

1.1.2.3.1 Cave reservoirs

The cave reservoirs are the most important ones in the Tarim Basin, with caves (diameter >100 mm) serving as the accumulation spaces. These types of reservoirs are concentrated on the top of the Yingshan Formation and in active faulting zones, and are controlled by tectonic stress and unconformable karst. They are the major contributor to oil and gas. In case of drilling breaks or fluid leakage, the cave reservoir reflects dark bands sandwiched with light blocks locally, or totally black plates on the logging images, and bead-like responses on the seismic profile (Fig. 1.13A).

Fig. 1.12 Coherence map of local limestone top in the center-western Tazhong region.

1.1.2.3.2 Vuggy reservoirs

The vuggy reservoirs are also developed well, with the accumulation space consisting of dissolved vugs (diameter < 100 mm) formed by primary pores due to dissolution; fractures are underdeveloped as a result of penecontemporaneous leaching of meteoric fresh water. The matrix porosity is generally <2%; in the intervals with dissolved vugs, it reaches 4%–6%, or even exceeds 10% locally. The dissolved vugs can be observed on the FMI images, generally presenting as irregular dark spots (Fig. 1.13B).

1.1.2.3.3 Fractured-vuggy reservoirs

The fractured-vuggy reservoirs are the most dominant ones in the Tarim Basin, where vugs act as the major accumulation space, and fractures can partially function as accumulation space, but serve more as flow pathways. Compared with vuggy or fractured reservoirs, the fractured-vuggy reservoirs have higher storage and flow capacities due to the coexistence of vugs and fractures. They reflect as connected black spots and vertical black bands on dynamic FMI images (Fig. 1.13C).

1.1.2.3.4 Fractured reservoirs

The fractured reservoirs contain abundant fractures, scarce vugs, and no matrix pores, with the vug porosity of <1.8% and the fracture porosity of

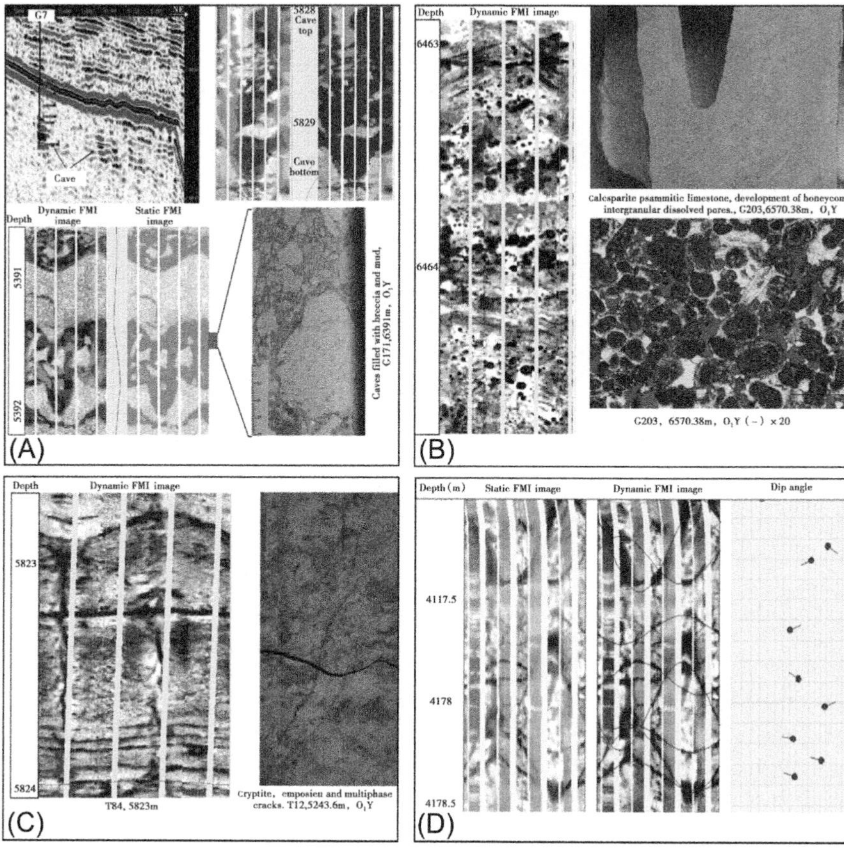

Fig. 1.13 Responses of Lower Ordovician Yingshan Formation karstic reservoirs, the Tazhong region. (A) Responses of cave reservoirs. (B) Responses of vuggy reservoirs. (C) Responses of fractured-vuggy reservoirs d. Responses of fractured reservoirs.

>0.04%. Fractures act as both flow pathways and major accumulation spaces. With low (matrix) porosity and high permeability, the fractured reservoirs have storage and flow capacities controlled by fracture distribution and development degree (Fig. 1.13D).

1.1.2.4 Strong heterogeneity of reservoirs with complex and diverse inner-structures

The carbonate reservoirs in the Tarim Basin have strong heterogeneity. The accumulation spaces are of complex and diverse origins, with the pores, caves, and fractures varying in size and connectivity. With various combinations of accumulation spaces, the reservoir types have greatly distinct connectivity. Generally, the carbonate reservoirs are characterized by strong heterogeneity and complex and diverse inner-structures.

Well D112 in the ZG8 well block of Tazhong I gas field, the Tarim Basin, was drilled to the bottom of the beads without any engineering anomaly (Fig. 1.14). After imaging logging and remote acoustic logging analysis, re-sidetracking was implemented. Leakage occurred at the position 32 m away from the original hole and at the same depth. The separator flare height was >10 m, hydrocarbons contents of gas logging >99%. The well testing revealed a high yield and a stable tubing pressure.

Based on the drilling and logging data, and later production testing analysis, it was concluded that the bead-like responses of Well D112 don't represent cave reservoirs, but an aggregation of multiple small fractured vuggy reservoirs. The beads are less connected within the range of about 30 m.

Well D11C2 failed because water breakthrough occurred when its vertical hole was drilled to the bottom of beads (Fig. 1.15). Instead, sidetracking was conducted to the high position of beads in the Yingshan Formation, which was successful. The wellhead sampling results revealed the H_2S content of 2000 ppm, the gas-oil ratio of 2172 m^3/m^3, oil viscosity of 1.147 mPa·s, and density of 0.8082 g/cm^3. Based on the inner-structures of beads identified, Well D11C2 was deployed for drilling the top of beads in the overlying Lianglitage Formation, but leakage occurred during the drilling. The liquid level data revealed no pressure loss, suggesting no connection with the underlying producing beads. The wellhead sampling results revealed the H_2S content of 50 ppm, the gas-oil ratio of 13,134 m^3/m^3, oil viscosity of 0.8042 mPa·s, and density of 0.7807 g/cm^3. This confirms that there are no connections between upper and lower beads. The beads at the high position suffered a more serious gas cut later, and both upper and lower beads have complex inner structures and poor connectivity.

The strong heterogeneity and complex and diverse inner structures lead to the disconnection between accumulation spaces of the reservoirs in a producing formation. Each accumulation space has its own oil-gas-water contacts and initial formation pressure. The accumulation spaces are mostly fractured vuggy systems controlled by a single well. New fractured vuggy systems with initial formation pressure can still be encountered in the vicinity of the depleted fractured vuggy systems. Neglect of a fractured vuggy reservoir on the drilling section may indicate a missing oil and gas reservoir, which is unlikely to be discovered by additional drilling.

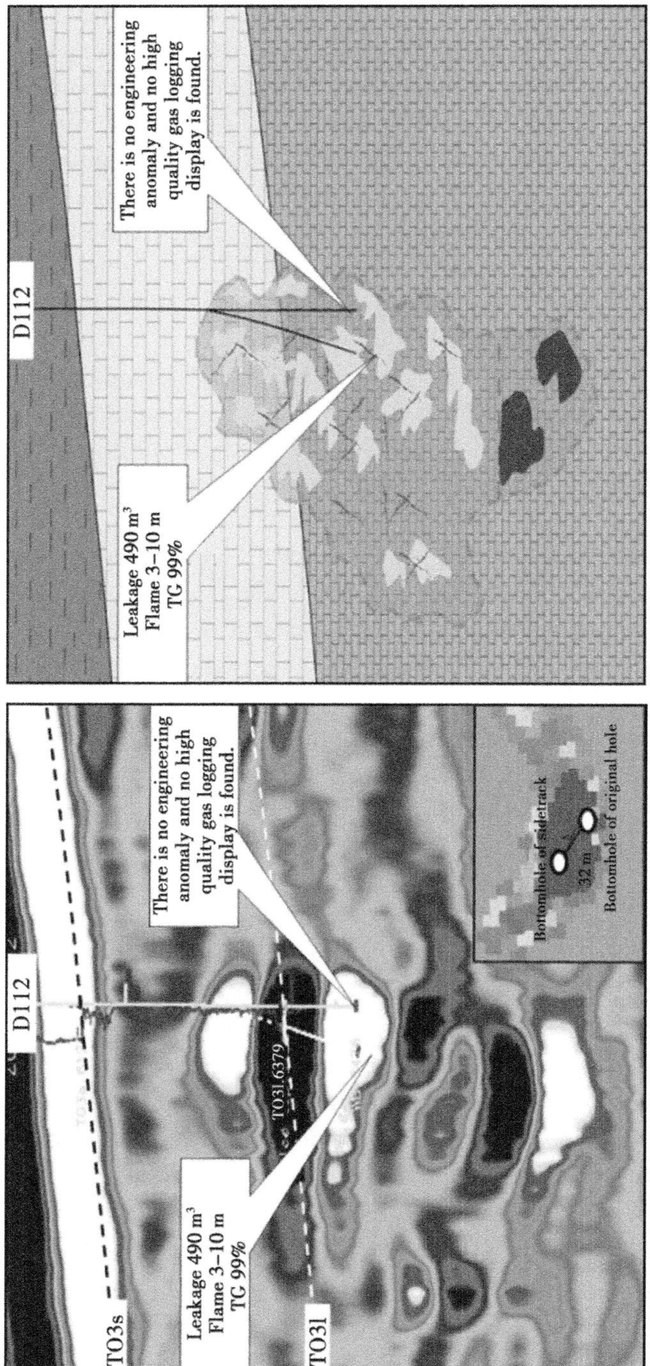

Fig. 1.14 Seismic profile and pattern of Well D112.

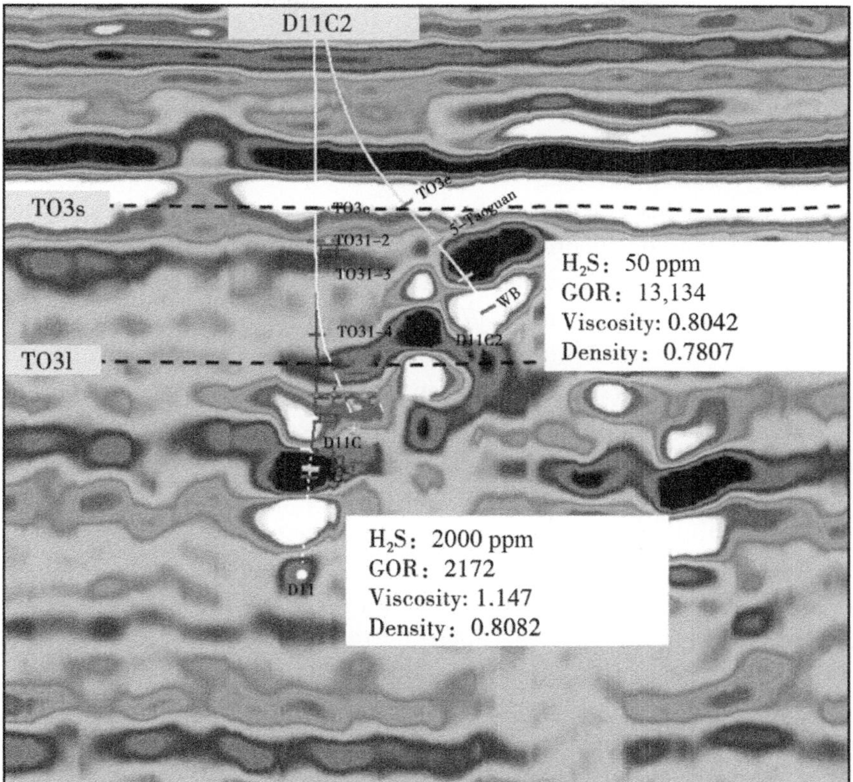

Fig. 1.15 Seismic profile of Well D11C2.

1.2 Typical characteristics of gas reservoir development

1.2.1 Complex flow mechanism

A conventional clastic rock reservoir is a typical porous medium, with multiple micrometer-size pores, while the fractured vuggy reservoir contains micrometer-size dissolved secondary pores, structural microfractures and dissolved fractures, and meter-size karst caves, which vary remarkably in size. The diverse accumulation spaces of the fractured vuggy reservoirs determine the complex flow pattern of internal fluid; namely, the coexistence of aperture flow in large fractures, cave flow in large unfilled caves, and pore flow in small and medium fractures, dissolved pores, and partly filled caves.

The flow pattern of fluids in fractured vuggy reservoirs is controversial. Some believe it belongs in the category of engineering fluid mechanics, and others classify it in the category of traditional percolation mechanics.

1.2.2 Large difference in productivity between wells

The well productivity varies significantly due to fault zone scale, structural location, and reservoir development. For example, in the Tarim Basin, the Halahatang oil field has a number of NE- and NW-trending strike-slip faults, which represent a main controlling factor for the fractured vuggy reservoirs, and also major pathways for hydrocarbon charging. The fault zones are very different in well productivity. For example, the Ha 9, Ha 7, and Ha 5 fault zones reflect the average well cumulative oil production of over 2.67×10^4 t, 1.71×10^4 t, and 1.17×10^4 t, respectively (Fig. 1.16).

Moreover, the initial productivity of wells varies greatly. For example, in the Lungu oil field, Well LG701 had an initial productivity as high as 400 t/d, and Wells LG7-5-2 and LG701-4 achieved initial productivity of 100–200 t/d; some wells such as Well LG7-16 recorded initial productivity below 10 t/d.

In a fault zone, well production varies across segments. For example, the ZG 10 fault in the Tazhong I gas field is divided into echelon, pull-apart, and horsetail segments, from south to north, where the average cumulative productions are $0.37 \times 10^8\ m^3$ gas and 1.9×10^4 t oil, $0.05 \times 10^8\ m^3$ gas and 0.25×10^4 t oil, and $0.2 \times 10^8\ m^3$ gas and 0.93×10^4 t oil, respectively. Obviously, the echelon and horsetail segments have significantly better production performance than the pull-apart segment (Table 1.2).

1.2.3 Less regularity of production performance

For the carbonate reservoirs, the matrix often has poor properties and contains no hydrocarbons, while hydrocarbons are mainly endowed in caves, vugs, and fractures, which are very distinct in the magnitudes of permeability and porosity, leading to serious heterogeneity and anisotropy in the reservoirs. The fractured vuggy carbonate gas reservoirs are unique for their accumulation patterns, which result in strong heterogeneity, spatially. Accordingly, the flow pattern is complex, such as pipe flow, seepage flow, or the coexistence of both, in most cases. This complexity brings about great differences in development characteristics. Production, water cut, GOR, and other development indexes are distinct between wells, and even in a well during its life cycle, showing less regularity represented by pressure fluctuation and rapid fluid changes among oil, gas, and water.

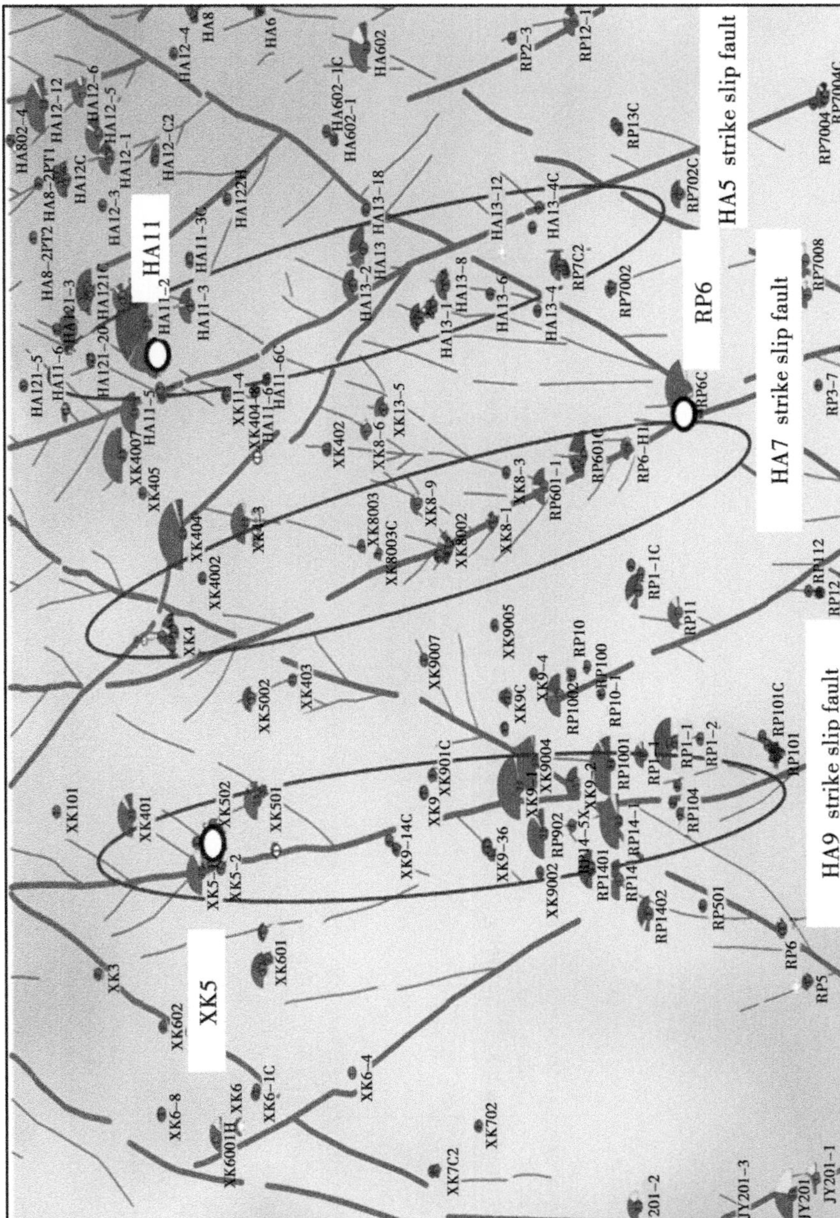

Fig. 1.16 Current production test in Harahatang oil field (Ha 9–Ha 5 fault zones).

Table 1.2 Well production performance in segments of ZG 10 fault zone, Tazhong I gas field

No.	Segment	Drilled wells	Successful wells	Success rate (%)	Average cumulative well production—gas (10^8 m^3)	Average cumulative well production—oil (10^4 t)
1	Echelon	8	8	100	0.37	1.90
2	Pull-apart	4	4	100	0.05	0.25
3	Horsetail	4	3	75	0.21	0.93

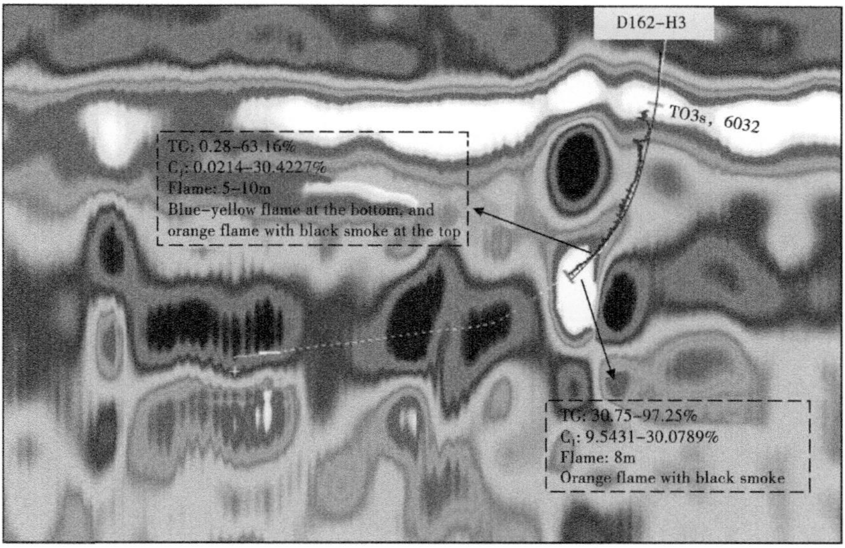

Fig. 1.17 Seismic profile of Well D162-H3.

1.2.3.1 Highly variable pressure with fluctuations

In the fractured vuggy carbonate gas reservoirs, the complex pore-cave-fracture combinations and distribution lead to variable flowing pressure at different positions, and also distinct flow patterns of fluids. Therefore, such reservoirs often show highly variable pressure with fluctuations. For example, Well D162-H3 in the Tazhong I gas field reflects bead-like strong responses on the seismic profile (Fig. 1.17). Its drilling was started on April 21, 2013, and finished on November 8, 2013, to 6414 m in the Lower Ordovician Langlitage Formation. In the drilling process, leakage started at 6366.11 m with a cumulative volume of 929.5 m^3, and overflow occurred at 6366.11 m with a volume of 0.7 m^3. In the target horizon, the gas logging results revealed 3 gas layers (29 m). The best show was observed in the 6384–6401 m interval, with the highest TG of 97.37%.

Well D162-H3 was put into production on November 18, 2013, with an initial tubing pressure of 36 MPa, daily oil flow rate of 42.81 t, the daily gas flow rate of 30,000 m^3, and no water. During the production test, the tubing pressure dropped rapidly in the beginning, and rose suddenly after reaching 1 MPa. Subsequently, the tubing pressure and production demonstrated the intermittent and irregular variation of ups and downs. Until the well was shut on November 20, 2017, it had been producing for 823 days, with a cumulative production of 3.9753×10^4 t oil and 0.2444×10^8 m^3 gas (Fig. 1.18).

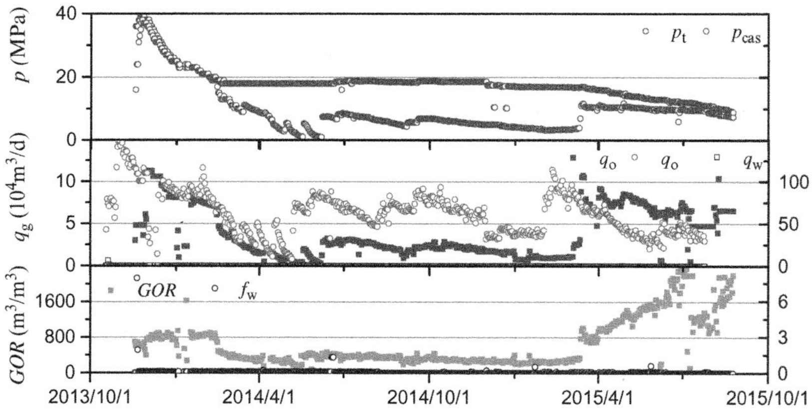

Fig. 1.18 Production test curves of Well D162-H3.

1.2.3.2 Highly variable fluid properties throughout the development phase

For the fractured vuggy carbonate gas reservoirs, the characteristics of multiphase accumulation, multiphase charging, and the existence of fault systems lead to complex horizontal and vertical fluids distribution. Furthermore, different vug-cave-fracture combinations and flow resistances create the product variation between oil, gas, and water. For example, Well M62-2 in the Tazhong I gas field reflects bead-like strong responses on the seismic profile (Fig. 1.19). Its drilling was started on May 7, 2005, and finished on July 11, 2005, to 4825 m in the Lower Ordovician Langlitage Formation. In the drilling process, leakage started at 6366.11 m with a cumulative volume of 1106.7 m^3, and overflow occurred at 4788.21 m with a volume of 2 m^3. In the target horizon, the gas logging results revealed 2 gas layers (29 m). The best show was observed in the 4779–4805 m interval, with the highest TG of 47.19%. Based on the well logging data interpretation, there is one Type-I fractured-vuggy reservoir, with a thickness of 2.5 m, and porosity of 6.8%, four Type-II fractured-vuggy reservoirs, with a thickness of 15.5 m and porosity 2.7%, and four Type-III reservoirs, with a thickness of 9.5 m and porosity of 1.64%.

Well M62-2 was put into production on December 22, 2005 after production testing, with an initial tubing pressure of 36.6 MPa, daily oil flow rate of 26 t, daily gas flow rate of 98,028 m^3, and no water. Up to March 14, 2018, the well had been producing for 3869 days, with a cumulative

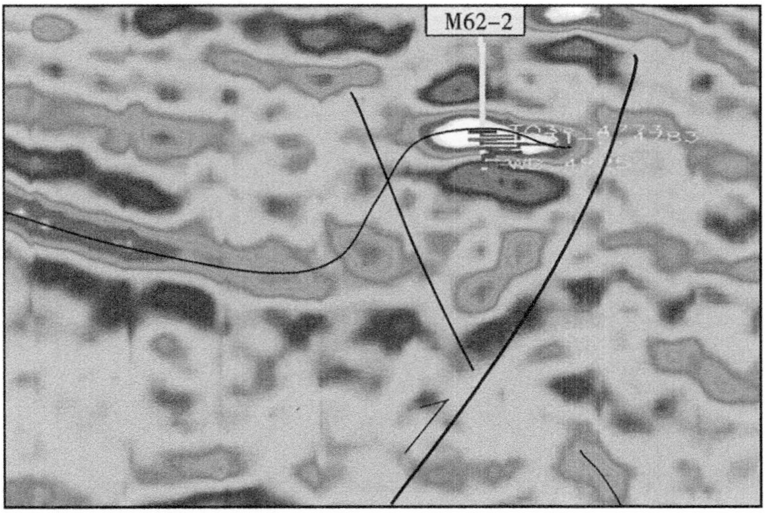

Fig. 1.19 Seismic profile of Well M62-2.

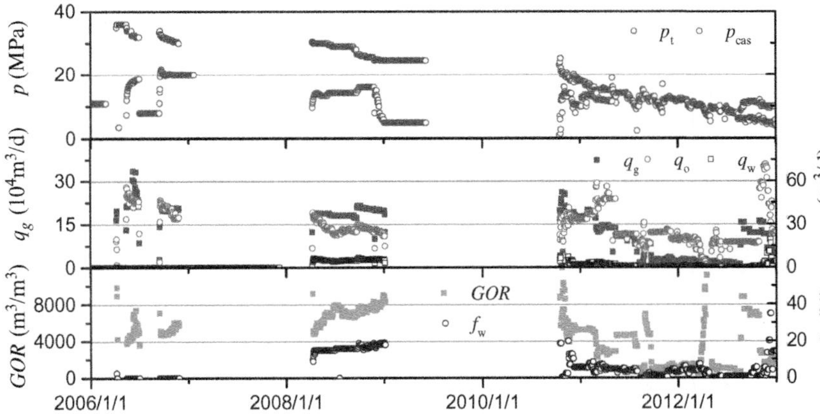

Fig. 1.20 Production test curves of Well M62-2.

production of 3.6961×10^4 t oil, 0.3867×10^8 m^3 gas and 0.2439×10^4 t water (Fig. 1.20). During the production test, the fluid property varied in different periods. The PVT test data obtained on August 18, 2005 showed that the formation was condensate gas reservoirs with a medium condensate oil content. Additional PVT test data obtained on August 15, 2011 showed crude oil, indicating a significant change in the hydrocarbon phase (Fig. 1.21).

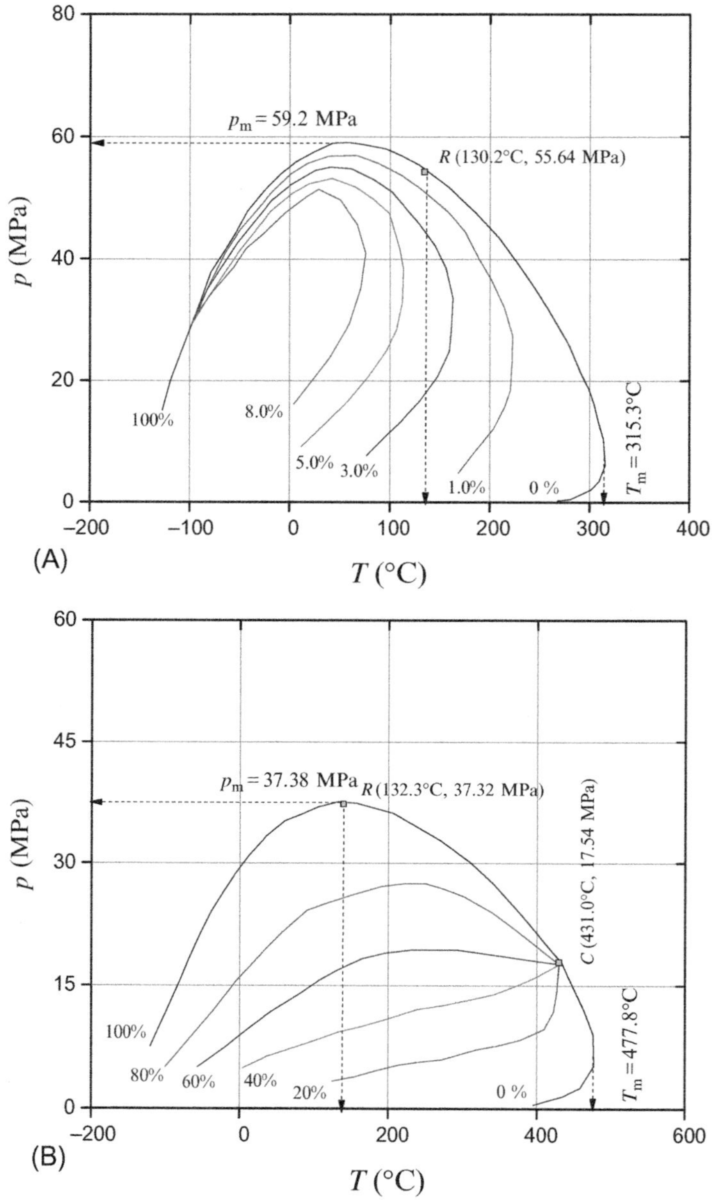

Fig. 1.21 Formation fluid phases of Well M62-2 in different periods. (**A**) August 18, 2005. (**B**) August 15, 2011.

Due to diverse types and distinct well production performance, the characteristics of fractured vuggy reservoirs are closely related to the accumulation space type, fracture-vug combination, reservoir size, connectivity with water, and water size. The wells vary greatly in production performance, but those wells encountering the same type of accumulation space suggest a similar trend of production.

1.2.4 Challenges in gas reservoir description

1.2.4.1 Increasing precision in reservoir description

Carbonate rocks, a kind of important reservoir rock, are complex in deposition, diagenesis, and evolution, so the carbonate reservoirs usually present dotted distribution, strong heterogeneity, and complicated spatial structures with the development of fractures and caves affected by numerous factors, which ultimately leads to great complexity in the formation, distribution, and enrichment of carbonate oil and gas fields. In the Tazhong region of the Tarim Basin, the carbonate reservoirs are observed with the coexistence of vugs, caves, and fractures, of which vugs and caves comprise the main accumulation spaces, and fractures act as the flow pathways.

In carbonate reservoirs, the development and distribution of caves are extremely heterogeneous, the filling degree and types are variable, and fluid properties are complex. Moreover, reservoirs at and near wells are largely distinct. Thus, the core analysis and logging evaluation of carbonate reservoirs are extremely challenging. Logging techniques are limited in the measurement range. Conventional logging techniques only involve dozens of centimeters around the borehole. A remote acoustic logging technique has emerged rapidly in recent years, with its measurement extending to more than a dozen, to dozens of meters, which, however, is still insufficient for the description of accumulation space in a carbonate reservoir. In addition, the logging interpretation based on a homogeneous geological model often provides a smaller reservoir porosity, which is contrary to the case of heavy fluid leakage, making it impossible to reflect the reservoir properties using logging techniques.

The advancement of seismic acquisition and processing technology and seismic attribute technology enables the description of carbonate reservoirs based on seismic attributes. So far, seismic attributes have been satisfactorily applied in predicting carbonate reservoirs. However, the strong anisotropy and various fracture-vug combinations deduce complex and variable seismic wave responses. It is difficult to obtain accurate and quantitative reservoir parameters through a single seismic attribute analysis.

1.2.4.2 Increasing accuracy in reserves estimation

The carbonate reservoir spaces are mainly composed of caves, vugs, and fractures, which are different in size and complex in distribution. The carbonate reservoirs with extremely strong heterogeneity are usually classified as cave reservoirs, vuggy reservoirs, and fractured reservoirs. Different from clastic rock reservoirs, the carbonate reservoirs are generally characterized by quasi-layered distribution, oil abundance, and local enrichment; and the carbonate reservoirs at and near wells are largely distinct, so the quantitative description of reservoir thickness, porosity, and oil-water contact (OWC) is very difficult. Practices have demonstrated that the well control technique for estimating reserves in carbonate reservoirs often suggests a large hydrocarbon-bearing area, large geological reserves, small recoverable reserves, and a low producing degree of reserves, and the methods of reserves estimation for homogeneous sandstone reservoirs are not suitable for the highly heterogeneous carbonate reservoirs.

For the fractured vuggy carbonate reservoirs with strong heterogeneity and complexity, the volumes of caves, vugs, and fractures in the reservoirs can be described quantitatively by using the fracture-vug spatial characterization, connectivity analysis, and spatial description techniques, together with the geological modeling and attribute volume integration technique. The resulting "fracture-vug characterization based volumetric method" for reserves estimation is a valuable test for carbonate gas reservoirs. Specifically, based on the reservoir characterization, the fracture-vug development zone, fracture-vug belt, and fracture-vug system are divided, and the estimation units for possible, probable, and proved reserves are determined depending on the well control degree and the producing degree of reserves. Then, the effective thickness and porosity of caves, vugs, and fractures in a single well are determined according to the reservoir type, porosity, and seismic facies, and the recovery factor of the concerned block is calculated using the analogy method (both static and dynamic) and the well performance method. Finally, the reserves estimation is completed by the gridding integral method.

The fracture-vug characterization based volumetric method is used for caves, vugs, and fractures. The quantity of reserves derived by this method is 40% of that by the traditional volumetric method, so this method is more effective in reflecting the reservoir heterogeneity. This method integrates the quantitative reservoir characterization. According to the geometric characterization of fractures and vugs, and the division of fracture-vug belts and systems, the distributions of fractured-vuggy bodies and sweet spots are clarified. The reserves estimated by this method are the reserves in caves and associated vugs and fractures. When the reserves in caves are recovered, the reserves in associated vugs and fractures can also be recovered. All these reserves are movable and can be directly considered as a reference in development planning. This method is more applicable to heterogeneous carbonate reservoirs with respect to oil-bearing area, estimation units division, and recovery demonstration.

1.2.4.3 Increasing accuracy in performance forecasting

In the Tarim Basin, the discovered carbonate reservoirs are deep (6000–7000 m average) and covered by desert, gobi, and eluvial soil on the surface. Disrupted by the intense attenuation by absorption and scattering, the signal-to-noise ratio (SNR) is low. The earth-filtering effect and surface attenuation seriously attenuate high-frequency components of seismic waves, thus restricting the seismic resolution and reducing the accuracy of reservoir identification. Therefore, it is extremely difficult to identify, predict, and depict the fractured-vuggy bodies by using seismic data, not to mention their inner structures and fillings. The complex size and connectivity of irregular fractured-vuggy bodies where oil and gas are accumulated bring great difficulties to subsequent 3D geological modeling.

Even if the modeling results are satisfactory, the existing numerical simulation methods and techniques based on continuous porous media are inapplicable to the field applications in fractured vuggy reservoirs with complex flow mechanisms (for the coexistence of aperture flow, cave flow, and pore flow). The equivalent multimedia and coupled numerical solution can be used, on the basis of coupling of continuous medium and discrete medium, to realize the numerical simulation of fractured vuggy carbonate reservoirs, but the calculation results are very uncertain.

For the purpose of effective development of the fractured vuggy reservoirs, it is essential to establish appropriate key development index prediction techniques in conjunction with advanced reservoir engineering methods, depending on the available recognition of geological features and the well production performance of such reservoirs.

1.2.4.4 Designing strategies for gas reservoir dynamic description

In this book, the dynamical description technique for fractured vuggy carbonate gas reservoirs based on the lifecycle well testing analysis and the advanced production decline analysis is proposed, as shown in Fig. 1.22. This technique, without the deficits of static description, can help realize the reservoir evaluation, reserves estimation, and productivity estimation of fractured vuggy carbonate gas reservoirs, ultimately contributing to the effective development.

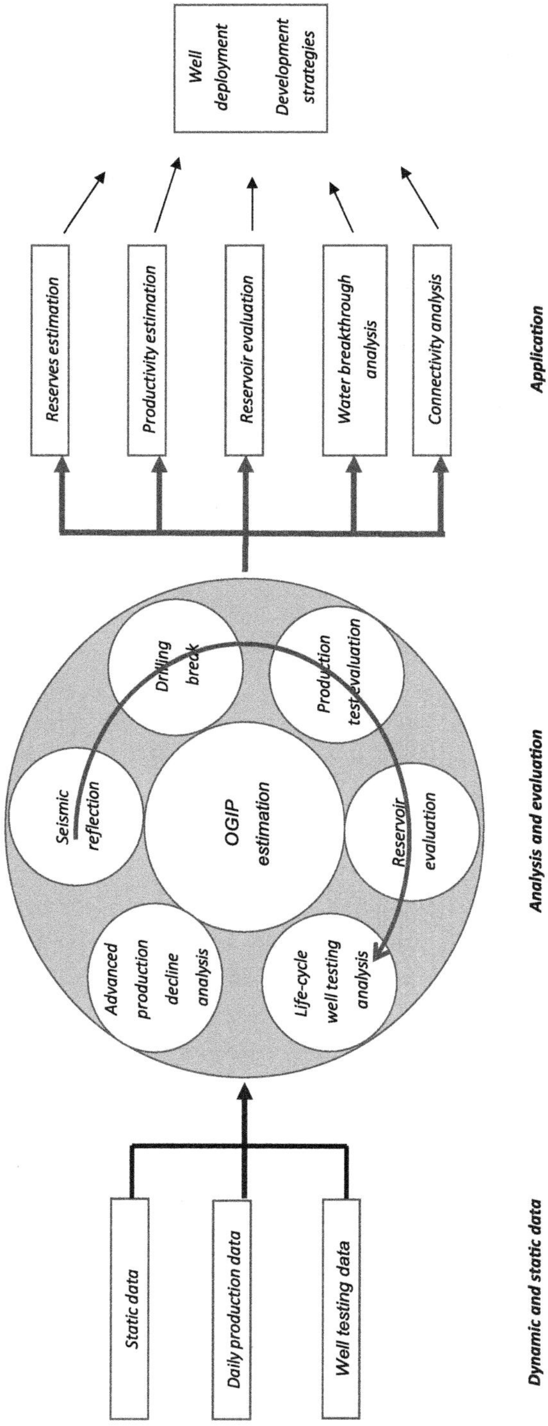

Fig. 1.22 Schematic diagram of dynamic description technique for fractured vuggy carbonate gas reservoirs.

1.3 Summary

In fractured vuggy carbonate reservoirs, unlike conventional sandstone reservoirs, the oil and gas distribution is controlled by the existence of fractured vuggy bodies instead of local structures. Generally, hydrocarbons exist throughout the reservoir, and are enriched locally. A single reservoir is small and has no uniform gas-water contact (GWC) macroscopically. Such reservoirs have strong heterogeneity and complex inner structures; the matrix contains no pores and reflects extremely poor properties; caves, vugs, and fractures created by secondary dissolution constitute the main accumulation spaces.

The diversity of accumulation spaces determines the complexity of flow patterns of fluids in the fractured vuggy carbonate reservoirs; namely, aperture flow, cave flow, and pore flow coexist. In practical production, such reservoirs are found to have poor performance regularity, variable well productivity, and rapid production decline.

Fine reservoir description, reserves estimation, and development index forecasting are essential for the effective development of fractured vuggy carbonate gas reservoirs. The proposed dynamical description technique for fractured vuggy carbonate gas reservoirs, without the deficits of static description, can help in performing the reservoir evaluation, reserves estimation, and productivity estimation of such reservoirs.

The following chapters will illustrate the dynamical description technique for fractured vuggy carbonate gas reservoirs and its application.

CHAPTER 2

Introduction to the dynamic description technique of gas reservoirs

Contents

2.1 Dynamic description technique of gas reservoirs

2.1.1 Status and role

A dynamic description of gas reservoirs refers to the process by which the gas reservoirs are characterized comprehensively and accurately to obtain the well and reservoir parameters, based on the dynamic data (e.g., pressure, production, and fluids) acquired during tests and development, and the static data (e.g., geology), by using gas reservoir engineering methods, including modern well test analysis methods and advanced production decline analysis methods (e.g., production analysis or PA, or rate transient analysis or RTA), as well as advanced computer software. The process is shown in Fig. 2.1.

Geophysical prospecting, well logging, and well test-based dynamic description are three major techniques used to characterize gas reservoirs. They are distinctive and noninterchangeable, with the range of descriptions as shown in Table 2.1.

For complex gas reservoirs in the preliminary evaluation phase, the geological study can clarify the regional depositional setting and reservoir

Dynamic Description Technology of Fractured Vuggy Carbonate Gas Reservoirs
https://doi.org/10.1016/B978-0-12-818324-3.00002-4

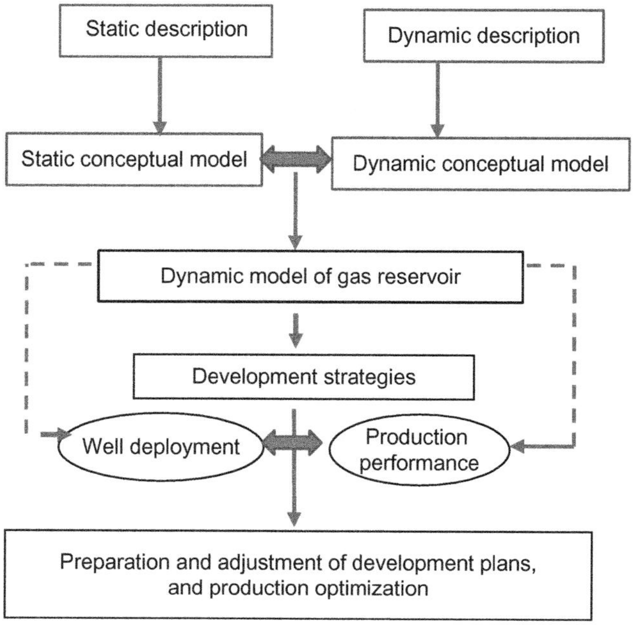

Fig. 2.1 Process of the dynamic description of gas reservoirs.

Table 2.1 Range of static and dynamic descriptions of gas reservoirs (Liu, 2008)

Static data	Geological study								
	Seismic survey								
	Core analysis								
	Well logging								
Dynamic data	Formation test								
	Lifecycle well test analysis								
	PA or RTA								
Range (m)		10^{-3}	10^{-2}	10^{-1}	10^{0}	10^{1}	10^{2}	10^{3}	10^{4}

development characteristics. However, due to insufficient resolution of seismic data and limited wells, the geological model of reservoirs primarily based on static data can be vividly regarded as a "static image" on which there are only a few clear well sites (for well logging). Thus, such a model is inadequate for efficient development of gas reservoirs.

Actually, the development of gas reservoirs needs a "dynamic image," which means the dynamic flow characteristics of underground fluids (gas, oil, and water) under the operating conditions; namely, the effective characteristics of flow units. A static description technique (e.g., geophysical prospecting and well logging) can identify the existence of faults and extensive reservoir bodies, but fails to address certain critical aspects, such as fault sealing and well drainage areas. In contrast, a dynamic description technique based on well testing eliminates the deficiency of the static description.

2.1.2 Methodology

The methods of dynamic description include wellbore node analysis, production logging analysis, wireline formation testing, well test analysis, advanced production analysis (or production data analysis), history matching analysis, and production performance forecasting. These methods are very different in the application scope, as shown in Fig. 2.2.

2.1.2.1 Wellbore node analysis

Wellbore node analysis is conducted to identify the fluid distribution characteristics based on static pressure (flow pressure) and static temperature (flow temperature) gradients tested. The static pressure gradient analysis can effectively reveal the structural features of gas reservoirs. It can be divided by the exploration phase and the development phase.

According to the static pressure gradient line in the exploration phase, we can judge which wells are drilled in a connected unit. In a suspected connected unit, the nonalignment of static pressure points of wells on one "static pressure gradient line of natural gas" is often found in two cases.

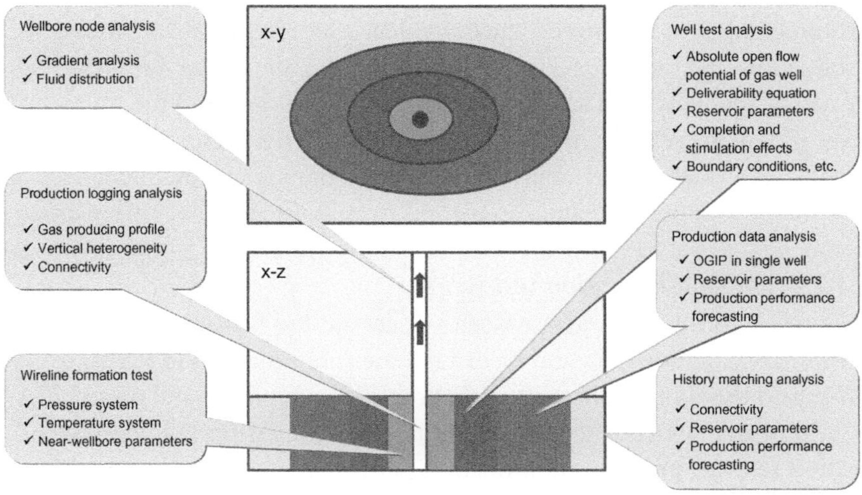

Fig. 2.2 Methods and application scopes of the dynamic description of gas reservoirs (Oliver et al., 2013).

In one case, the static pressure points are aligned on two or three "static pressure gradient lines of natural gas," which means that the wells may be deployed in two or three independent gas pools. In the other case, the static pressure points are aligned on an approximate "static pressure gradient line of hydrostatic column," which suggests that the reservoirs were covered in the same hydraulic system when gas was migrated and accumulated, but natural gas therein does not exist in the same connected gas pool.

Are the wells in a gas pool considered connected if their static pressure points are aligned on one "static pressure gradient line of natural gas" in the exploration phase? The answer is indefinite. In some gas fields of the Sichuan Basin (especially the southern Sichuan Basin), China, the wells had roughly the same initial pressure, but reflected significantly different formation pressure after producing for a period of time according to the shut-in pressure tests. This implies that the areas controlled by the wells are independent, although they are not necessarily separated completely.

2.1.2.2 Production logging analysis

Production logging analysis is conducted to measure the flow profiles of fluids in producing wells and injection wells, including the velocity (flow rate), density, water holdup, temperature, and pressure of fluids. Then, the properties and flow rates of fluids produced or absorbed in perforation intervals are identified to evaluate the occurrence of gas wells and the exploitation characteristics of gas reservoirs.

For gas wells with multiple perforation intervals, the production profiles acquired can be used to derive the individual layer production, including gas production and water production, and analyze the water-breakthrough horizons, the individual layer productivity contribution, and the individual layer producing reserves, thus providing a basis for the evaluation of gas reservoirs, preparation and adjustment of development plans, detection of downhole complexities, and stimulation treatments and effect evaluation.

2.1.2.3 Wireline formation test (WFT)

A wireline formation tester can recover certain fluid samples from a short well interval (about 1 m), and record the pressure responses in the recovery phase and the shut-in phase. It is applicable to open holes and cased holes.

Compared with conventional pressure transient testing, the wireline formation test can provide the formation permeability and pressure response within multiple small-scale ranges in the test intervals, and collect multiple fluid samples at different depths.

Because only a few samples are taken, the wireline formation test is limited in access depth, and can hardly investigate the lateral boundary of reservoirs. Generally, the small-scale permeability obtained in the formation test will be scaled up before being used in production performance forecasting. However, the skin factors obtained in the wireline formation test cannot be used in the production performance forecasting.

Wireline formation testing is conducted to collect bottom hole fluid samples, estimate the changes of formation pressure and permeability with the depth, identify the depleted formation, and obtain the vertical permeability. It follows the same data interpretation theory as a conventional well test analysis. The permeability obtained can also be used to establish the correction relation of a permeability derived from wireline logging interpretation.

2.1.2.4 Well test analysis

A "well test" refers to the whole process during which a pressure gauge is run in a hole to measure the changes of pressure and/or flow rates over time, and the test data are analyzed and processed. It includes two parts: field testing and data analysis and processing. Well testing is a critical technique to understand oil and gas reservoirs and evaluate the dynamic reservoir characteristics, well completion efficiency, and stimulation effects. The data obtained in the well test is the only data collected under the fluid flowing conditions, so the analysis result can best represent the dynamic characteristics of reservoirs.

2.1.2.4.1 Types of well test

A well test can be categorized by various criteria—for example, a single-well test or multiwell test by the number of wells involved, or a drawdown test or buildup test by the well status (producing or shut-in).

(1) Deliverability test and transient test

The deliverability test of a gas well is conducted to obtain fluid samples, determine the well deliverability and stable productivity, and obtain the inflow performance relationship (IPR) curve necessary for the systematic analysis of gas reservoirs. The deliverability test includes a back pressure test, isochronal test, modified isochronal test, and a single-point test. The deliverability equation of a gas well can be established according to the data of deliverability test and the deliverability evaluation methods to obtain the absolute open flow potential and IPR curve for guiding the reasonable production allocation.

The transient test is conducted to determine the gas phase permeability under formation conditions, calculate the average formation pressure in the drainage area, evaluate the stimulation effects, and estimate the drainage area and/or dynamic reserves. Generally, it refers to the buildup test.

(2) Buildup test and drawdown test

The buildup test, or pressure buildup test, refers to the process that the well is shut in after producing for a period of time to observe the changes in flowing pressure.

The drawdown test, or pressure drawdown test, refers to the process during which the changes of flowing pressure are observed when new wells or wells that maintain stable formation pressure after a long period of being shut-in produce at a constant rate.

(3) Single-well test and multiwell test

The single-well test refers to the process during which a well is monitored for its pressure response by changing its production rate. It is used to quantitatively describe how the reservoir is damaged or stimulated, and estimate the formation permeability, drainage area, well-controlled reserves, and average formation pressure in the drainage area.

The multiwell test refers to the process during which the production rate of a well (active well) is changed to monitor the pressure of the surrounding one, or other wells (observation wells). The multiwell test is more complicated than the single-well test, and it is often used to quantitatively describe the interwell connectivity or estimate the permeability in a direction.

(4) Exploration-well test and development-well test

The exploration-well test focuses on the evaluation of the overall properties of reservoirs. It is mainly used to take fluid samples and estimate the initial formation pressure, gas well deliverability, and the distance between the well and boundary, and well-controlled reserves. The exploration-well test is usually performed when the drilling rig is still at the well site, so its cost may be very high. The data obtained in the test are valuable references for making major investment decisions.

The development-well test focuses on the evaluation of properties of a single well and areas near the wellbore. It is mainly used to estimate the average formation pressure, evaluate the stimulation effects, quantify the near-wellbore reservoir damages, and determine the reservoir permeability. The development-well test is more economical than the exploration-well test, and the data obtained in the test are mainly used for making small investment decisions, such as the decision on reservoir stimulation.

2.1.2.4.2 Purpose of well test

The well test plays an important role in the lifecycle of a gas well, including the reservoir exploration, preliminary evaluation, and development phases.

(1) Exploration phase

A well test is critical in the exploration phase. The data obtained are necessary for major investment decisions. In this phase, a DST must be implemented to confirm the existence of a gas layer, and preliminarily evaluate the reservoir permeability and the possibility of formation damage in the drilling process.

A well completion test is needed to further verify the scale and deliverability of a gas reservoir. Once the drilling is finished after the target horizon is encountered, the exploration well is completed by the casing or other techniques, and the gas test is performed layer by layer, so as to derive the absolute open flow potential of a gas well. The absolute open flow potential is the most convincing parameter to measure the well deliverability.

Reserves estimation will be initiated immediately after the data provided by an exploration well; including production, reservoir pressure, and permeability; prove the existence of gas reservoirs. In the exploration phase, well test data cannot be used to directly calculate the reserves, but only supplement or correct the reserves estimation results to a certain extent. Specifically, the well test can provide the data of deliverability for reserves estimation, characteristic coefficients for the stable productivity of dual-porosity reservoirs, and plane distribution of reservoir and initial pressure data for reserves estimation (Table 2.2).

(2) Preliminary evaluation phase

In the preliminary evaluation phase, main activities are carried out to deepen the geological cognition on gas reservoirs, estimate the reserves, ascertain the deliverability, select the principal development technologies, and prepare the development conceptual design or development plans, so as to achieve scientific gas field development.

The systematic well test is performed for conventional gas reservoirs, and modified isochronal well testing is performed for low-permeability gas reservoirs; then, the deliverability equation of the gas well is established to calculate the absolute open flow potential. A drawdown test and buildup test are performed for representative wells to derive the reservoir parameters. A multiwell interference test is mainly conducted to evaluate the connectivity of reservoirs or fracture systems and calculate the interwell reservoir parameters.

Table 2.2 Role of the well test in the exploration and development of gas fields (Zhuang, 2013)

Phase/item		Purpose of well test											
		Identify the gas-bearing potential of reservoirs	Test the reservoir pressure	Determine the absolute open flow potential through a deliverability test	Interpret the reservoir permeability by a transient test	Evaluate the quality of well drilling and completion according to skin factor	Determine the length and flow conductivity of induced fractures	Determine the parameters of fractured dual-porosity reservoirs	Provide the turbulence coefficients during the production of gas wells	Determine the impermeable boundary distribution of reservoirs	Clarify the lateral connectivity of the interference test	Derive the gas well-controlled dynamic reserves	Verify the dynamic reserves of gas reservoirs
Exploration phase	DST during drilling of exploration wells	★	★	☆	★	★							
	Completion gas testing of exploration wells	★	★	★	★	★	☆						
	DST and completion gas testing of detailed exploration wells	★	★	★	★	★	☆	☆					
	Reserves estimation of gas-bearing blocks	■	■	■	■	■	□	□					

Preliminary evaluation phase	Deliverability test and other transient tests of appraisal wells	★	★	★	★	★	★	★	★	☆	☆	☆	
	Acid fracturing stimulation		★	★	★	★	★	☆	★	☆			
	Production test and extended well test of appraisal wells		★	★	★	★	★	★	★	★	★	★	★
	Verification of reserves in gas reservoirs		■	■	■	■	□	■		■	■		■
	Numerical simulation of gas reservoirs, and preparation of development plans		■	■	■	■	■	■	■	■	■	□	■
Development phase	Dynamic monitoring of gas reservoirs		★	☆	★	★	☆	☆	☆	☆	☆		★
	Completion gas test of adjustment wells	★	★	★	★	★	★	★	★	☆			★

Note: ★, Mandatory item; ■, Mandatory parameter; ☆, Contingent item; □, Contingent parameter.

(3) Development phase
In the whole development process, conventional well test methods can be applied to monitor the gas reservoir performance. For a gas well in normal production, however, it is apparently impracticable to conduct a well test by frequently tripping out the pressure gauge while the well is opened and shut in; otherwise, a permanent pressure gauge is installed. According to the needs of development, the transient test should be performed in the early stage, when new wells are put into production when great changes occur in the production rate or pressure of producing wells, or before and after stimulation. For key wells, a deliverability test and buildup test should be performed regularly, and an interference test can be arranged when necessary. In the production process, the wellhead pressure and flow rate should be recorded comprehensively and accurately, and the performance should be monitored.

As the research activities are advanced and the open time of gas wells extends, the radius of investigation increases. Now, it is necessary to perform the extended buildup test. Some parameters that cannot be collected in the early stage, such as transient and stable deliverability, boundary distance, and shape, block size, dual-porosity reservoir parameters, dual-permeability formation parameters, composite formation parameters, non-Darcy flow coefficient, reservoir connectivity and dynamic reserves of blocks can be obtained from the extended well test analysis. On this basis, the dynamic model of the gas reservoir can be established and effectively used in the analysis of gas reservoir performance.

2.1.2.4.3 Forward and inverse problems

(1) Forward problem
The pressure response is determined under the condition that the flow rate history and the reservoir model are known, as shown in Fig. 2.3. If the flow rate history is given, then a pressure response will always be obtained for any given reservoir model. In other words, the solution to the forward problem is unique.

(2) Inverse problem
The unknown reservoir model is determined under the condition that the flow rate history and the pressure response are known, as shown in Fig. 2.4. If the flow rate history is given, then the pressure response given by many different reservoir models may be the same. In other words, the solution to the inverse problem is not unique.

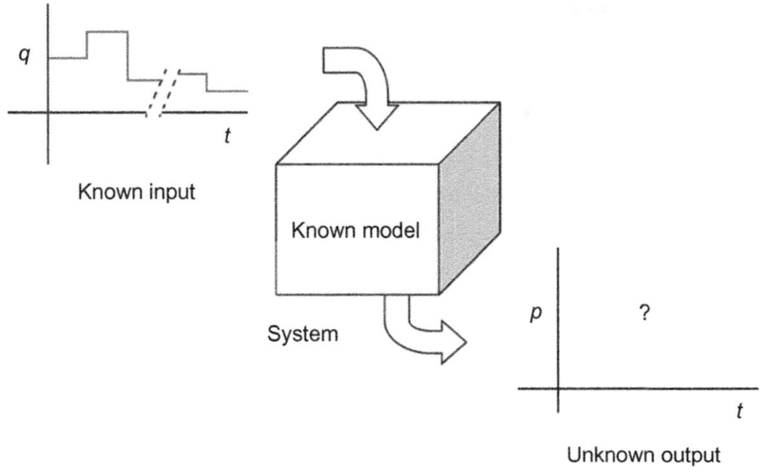

Fig. 2.3 Schematic diagram of a forward problem (well test design).

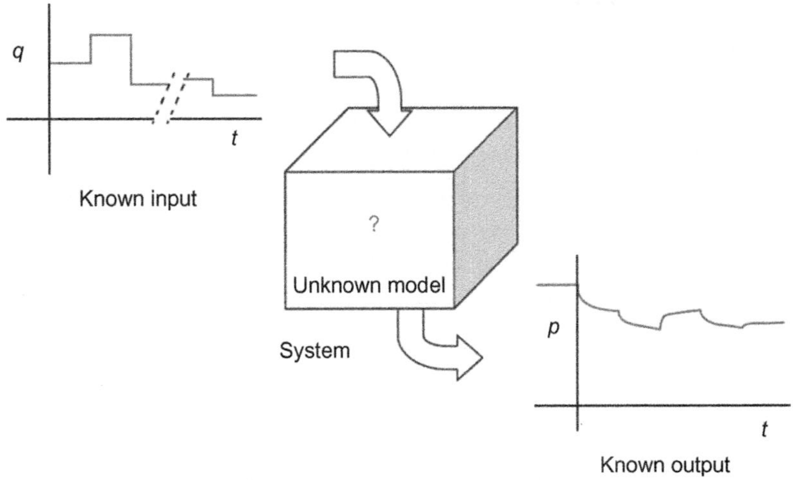

Fig. 2.4 Schematic diagram of an inverse problem (well test interpretation).

The forward problem has a unique solution; in other words, the output is constant for a given input and model. The inverse problem does not have a unique solution; in other words, different models can produce the same output for a given input.

In nature, the well test interpretation belongs to an inverse problem. Two core problems of the well test are to identify the unknown reservoir model and to solve the unknown parameters of this model. Therefore, it is necessary to reduce the multiplicity of interpretation results wherever possible during interpretation.

2.1.2.4.4 Well test analysis methods

In the early stage of gas reservoir development, the gas well performance is primarily monitored by collecting and analyzing the deliverability test and transient test data.

Several publications focus on the gas well test theories and analysis, and numerous papers in journals and academic conferences also deal with the well test. These literature illustrates the establishment of theoretical models, the solution of mathematical equations, and field applications. All quantitative well test interpretation methods can be categorized into three types: straight-line segment analysis, type curve analysis, and simulation/history matching. The well test analysis has become an effective means for the precise description of reservoirs and the performance analysis of gas reservoirs, rather than a tool of single-well evaluation originally.

Currently, the transient test analysis technique features the integration of an analytical solution and numerical solution. As compared with the well test, simply based on the analytical solution, the well test based on the numerical solution can provide a geological model that is closer to the actual reservoir shape and parameter distribution, and can effectively consider the influence of multiwell production. Thus, the latter is more practical for reservoir description and production performance forecasting of gas reservoirs. The numerical well test method has been successfully merged into commercialized well test analysis software overseas. Internationally representative well test software includes Saphir (KAPPA), PanSystem (EPS), and Welltest200 (Schlumberger). In recent years, these software programs have introduced the numerical well test module, and absorbed the traditional well performance analysis.

The well test interpretation is a process to solve the inverse problem, and it is still suffering from multiplicity. Especially for low-permeability tight fractured gas wells, actual well tests often fail to derive the pseudo-radial flow, leading to great uncertainty in the calculation of reservoir permeability and induced fracture half-length.

2.1.2.5 Production data analysis

2.1.2.5.1 Advanced production decline analysis methods

Advanced production decline analysis (PA or RTA) is performed to obtain the characteristic parameters of reservoirs based on the production rate and pressure data acquired in the production process, similar to the pressure transient analysis. Under normal conditions, this method can be used to estimate the original gas in place (OGIP), formation permeability, skin factors, or fracture half-length.

The advanced production decline analysis methods include the traditional Arps type curve method (Arps, 1945), the classical Fetkovich type curve matching method (Fetkovich, 1980), the Blasingame (Palacio and Blasingame, 1993) and Agarwal-Gardner (1998) type curve matching analysis method, and the Mattar flowing material balance reservoir engineering method (1998).

The Arps type curve method can better describe the production decline law of producing wells under the conditions of constant flowing pressure and fully boundary-dominated depletion. Most important, this method needs no formation parameters. However, it is not applicable to the data analysis in the transient production phase.

The Fetkovich type curve matching method organically combines the transient production decline curve and the Arps production decline equation by using the concept of equivalent wellbore radius to show the production decline law and the influence of boundary. The biggest advantage of this method is that it can reliably determine whether the production period is in the transient-dominated stage or in the boundary-dominated stage.

The Blasingame (1993) and Agarwal-Gardner (1998) type curve matching analysis methods are used to identify the influences when variable flowing pressure, variable production rate, and PVT property of natural gas change with the pressure, by introducing the pseudo-time (or material balance pseudo-time), and the production normalized pseudo-pressure (or pseudo-pressure normalized production). The type curves on production integral, production integral derivative, cumulative production-time, and production-cumulative production are added, respectively, in these two methods as assisted matching analysis curves, to reduce the multiplicity of interpretation results.

Mattar and McNeil (1998) proposed to analyze the production performance data using the "flowing material balance" method and discussed the calculation of material balance time in detail. Mattar and Anderson (2003) believed that no production analysis method can meet all types of gas reservoirs, and the best way to reduce analysis errors is to combine various analysis methods and make full use of the flowing pressure data.

2.1.2.5.2 Similarities and differences between advanced production decline analysis and modern well test analysis

The advanced production decline analysis and modern well test analysis are two major techniques for a dynamic description of gas reservoirs. They have independent strong points, and also common features. Both methods can be combined and balanced organically in the reservoir parameter evaluation to reduce the uncertainty of interpretation results. Their similarities and differences are shown in Table 2.3.

Both methods are based on the classical filtration theory and use the type curve matching method to obtain parameters. Under the conditions of the

Table 2.3 Similarities and differences between advanced production decline analysis and modern well test analysis (Sun, 2015)

Item	Modern well test analysis		Advanced production decline analysis
Theory	Filtration theory		
Time range	Traditional pressure buildup analysis	Hours, days, and weeks	Days, months, and years
	Lifecycle analysis	Hours, days, weeks, months and years	
Analysis stage	Traditional pressure buildup analysis	Shut-in stage	Production phase
	Lifecycle analysis	Lifecycle	
Data source	Buildup test, wireline formation test, and permanent pressure gauge		Wellhead metering or permanent metering
Detection range	Detection range in the shut-in period		Drainage area of well or clusters
Development phase	Early manual matching	MDH, Horner	Arps
	Classical curve matching	Ramey	Fetkovich
	Modern curve matching	Bourdet	Blasingame, A-G, NPI
Data diagnosis	Requirement	Radial flow	Pseudo-steady state, and boundary-dominated flow
	Capacity	High	Moderate
Analysis results	Traditional pressure buildup analysis	Formation flow capacity, skin factor, and boundary condition	Well drainage area and well-controlled reserves, formation flow capacity, and skin factor
	Lifecycle analysis	Well-controlled reserves can be calculated	

Table 2.3 Similarities and differences between advanced production decline analysis and modern well test analysis (Sun, 2015)—cont'd

Item	Modern well test analysis		Advanced production decline analysis
Reliability of results	Traditional pressure buildup analysis	Relatively high	Relatively high
	Lifecycle analysis	High	

complicated boundary, multiphase flow, and multiwell interference, the numerical solution method can be used through the establishment of a model. However, the two methods adopt data with different levels of precision. The advanced production decline analysis method only requires daily production and pressure, while the modern well test analysis method requires high-precision pressure transient test data. As compared with the pressure transient analysis, the advanced production decline analysis method is superior because of the lower cost of data collection (no special equipment, no shut-in, and without hindering normal production) and larger detection range. By virtue of this advantage, the advanced production decline analysis method has become one of the most important dynamic description techniques for low-permeability reservoirs.

The pressure transient data are acquired under a controlled condition exclusively for estimating characteristic parameters of gas reservoirs, while the data used in the advanced production decline analysis are collected under a routine production condition. Therefore, the latter data reflect a bigger noise and are more sensitive to production conditions. Accordingly, the early characteristics of the latter data cannot be used to reliably identify whether the well is damaged. Similarly, the latter method is inadequate to trace the changes of skin, effective permeability, and pressure in the drainage area with the time.

2.1.2.6 History matching analysis

The log-log curve matching results of modern well test analysis, or advanced production decline analysis, are used as the initial values for history matching, and the analytical solution or numerical solution is used to calculate the pressure response of the gas reservoir. The parameters in the model include those that are known or unknown. The values of the unknown parameters are adjusted repeatedly until the pressure response calculated through the model matches the measured response. The matching process can be achieved by manually adjusting the parameters, or completed automatically by computer nonlinear regression matching.

The biggest advantage of the history matching analysis is that multiple phenomenon characteristics can be considered comprehensively during analysis, and all data are used in the overall analysis to the utmost extent, so as to ensure the highest conformity of interpretation results to all data.

2.1.3 Single well dynamic description

The core of single well dynamic description is the evaluation of deliverability and stable productivity for the purpose of single well production performance forecasting. Various dynamic parameters of gas reservoirs can be obtained through such dynamic description. By functions, these parameters can be categorized into five types: (1) reservoir flow characteristic parameters, such as permeability, storativity ratio, and interporosity flow coefficient; (2) reservoir boundary characteristic parameters, such as boundary characteristics and distance, radius of complex model, and drainage area; (3) parameters for effect evaluation of stimulation treatments, such as skin factor, half-length of hydraulic fractures, flow conductivity, and fracture skin factor; (4) gas reservoir (well) parameters, such as formation pressure, absolute open flow potential, and dynamic reserves; and (5) parameters for evaluation of stable productivity, such as reasonable production allocation and cumulative production. Some dynamic parameters are detailed in Table 2.4.

2.1.4 Gas reservoir dynamic description

The dynamic description of gas reservoirs (clusters) refers to the description of gas reservoirs (clusters) from points to surfaces on the basis of the single well dynamic description, including pressure system analysis (gradient test), fluid distribution analysis (RFT, FMT, MDT, etc.), reservoir distribution characteristics (multiwell buildup test, interference test, etc.), influence of fractures on flow (buildup test, interference test, etc.), well-type adaptability analysis (numerical well test), dynamic reserves estimation (buildup test, advanced production decline analysis, material balance analysis), and gas reservoir performance forecasting (buildup test, advanced production decline analysis, numerical simulation). Especially, the pressure system analysis, evaluation on effective reservoir distribution, and performance forecasting of gas reservoirs (clusters) are the core tasks.

Table 2.4 Main parameters of single-well dynamic description

Single-well dynamic description technique	Flow characteristic parameters		Reservoir boundary parameters		Stimulation evaluation parameters		Gas reservoir (well) parameters			Stable productivity evaluation		
	K	ω/λ	L/R	A	S	x_f	p_i	q_{AOF}	q_i	G	q	G_p
Deliverability test								+				
Production logging									+			
Formation test	+						+					
Buildup test	+	+	+	+	+	+	+			+	+	+
Advanced production decline analysis	+	+	+	+	+	+				+	+	+
History matching	+	+	+	+	+	+	+			+	+	+

Note: +, the parameters that can be evaluated by each technique.

2.2 Dynamic monitoring of gas reservoir development

2.2.1 Significance and role

Dynamic monitoring and performance analysis are conducted throughout the process of gas reservoir development. They are invaluable techniques for obtaining intensive knowledge on the characteristics and dynamic changes of gas reservoirs and gas wells, supporting the decision-making for production optimization, ensuring the normal operation of production systems, evaluating the implementation effects of stimulation and development plans, and providing the basis for development adjustment and potential tapping. The quality of such work directly affects the level and effect of gas reservoir development. The requirements for dynamic monitoring and performance analysis are different depending on the work objectives and tasks for the stages of gas reservoir development (Table 2.5).

Dynamic monitoring and performance analysis are complementary to eachother, but focus on different aspects. Dynamic monitoring is used to acquire the basic data in development stages and special tests, such as single-well pressure, temperature, productivity, and fluid properties, and to conduct a brief analysis to understand the basic performance behaviors of gas reservoirs/wells and discover any abnormality in a timely manner. Performance analysis is conducted to raise the technical requirements for data acquisition through dynamic monitoring according to the development needs and engineering theory of gas reservoirs, and thoroughly investigate and identify the characteristics and production performance of gas reservoirs/wells based on the data acquired through dynamic monitoring.

2.2.2 Contents and methodology

2.2.2.1 Principles

The dynamic monitoring of gas reservoir development mainly includes preparation of a monitoring plan, the establishment of the monitoring system, and acquisition of monitoring data. It follows the principles of "systematic, accurate, and practical"—the monitoring objects are relatively comprehensive, the data is acquired to meet the technical requirements for dynamic analysis, and the monitoring process is technically and economically feasible, as shown in Table 2.6.

2.2.2.2 Purpose and contents

The dynamic monitoring of gas reservoir development is involved in reservoir engineering, production engineering, surface gathering and

Table 2.5 Requirements for dynamic monitoring and performance analysis in different development stages (Li et al., 2016)

Stage	Objectives	Main tasks					Key research objects
Predevelopment evaluation	Complete the conceptual design of gas reservoir development and deliver the proved reserves	Get deeper knowledge on geological characteristics and development laws of gas reservoirs	Trace and analyze gas well deliverability	Trace, analyze, and evaluate the reserves of gas reservoirs	Establish and improve the description mode of gas reservoirs	Put forward the requirements for static and dynamic data acquisition and deploy the appraisal wells	Initial formation pressure of the gas reservoir; the relationship between pressure, temperature, and fluid distribution; flow characteristics of the reservoir; absolute open flow potential of a gas well; contamination or improvement of the gas well
	Complete preparation of the gas reservoir development plan					Make production test, optimize the development method, divide the development layer series and units, and determine the well placement, and production allocation of the gas well and the gas recovery rate of the gas reservoir	Formation pressure, permeability and other flow characteristic parameters; deliverability equation of gas well, interwell, and interlayer connectivity; recoverability of reserves, contamination, or improvement of gas well
Productivity construction	Achieve the designed production capacity					Put forward requirements for additional data acquisition, and	Difference between actual deliverability and forecasted deliverability, factors

Continued

Table 2.5 Requirements for dynamic monitoring and performance analysis in different development stages (Li et al., 2016)—cont'd

Stage	Objectives	Main tasks	Key research objects
	according to the development plan	optimize the production allocation of a gas well, well location, and drilling sequence of development wells to be drilled	influencing the deliverability, and reasonable production of gas well
Stable production	Improve the stable production capacity of the gas reservoir and prolong the stable production period	Maintain the normal production of the gas reservoir, optimize the production regime of a gas well, and carry out such targeted treatments as stimulation for abnormalities, control water in waterflooding reservoir, and supplement development wells if necessary	Pressure, deliverability, flow characteristics, connectivity, contamination or improvement status, water invasion performance, well-controlled reserves, distribution of remaining reserves, etc., and their change laws
Production decline	Mitigate the production decline of the gas reservoir	Deploy development adjustment wells and infill wells if necessary to increase the producing degree	Pressure, deliverability, flow characteristics, connectivity, contamination or improvement status, water invasion

						of reserves; control water in water-flooding reservoirs and carry out the treatments such as stimulation in case of abnormalities	performance, well-controlled reserves, distribution of remaining reserves, etc., and their change laws, and production decline law
Low production	Enhance the ultimate recovery of the gas reservoir					Reduce the abandonment pressure of gas reservoirs as much as possible, prolong the production duration of gas wells, and tap the development potential of gas reservoirs	Deliverability, distribution of remaining reserves, recovery ratio, etc.

Table 2.6 Principles of dynamic monitoring

Basic principles	**Systematic, accurate, and practical**
Specific principles	Dynamic monitoring should be conducted depending on the development characteristics of various gas reservoirs Dynamic monitoring should be in accordance with performance analysis of gas reservoirs in different development stages Dynamic monitoring should consider both local and general points and highlight key aspects Wells monitored should be representative and pertinent Data monitored should be continuous and comparable Dynamic monitoring should be conducted, especially when an abnormality occurs in production, pressure, water-gas ratio, and other indicators Dynamic monitoring should be conducted carefully to provide high-quality dynamic data for the preparation of a development plan and the development decision-making

transmission engineering, and other aspects. The dynamic monitoring of reservoir engineering is conducted to understand the characteristics and development law of gas reservoirs and the deliverability of gas wells to provide the basis for creating relevant development strategies. The dynamic monitoring of production engineering is conducted to evaluate the casing program, assess the adaptability and safety of related processes, or test new processes, so as to support the selection of optimal gas production technology. The dynamic monitoring of surface gathering and transmission engineering is performed to establish the operation of the gathering and transmission pipelines and equipment to make sure that the production system is working safely and is stable, and to perform optimization and make adjustments when necessary. See Table 2.7 for the purpose and contents of dynamic monitoring of reservoir engineering.

2.2.2.3 Methodology

The dynamic monitoring is divided into downhole monitoring and wellhead monitoring. Generally, downhole monitoring is not very sensitive to abnormal factors, and can provide the results with high accuracy; however, the downhole monitoring is costly and difficult to implement. In contrast, wellhead monitoring is simple and operable, but less accurate and applicable in some complex cases. For example, when a gas well produces

Table 2.7 Purpose and contents of dynamic monitoring of gas reservoir engineering

Purpose	Provide necessary basic data for production management and performance analysis	Evaluate the deliverability and production stability of the gas well Review and estimate reservoir reserves Identify the flow characteristics of the reservoir Describe formation pressure and fluid distribution Identify the interwell and interlayer connectivity Identify water invasion Analyze the decline law Diagnose gas well contamination or improvement Evaluate the effect of the development plan Determine abandonment conditions and forecast recovery ratio
Contents	Routine monitoring	Wellhead pressure and temperature Well production or injection rate Components of produced fluids Salinity and ion content of produced water
	Special monitoring	Wellbore pressure/temperature gradient and flowing pressure/temperature Well test Production logging Sampling of produced fluids for PVT analysis

a high-yield of water, the pressure data monitored at the wellhead cannot accurately reflect the change law of flowing pressure, and cannot be used to accurately and quantitatively analyze the characteristics of gas reservoir development and gas well production.

According to the purposes and contents, the dynamic monitoring can be classified into six categories; that is, production data monitoring, well-testing data monitoring, gradient monitoring, fluid sample monitoring, production logging data monitoring, and connectivity monitoring.

2.2.2.3.1 Production data monitoring

Monitoring of production data is a routine dynamic monitoring of gas reservoir development. Based on the data of wellhead tubing/casing pressure, production, or injection acquired, the daily production report is completed, the gas production curve for a production well or the injection curve for an injection well, as well as the gas production curve for a gas reservoir, are prepared, and relevant fundamental analysis is conducted. Depending on actual situations, the online monitoring of H_2S content in gas producing wells, the salinity of the produced water, and flowing pressure (acquired from the permanent downhole pressure gauge) can also be incorporated in the monitoring and analysis of production data.

The fundamental analysis of production data, which is produced in combination with the gas production curves of production wells, includes the evaluation of production stability, calculation of wellhead pressure decline rate, analysis and forecasting of gas-water ratio, and identification of liquid loading in the wellbore for the annular packer-free wells based on the changes of casing-tubing pressure difference.

2.2.2.3.2 Well testing data monitoring

A well test is a typical high-quality dynamic monitoring and performance analysis technique in gas reservoir engineering, and also one of the most effective tools to profoundly reveal the characteristics and production performance law of gas reservoirs/wells. Because wellhead pressure is susceptible to wellbore fluid density and temperature changes, the well testing data acquired at the wellhead are unsatisfactory for well test analysis. Therefore, even for pure gas wells, the downhole pressure test is recommended. Well testing methods should be selected depending on purposes, as shown in Table 2.8.

For new appraisal wells that have revealed oil/gas shows, if the well testing shows the stable result, a buildup test should be conducted in a timely manner to obtain the initial formation pressure, temperature, reservoir damage, and physical property parameters of the reservoir. For some new development wells that have revealed oil/gas shows, a buildup test can be conducted if no pressure data have been acquired in the corresponding

Table 2.8 Well testing methods

Purpose	**Well testing methods**
Evaluate contamination or improvement of gas well Diagnose dynamic characteristics of gas reservoir/well Calculate flow parameters and detect boundaries	Shut-in pressure buildup well test after production at a constant rate is preferred
Evaluate gas well deliverability	The systematic well test is usually selected Modified isochronal well test is used for low-permeability reservoir or gas well, which requires a long period of time to reach stable production
Analysis of interwell connectivity	Interference well test is usually selected A pulse well test can be conducted for high-permeability reservoirs or active gas wells with large production and small well spacing, in order to effectively identify pressure interference in different directions in the multiwell system
Analysis of water invasion performance	Results of buildup well tests in different phases are compared to diagnose the behavior of water invasion

fractured vuggy units. For the new development wells that are proved with stable production through well testing, a buildup test can be conducted if no pressure buildup data have been acquired in the corresponding fractured vuggy system. For the oil and gas wells that show energy voidage in well testing and are suspected to be in connection with adjacent wells according to static data, a buildup test can be conducted to find out whether there is interwell interference. For the purpose of research, some old wells with stable and high production can be selected for a buildup test once a year. The buildup testing data should be acquired above the dew point pressure to ensure accurate reservoir parameter evaluation.

2.2.2.3.3 Gradient monitoring

Monitoring of wellbore pressure/temperature gradient can help in investigation of the distribution of wellbore fluid, diagnosis of the liquid loading in the wellbore, identification of the liquid level, and in obtaining the flowing pressure and temperature data, to provide the basis for calculating the wellbore pressure and temperature at different depths.

For condensate gas wells in fractured vuggy reservoirs, which are usually 6000 m, and contain the multiphase flow, the conventional conversion method may produce significant calculation errors in the flowing pressure. Therefore, the monitoring of the flowing pressure gradient should be enhanced to restrict the converted flowing pressure, to ensure the accurate estimate of dynamic reserves.

For the wells with stable tubing pressure and production after they are put into production, or the high-productivity wells (oil $\geq$40 t/d or gas $\geq$80,000 m^3/d) with a rising tubing pressure, the flowing pressure gradient test should be conducted every 3 months. For the medium- and low-productivity wells, the flowing pressure gradient test should be conducted once every half a year. For the wells with abnormalities such as water breakthrough and production fluctuation, the flowing temperature and flowing pressure gradient tests should be conducted duly.

For new development wells that have revealed oil/gas shows, the static pressure gradient test should be conducted in a timely manner if no buildup test or other special test is required. For high-productivity wells, the static temperature/static pressure gradient test should be conducted once a year. For stimulation wells, the static pressure gradient test should be conducted prior to the job if the testing conditions are available.

2.2.2.3.4 Fluid sample monitoring

Fluid sample monitoring is an important technique to obtain the fluid properties and PVT parameters of gas reservoirs, the distribution characteristics and changes of formation fluids, the interwell connectivity, and the signs of water breakthrough in gas wells. Monitoring results in the gas reservoir engineering scope, such as H_2S and CO_2 contents in natural gas and pH value of formation water, are also a necessary reference for material selection and safety design of downhole pipe string, gas producing wellheads, and surface production facilities. The fluid sample monitoring methods include downhole sample monitoring and wellhead sample monitoring. The wellhead sample monitoring is sufficient for the conventional gas reservoir, while the downhole sample monitoring is necessary for the special gas reservoirs with fluid phase transition, such as condensate gas reservoirs and high-sulfur gas reservoirs.

2.2.2.3.5 Production logging data monitoring

Production logging for dynamic monitoring of gas reservoir engineering can be conducted to acquire multiple parameters such as flow rate, pressure, temperature, fluid density, and a gas holdup at a given depth in the wellbore. The production logging results can be used to monitor and analyze: (1) the fluid production profile, layer water cut, and flow pattern of water-carrying gas in the wellbore of the gas wells with water breakthrough; (2) the water production profile of the water wells and the layer water absorption of the gas-field water reinjection wells; and (3) the single-layer production contribution of multilayer commingled production gas wells, or the gas wells with strong vertical heterogeneity of reservoirs, to judge how difficult it will be to recover the reserves in each reservoir interval. Production logging can also be used as a special technique for acquiring data for well test analysis. The production logging based on well test design is conducive to identifying the dynamic characteristics of water breakthrough under different production pressure differences, and the production contribution changes of low-permeability reservoirs.

After being put into production, one or two wells at the low positions should be selected from each fractured vuggy system for fluid production profile testing once a year, or once semiannually, in the case of wells with medium/high water content. Given the working system is unchanged, if the production behaviors change obviously, the production profile test should be conducted in a timely manner. For the oil wells with high water content or abnormal water breakthrough, the water detection test should be conducted to find out the main interval(s) with water breakthrough.

2.2.2.3.6 Connectivity monitoring

Understanding the fractured vuggy units is fundamental for the development of fractured vuggy carbonate reservoirs. Currently, it is realized by way of a trace test or interference test, because the connectivity of such reservoirs is extremely complex. Nevertheless, for the purpose of future comprehensive adjustment of the fractured vuggy systems/units, it is necessary to fully understand the fractured vuggy units and clarify the interwell connectivity through tracer testing and interference testing of well clusters, especially in the key production areas.

2.2.2.3.7 Selection of monitoring technique

For the carbonate reservoirs in the Tarim Basin, which universally contain H_2S and are characterized by high pressure, high production, and multiphase production, the applicable dynamic monitoring techniques include wireline-suspended dynamic monitoring, and wireline-conveyed dynamic monitoring, both of which should be highly resistant to high H_2S.

The wireline-suspended dynamic monitoring technique presents no requirement for the phase state of the fluid in the wellbore, and can be used for all dynamic monitoring applications. The pressure data obtained are used to analyze reservoir properties, the wellbore phase state, and production performance; and to correct the wellhead pressure data. In the case of highly corrosive media, the technique is only used for testing short-term parameters, such as static pressure/temperature and gradient, and flowing pressure/temperature and gradient.

The wireline-conveyed dynamic monitoring technique is used for long-term monitoring of flowing pressure/temperature (e.g., shut-in buildup and deliverability tests) and accurate recording of formation pressure behaviors by landing/lifting the electronic pressure gauge into/out of the production string via a wireline. The wellhead is only occupied temporally at the time of landing/lifting, so the wellhead seal risk is reduced, and the wireline corrosion is mitigated. This technique can be applied to the short-, medium-, and long-term continuous monitoring of pressure history in the case of complex well conditions.

2.3 Summary

The connotation, status, role, and methodology of the dynamic description technique of gas reservoirs are illustrated in this chapter. The dynamic description of gas reservoirs refers to the process by which the gas reservoirs are characterized comprehensively and accurately to obtain the well and reservoir parameters, based on the dynamic data (e.g., pressure, production, and fluids) acquired during tests and exploitation, and the static data (e.g., geology), by using the gas reservoir engineering methods, including modern well test analysis methods and advanced production decline analysis methods (e.g., Production Analysis or PA, or Rate Transient Analysis or RTA), as well as advanced computer software.

Geophysical prospecting, well logging, and well test-based dynamic description are three major techniques used to characterize the gas reservoirs. They are distinctive and noninterchangeable. The methods of dynamic description include wellbore node analysis, production logging analysis, wireline formation test, well test analysis, advanced production analysis (or production data analysis), history matching analysis, and production performance forecasting. These methods are very different in the application scope.

The core of single well dynamic description is the evaluation of deliverability and stable productivity for the purpose of single well production performance forecasting. Two elements must be considered for the gas well test analysis. One is the physical property of gas as the function of pressure, and the other is the skin factor relevant to production rate or non-Darcy flow. When the pressure drawdown is small, the liquid phase diffusion equation can be used to describe the gas phase diffusion equation, and the solution is in the form of pressure. When the pressure drawdown is large, the square of pressure, pseudo-pressure, and normalized pseudo-pressure can be introduced, and the diffusion equation can be linearized; the solution is in the forms of pressure square, pseudo-pressure, and normalized pseudo-pressure.

CHAPTER 3

Well test analysis methods of fractured vuggy carbonate gas reservoirs

Contents

3.1 Challenges in well test analysis

3.1.1 Diversity of well test curve

Fractured vuggy carbonate reservoirs are highly heterogeneous, and log-log curves are complex and diversified, and they mainly fall into five types: pseudo-homogeneous model, dual-porosity model, composite model, fractured model, and fractured vuggy model. Here, these models are introduced, with the Tazhong I gas field as an example.

Dynamic Description Technology of Fractured Vuggy Carbonate Gas Reservoirs
https://doi.org/10.1016/B978-0-12-818324-3.00003-6

3.1.1.1 Reservoir models

3.1.1.1.1 Pseudo-homogeneous model

Homogeneous formation is a flow medium that is evenly distributed. The pseudo-homogeneous formation is identical to the homogeneous formation in the shape and trend of characteristics curves, but different in the medium where the fluid is stored and flows. This formation is characterized by certain natural deliverability and fast shut-in pressure buildup. It presents the derivative curve with an obvious radial flow segment on the log-log plot, thus reflecting the flow behaviors inside the formation and suggesting good reservoir conditions. Furthermore, it shows the best curve of stimulation performance, suggesting a high treatment result.

Using Well T621 as an example, the interpretation of logging data acquired from the 4851–4885 m interval (the first and second members of Lianglitage Formation, or Liang 1 and Liang 2 Members) reveals that the porosity is 2.66%, the lithology is biocalcarenite, acicular vugs and small vugs are developed, and the reservoirs are mainly Type II and III, and evaluated to be the vuggy type. During well testing, the flow pressure curve was similar to a horizontal stabilization line, suggesting the characteristics of the dry layer. The initial shut-in pressure built up fast, indicating good formation permeability. After fracturing, the well was produced with a 4.9 mm nozzle to contribute 103.87 m^3 oil per day and 47,320 m^3 gas per day, implying perfect stimulation results. The log-log curve shows the typical characteristics of a homogeneous formation, as shown in Fig. 3.1. The extensive caves with fewer fillings provide good flow pathways in the reservoirs, with permeability of 270 mD and high initial deliverability according to the well test interpretation. However, stable production depends on the elastic space of caves and the storage-seepage capacity of reservoir bodies communicating with the caves. During the period from December 8, 2004, to March 10, 2011, the cumulative oil production was 9.9×10^4 t and the cumulative gas production was $4400 \times 10^4 m^3$.

3.1.1.1.2 Dual-porosity model

The dual-porosity model is a key type of carbonate reservoir in the Tazhong I gas field. Although the porosity and permeability of the matrix are extremely low, a flow medium with characteristics similar to that of dual-porosity media is formed between high-conductivity fractures/vugs and microfractures in the effective storage-seepage bodies. This feature also indicates that the vugs and microfractures are relatively developed in the fractured-vuggy unit, which provides a certain accumulation capacity.

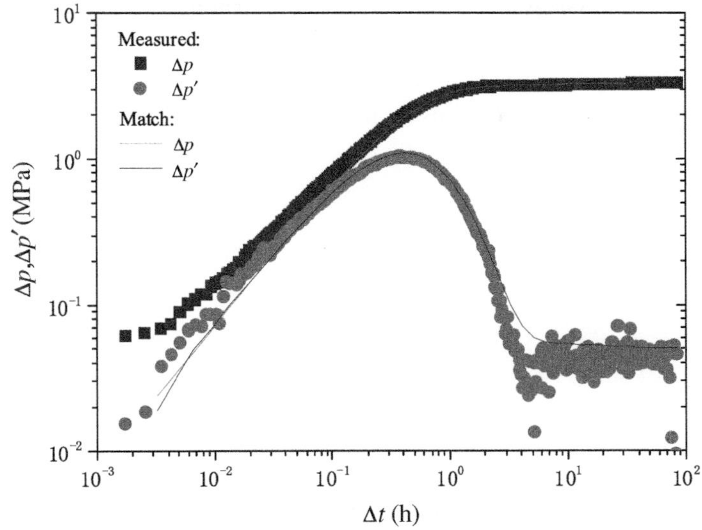

Fig. 3.1 Log-log matching plot for pseudo-homogeneous model of Well T621.

For Well T82, according to the logging data interpretation, in the 5430–5487 m interval, the porosity is 1.35%; the sedimentary subfacies is a lime-mud mound, while the microfacies are mound core and mound flat. In 5430–5450 m, the microfacies are mound flat, and the lithology is grayish white sand-clastic micrite-limestone with drusy vugs, which are generally fully-filled, and locally partly-filled. In 5450–5490 m, the microfacies are mound core, and the lithology is dark gray crypto algal micritic limestone, with a large number of drusy vugs filled with white calcites at the lower part. The dual-porosity model is used for well test interpretation (Fig. 3.2). The well was put into production testing on September 17, 2005, and shut in at the end of July 2006 due to high GOR. During this period, the cumulative oil production was 1.7×10^4 t and the cumulative gas production was $4016 \times 10^4 m^3$.

3.1.1.1.3 Fractured model

The fractured model is also a key type of carbonate reservoir in the Tazhong I gas field. For this type of well, the poor matrix porosity and permeability and insufficient energy supply lead to low deliverability and fast decline. For the Liang 2 Member in the 5595–5603 m interval in Well T828, the depositional setting is mainly composed of grain bank, reef mound subfacies, reef flank, and medium-high-energy sand-clastic bank microfacies. The lithology is bioclastic calcarenite limestone and stromatoporoid framestone, with

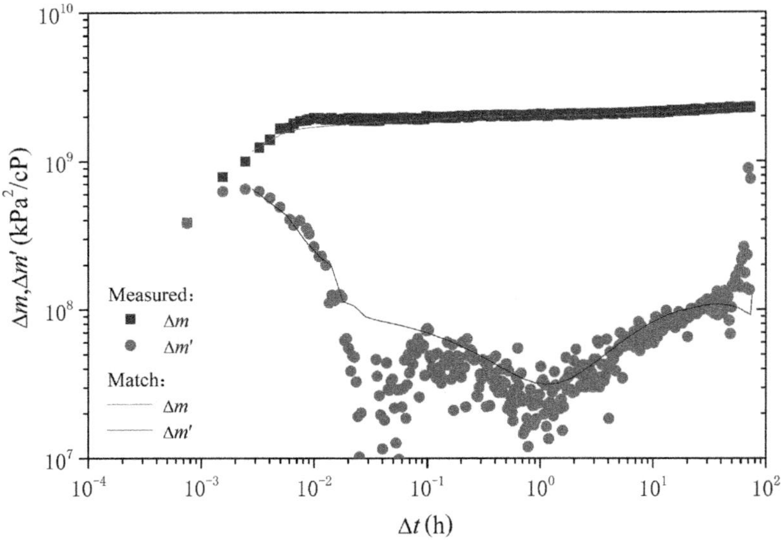

Fig. 3.2 Log-log matching plot of the dual-porosity model for 5430–5487 m interval in Well T82.

massive spines, caves, gray bioherm limestone, high-angle structural fractures, and cellular vugs. The average porosity is 1.8%, and the reservoirs are mainly Type I and Type II. During well testing, the flowing pressure curve was similar to a stabilization line. During the open flow, there was no production at the surface, which shows that the formation has no flowing capability. The shut-in pressure buildup was slow, revealing a low-permeability reservoir. The second shut-in derivative curve is abnormal in the late stage, which is affected by a phase change in the wellbore (Fig. 3.3), and only the data within 10 h are available for well test interpretation.

The well was tested for a short period from June 8 to August 10, 2008. Before the shut-in, the cumulative oil production was 220 t and the cumulative gas production was $39 \times 10^4 m^3$. The tubing pressure dropped from 7.5 MPa to 0. The result of an advanced production decline analysis also shows that the drainage radius is 70 m and the OGIP are only $70 \times 10^4 m^3$.

3.1.1.1.4 Fractured-vuggy model

The fractured-vuggy model is a major type of carbonate reservoir in the Tazhong I gas field. Although the matrix porosity and permeability are poor and the deliverability is low, this type of gas well has a certain period of stable production attributing to the energy supply.

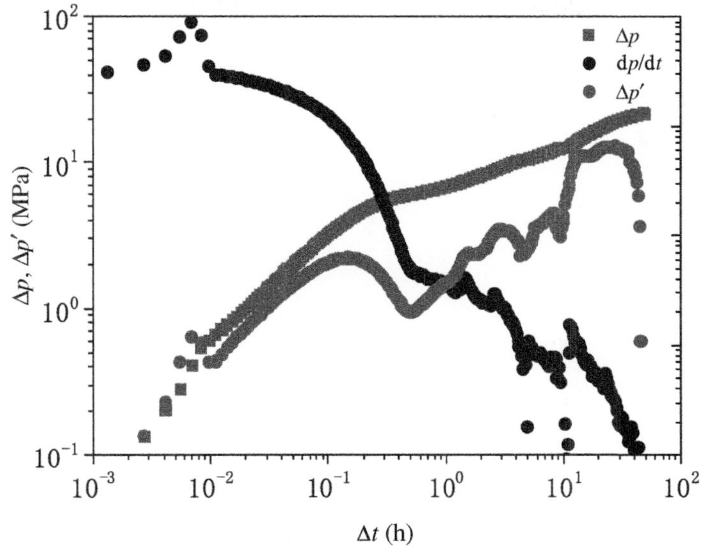

Fig. 3.3 Log-log matching plot for 5595–5603 m interval in Well T828.

For Well T62, in the Liang 2 Member in the 4704.5–4753.5 m interval, the depositional setting is mainly composed of the sand-clastic bank, and the stromatoporoid-sponge framework reef and the sedimentary microfacies are medium-high-energy intraclastic bank, reef top-reef flat, and reef flank-reef core. The lithology is mainly light gray crypto algae micrite limestone, micrite and sparite bioclastic intraclastic limestone, coral trellis limestone, with vugs and high-angle fractures filled with calcites. The porosity is 1.84%–6.2%, and Type I, II, and III reservoirs are all developed. The well was put into production testing on March 23, 2004, and into actual production after acid fracturing. The log-log curve shows obvious bead-like characteristics (Fig. 3.4). By January 5, 2009, when the well was shut in, the cumulative oil production was 1.4×10^4 t and the cumulative gas production was $3125 \times 10^4 m^3$, revealing the stable production in a stepped manner.

Fractures, pores, and vugs in carbonate reservoirs are sometimes distributed in separate zones locally, which are communicated by permeable but extremely narrow channels to form a bead-like configuration. As shown in Fig. 3.5, if a well is drilled in Zone A, it usually achieves very high deliverability. If a well is drilled in the smaller Zone B, the flow may be sustained for a short time; however, it is still possible to realize stable production if the zone communicates with a well-developed fracture zone. If a well is drilled in Zone C, acidizing results in a highly conductive and possibly extremely

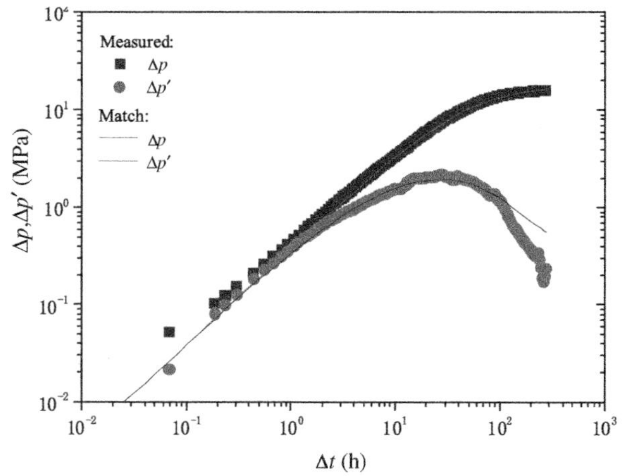

Fig. 3.4 Log-log matching plot for 4704.5–4753.5 m interval in Well T62.

Fig. 3.5 Structural diagram of bead-like fracture zones.

narrow zone nearby, and communicating with the wellbore; the communication channel can facilitate the gas drainage into the wellbore in spite of the poor permeability.

3.1.1.1.5 Composite model

The composite model is one of the main types of carbonate reservoirs in the Tazhong I gas field. This type of well is mostly characterized by poor peripheral conditions and a small inner control area, and it shows a rapid drop of flowing pressure and no stable production period in view of production.

For the Yingshan Formation reservoir in the 5666–5684 m interval in Well T83, the depositional setting is mainly composed of grain banks and

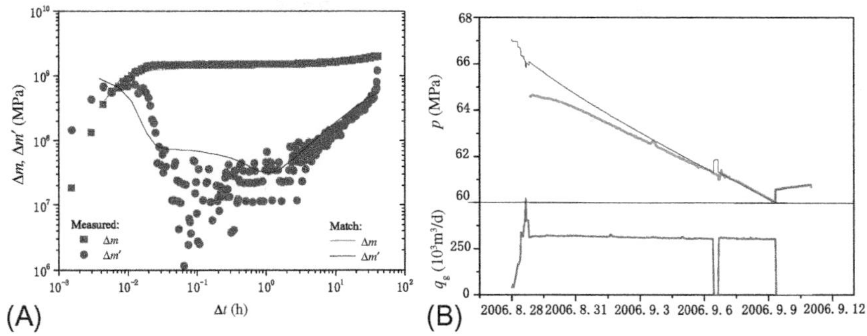

(A) (B)

Fig. 3.6 Well test interpretation matching plot for 5666–5684m interval in Well T83. (A) Log-log and (B) history match.

marl subfacies, and the sedimentary microfacies are mainly medium-energy sand-clastic bank and mound core. The lithology is dominantly crypto algal micritic limestone, in which argillaceous stripes are filled with black and green shale. The porosity is 0%–3.8% and the reservoirs are mainly Type III, intercalated with Type IV. The production test after acid fracturing shows that the production with a 7mm nozzle is essentially stable; but the flowing pressure drops rapidly by about 4.0MPa. The well test curve shows the characteristics of poor peripheral conditions of the composite model (Fig. 3.6), which demonstrates that the drainage area of a single well is very limited.

3.1.1.2 Well test characteristics of a fractured-vuggy carbonate gas reservoir

The well test characteristics of the Tazhong I gas field are diverse, mainly incorporating the pseudo-homogeneous model, dual-porosity model, fractured model, fractured-vuggy model, and composite model. Generally, the planar connectivity is poor and the permeability is low; however, there is the possibility of local connectivity.

The well test characteristics are closely related to the deliverability characteristics. In the case of pseudo-homogeneous and dual-porosity models, the production test reveals high and stable production with a slow decline. In the case of the fractured-vuggy model, the production is low, but stable. In the case of a fractured model and composite model with poor peripheral conditions, the production declines rapidly, and the pressure depletion is fast; moreover, there is almost no stable production period, as shown in Table 3.1.

Table 3.1 Correlation between well test characteristics and deliverability in the Tazhong I gas field

Reservoir type	Flow pathway	Physical model	Log-log curve	Production performance
Pseudo-homogeneous	Microfractures, high-permeability vugs	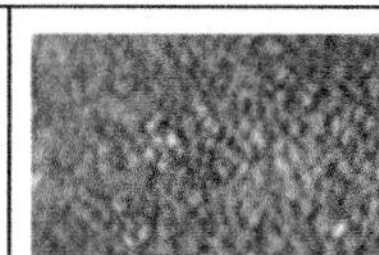	Fig. 3.1	Moderate decline
Dual-porosity	Fractures		Fig. 3.2	Stable deliverability
Fractured	Fractures, microfractures	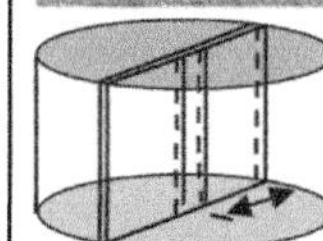	Fig. 3.3	Fast pressure depletion, almost no stable production period
Fractured-vuggy	Fractures and vugs	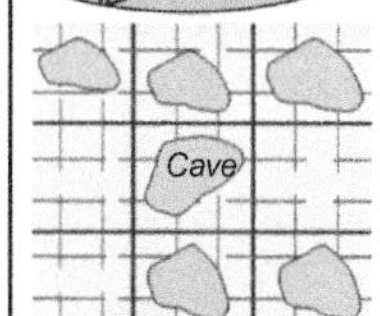	Fig. 3.4	Stable production, slow decline
Composite, with good inside conditions and poor outside conditions	High-permeability vugs, microfractures	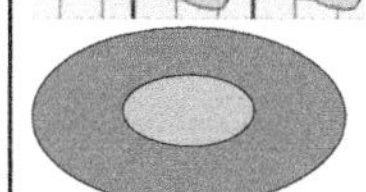	Fig. 3.6	Fast decline, almost no stable period

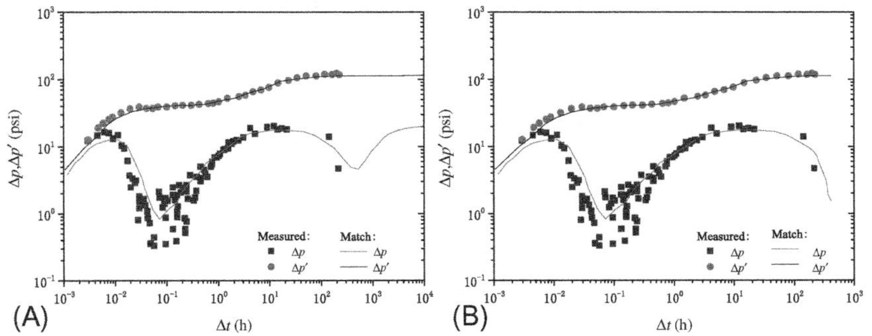

Fig. 3.7 Log-log plots of two interpretation models for Well T313 (Wu et al., 2007, Permission to publish by the SPE, Copyright SPE). (A) Tri-porosity model and (B) dual-porosity model.

3.1.2 Ambiguity of interpretation results

For Well T313 and Well TK609 in the Tahe oilfield, with TVD of over 5000 m, the drilling logs, core analysis, and geophysical characteristics all suggest that the wells encounter typical fractured-vuggy reservoirs. Interpretation of well test data for the two wells shows that both the tri-porosity model and dual-porosity model can be easily interpreted. For Well T313, the tri-porosity model is used for the interpretation, indicating that the pressure and pressure derivative are well matched; but the duration of pressure buildup is insufficient, resulting in the incomplete indication of the second derivative concave (the stage of flow from the matrix to the fractures). The dual-porosity model can also be used for this well. When the pressure derivative decreases in the late stage, a constant-pressure boundary model is applied, as shown in Fig. 3.7.

3.1.3 Complexity of the well test model

Even the simplest tri-porosity model has 8 parameters (Fig. 3.8). Some researchers have even proposed the tri-porosity model with more than 10 parameters.

To sum up, the well test of fractured-vuggy carbonate gas reservoirs is characterized by the diversity of curves, the ambiguity of results, and complexity of models. It is difficult to obtain reliable interpretation results only by relying on short-term pressure buildup data. A large number of well test interpretation experiences in gas wells reveal that the lifecycle well test analysis in a combination of short-term well test data with long-term production performance data is an effective method to reduce the ambiguity of well test interpretation results.

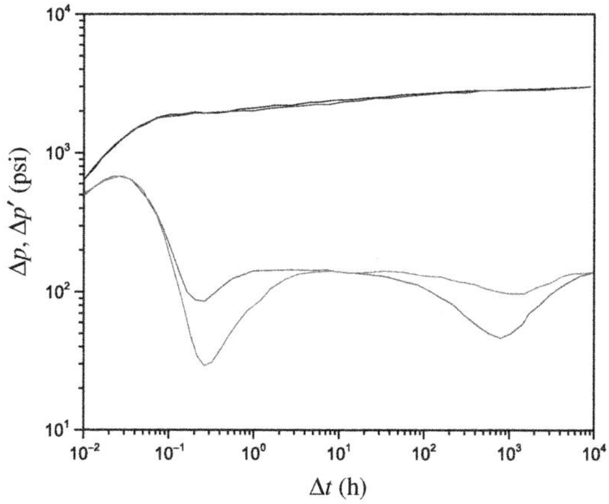

Fig. 3.8 Log-log plot of the tri-porosity model.

3.2 Well test analysis method

3.2.1 Progress of tri-porosity well test

As more and more fractured-vuggy carbonate reservoirs in the Tahe and Tazhong regions are developed, it is important to understand the well test behaviors for these reservoirs. Here, the progress of well test analysis for fractured-vuggy carbonate reservoirs is reviewed according to the available literature around the world.

3.2.1.1 Physical model

The media with two or more pore-structures are called multiporosity media. Generally, the naturally fractured oil and gas reservoirs are classified into dual-porosity media and tri-porosity media. For the dual-porosity flow model, it is assumed that there are two systems in the reservoir; namely the fractures and the matrix. The matrix is the main accumulation space, while the fractures serve as the main flow channels. In the past 50 years, the well test of dual-porosity reservoirs was discussed extensively, and now this technique is mature and perfect. A large number of carbonate rock development practices around the world demonstrate that, in addition to matrices and fractures, caves exist in the carbonate reservoirs, suggesting the existence of tri-porosity media. Here, the well test models and characteristics of the tri-porosity reservoirs will be presented.

Tan (2002) gave specific definitions to caves, fractures, and matrices based on the actual conditions of the Tahe oilfield. The cave system consists of huge caves with a diameter larger than 50 mm, and fractures communicating with these caves; and the super/macrofractures with a width larger than 5 mm, and vugs communicating with these fractures. The fracture system consists of small and medium fractures (visible to the naked eye) with a width of 0.1–5 mm, and vugs communicating with these fractures with a diameter of 2–50 mm. The matrix system consists of a carbonate pore complex split by smaller fractures (width > 0.1mm) with greatly variable storage-seepage capacity and complicated distribution, where the primary and secondary pores are mainly composed of microfractures and ultra-microfractures with a fracture width <0.1 mm, and the vugs (mostly dense intergranular dissolved pores) with a diameter <2 mm.

3.2.1.1.1 Fracture + dual-matrix model

The fracture system communicates with the wellbore, serving as the flow channel. There are two types of matrices (by the permeability): Matrix 1 is well connected with the fracture system, but Matrix 2 is less connected with the fracture system. The model is layered or massive, as shown in Fig. 3.9.

For the layered model, the matrix system is horizontally split by the fracture system, and transient crossflow occurs between the two systems. For the massive model, the matrix system is split by a set of orthogonal fracture systems, and transient crossflow occurs between the two systems.

3.2.1.1.2 Matrix + dual-fracture model

Depending on the nesting behavior, the model is found in three scenarios: (1) the pseudo-steady state crossflow from matrix and microfractures to macrofractures, which is similar to the layered model; (2) the pseudo-steady state crossflow from matrix to microfractures, and then from microfractures to macrofractures, with the drainage through the macrofractures to the wellbore; and (3) the pseudo-steady state crossflow from matrix to microfractures, and then from microfractures to macrofractures, with drainage through the microfractures and macrofractures to the wellbore, as shown in Fig. 3.10.

3.2.1.1.3 Triple-continuum model

The triple-continuum model is mainly composed of fractures, various caves (directly or indirectly connecting with fractures through small fractures or matrix), and a matrix (possibly containing some small vugs that are locally connected with fractures). Fractures are flow pathways, and caves and

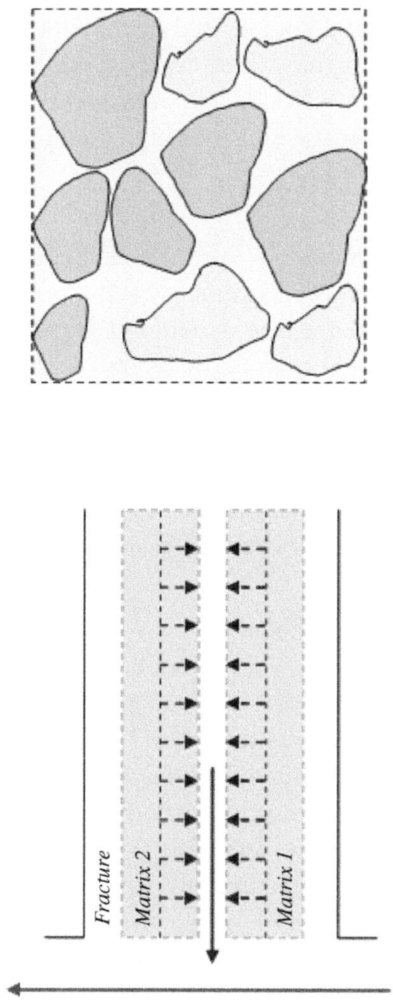

Fig. 3.9 Schematic diagram of fracture+dual-matrix model (Abdullah and Iraj, 1996, Permission to publish by the SPE, Copyright SPE).

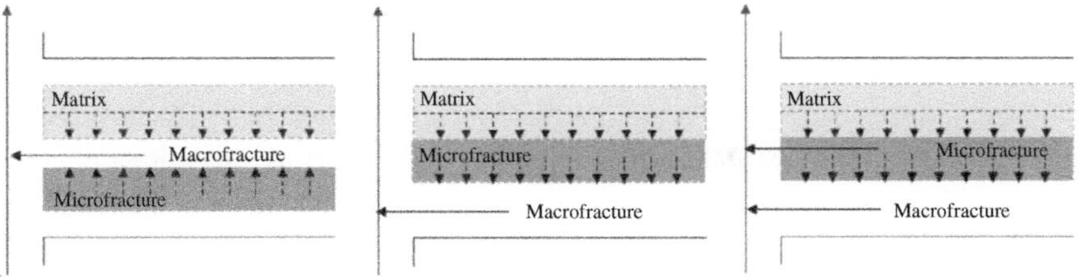

Fig. 3.10 Schematic diagram for dual-fracture+matrix model (Abdullah and Iraj, 1996, Permission to publish by the SPE, Copyright SPE).

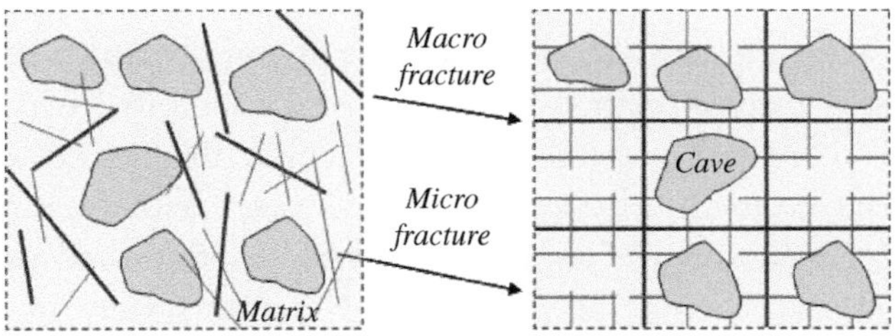

Fig. 3.11 Tricontinuum model (Wu et al., 2007, Permission to publish by the SPE, Copyright SPE).

matrices constitute the accumulation space. Fluid exchange can occur between matrices and fractures, matrices and caves, and fractures and caves (Fig. 3.11).

So far, the cognition on the fracture system, vug system, and matrix system have been gradually deepened, and the description of the relationship between them has been increasingly refined.

3.2.1.2 Well test model

Based on the dual-porosity model, Clossman (1975) first proposed the triporosity model. Later, Wu and Ge (1983) and Abdassah and Ershaghi (1986) performed in-depth studies on the model, and used it extensively, in many applications. Over the following decade, further research was conducted on tri-porosity media on the basis of their achievements. In the past 10 years, Abdullah and Iraj (1996), Camacho et al. (1987), Rodriguez (2004), Pulido (2006), Wu et al. (2007), Wang et al. (2007), Li et al. (2006b) and other scholars advanced the study of tri-porosity well testing to a new stage, and they successively proposed, developed, and improved the tri-porosity single-permeability and tri-porosity dual-permeability (nested) well test models for single-phase/multiphase fluid and different well types (vertical wells, partially perforated vertical wells, and horizontal wells). The crossflow between media is pseudo-steady state or transient state. The inner boundary conditions incorporate wellbore storage and skin effects, while the outer boundary conditions involve infinite, closed, constant pressure and other cases. Some models also take into account such factors as variable-wellbore storage, composite reservoirs, and stress sensitivity (Wang et al., 2007).

There are commonly three interpretation methods for tri-porosity well testing. First, the log-log analysis method is inherited. The Laplace transformation is used for numerical solutions, and the log-log curve matching is used for well test interpretation. In the matching process, in addition to the traditional LM least mean square algorithm, the evolutionary algorithm, and genetic algorithm are also applied, achieving ideal results. Second, the comprehensive analysis method for tri-porosity reservoirs based on Tiab's idea is applied in well test interpretation. Third, the boundary element numerical well test method has grown in recent years, with the advancement of computing technology. However, relatively sophisticated commercial well test software has not been developed. ECRIN only has a simple extended tri-porosity model.

3.2.1.3 Well test type curve

Fig. 3.12 shows the semi-log curves of vertical wells in fractured-vuggy reservoirs. It can be seen that the three parallel stabilization line segments are typical for the reservoirs, but they only appear within a certain range of parameters. The first segment corresponds to the radial flow in the fractures, the second one corresponds to the radial flow in the fractures and vugs, and the third one corresponds to the radial flow in the entire system. The first transition period is caused by the vugs (Matrix 1) with a larger crossflow factor and a smaller elastic storativity ratio, while the second transition period is caused by the matrix (Matrix 2) with a smaller crossflow factor and a larger elastic storativity ratio. Due to the influence of the wellbore storage effect, the first and second stabilization line segments may be hidden, and cannot be identified. Fig. 3.13 shows the semi-log curves of horizontal wells in fractured-vuggy reservoirs. It can be seen that there are five flow stages: (1) wellbore storage flow stage; (2) the first radial flow stage, reflecting the early radial flow perpendicular to the horizontal wellbore; (3) pseudo-steady state crossflow in vugs; (4) pseudo-steady state crossflow in the matrix; and (5) total radial flow in the system. Limited by the parameters, the radial flow stage between the two transition stages is not obvious.

Since the first tri-porosity model was built by Clossman in 1975, the studies of tri-porosity media have evolved in two stages. In the 1980s–1990s, the tri-porosity single-permeability model was mainly used for pressure analysis.

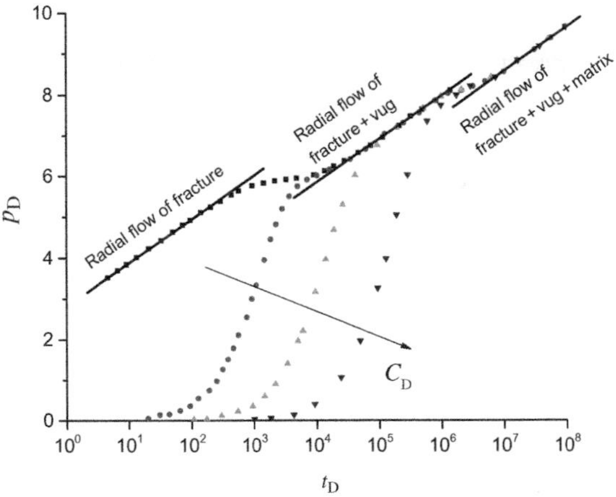

Fig. 3.12 Well test type curves of vertical wells in fractured-vuggy reservoirs and the influence of the well storage effect (Wu et al., 2007, Permission to publish by the SPE, Copyright SPE).

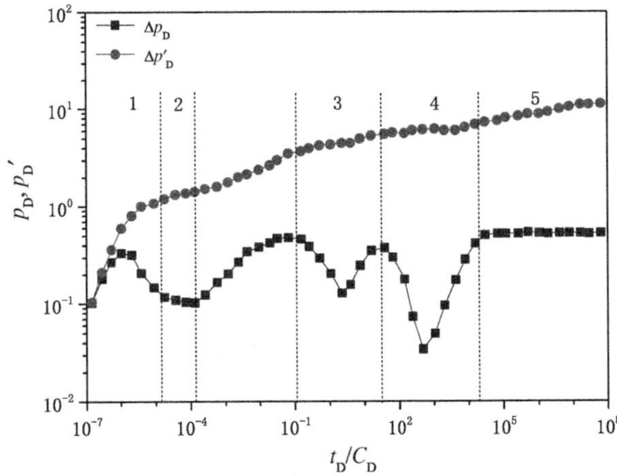

Fig. 3.13 Well test type curves of horizontal wells (Li et al., 2006b).

Since the advent of the 21st century, as knowledge of the fractured-vuggy reservoirs has increased, the tri-porosity dual-permeability model has triggered advances in the study of tri-porosity media. Although various tri-porosity models have been built, most of them are just theoretical, and relatively sophisticated commercial well test software has not yet appeared. The well test interpretation of tri-porosity media is much more complicated than dual-porosity media, with greater ambiguity. In some cases, it is difficult to distinguish between them, which is also one of the immediate concerns.

3.2.2 The multiwell well test analysis method

As discussed in Section 3.2, although the fractured-vuggy carbonate reservoirs are highly heterogeneous with less interwell connectivity, there is still the possibility of connectivity. The gas reservoir with high permeability and good interwell connectivity is essentially a multiwell system. After it is put into production, the pressure buildup data for testing wells may be affected by offset wells. The pressure derivative curve may show obvious "downwarping" or "upwarping" in the middle and late stages (Deng et al., 2015). The single-well test analysis method often interprets this phenomenon as being influenced by boundaries. The improper interpretation results may mislead production decision-making (Sun, 2016).

In order to correctly understand the well test characteristics of reservoirs with multiwell connectivity, and correctly identify the suspected boundary characteristics, the well test type curves are plotted for two scenarios: (1) offset wells producing simultaneously; and (2) offset wells shut in simultaneously. Accordingly, the multiwell buildup test analysis method is established. The theoretical analysis of a long-term asymptotic solution shows the results in two aspects. First, in both of the preceding scenarios, the pressure buildup derivative curves present a "stepped rise," with multiple stabilization lines of radial flow; and the ratio of each stabilization line-height to the first stabilization line height is the algebraic sum of the dimensionless production rate of the testing well and the affecting offset wells. Second, when the offset wells are producing continuously and interfere with the testing well, the pressure buildup derivative curve of the testing well presents "downwarping" in the middle and late stages.

3.2.2.1 The multiwell system in infinite homogeneous reservoirs

3.2.2.1.1 The well test model

It is assumed that N wells are producing at a constant rate in an infinite homogeneous reservoir, the influences of gravity and capillary force are neglected, and the influences of the skin effect and wellbore storage effect are considered for the testing well, but not considered for the offset wells. The definite solution for the testing well (which is represented by subscript 1) based on the effective wellbore radius model can be described as follows:

$$\frac{1}{r_D}\frac{\partial}{\partial r_D}\left(r_D\frac{\partial p_{1D}}{\partial r_D}\right)=\frac{1}{C_D e^{2S}}\frac{\partial p_{1D}}{\partial(t_D/C_D)} \tag{3.1}$$

$$p_{1D}(r_D,0)=0 \tag{3.2}$$

$$\frac{dp_{wD}}{d(t_D/C_D)} - \left(r_D\frac{\partial p_{1D}}{\partial r_D}\right)_{r_D=1} = 1 \tag{3.3}$$

$$p_{wD} = p_{1D}(1, t_D/C_D) \tag{3.4}$$

$$p_{1D}(r_D \to \infty, 0) = 0 \tag{3.5}$$

The diffusion equation, initial condition, and outer boundary condition for the definite solution of the remaining $N-1$ offset wells are the same as those of the testing well; but the inner boundary condition should be expressed as follows:

$$-\left(r_D\frac{\partial p_{jD}}{\partial r_D}\right)_{r_D=1} = q_{jD} \quad j=2,3\ldots,N \tag{3.6}$$

3.2.2.1.2 Model solving

In the Laplace space, the exact solution for bottom hole pressure of the testing well is as follows:

$$\bar{p}_{wD}(z) = \frac{1}{z}\left[\frac{K_0(\sigma) + \sum_{j=2}^{N} q_{jD}K_0(\sigma r_{jD})}{zK_0(\sigma) + \sigma K_1(\sigma)}\right], \quad \sigma = \sqrt{\frac{z}{C_D e^{2S}}} \tag{3.7}$$

In the Laplace space, the pressure response of an offset well at the testing well is

$$\bar{p}_{jD}(z, r_{jD}) = \frac{q_{jD}K_0(\sigma r_{jD})}{z\sigma K_1(\sigma)}, \quad \sigma = \sqrt{\frac{z}{C_D e^{2S}}}, \quad j=2,3\ldots,N \tag{3.8}$$

The real space solution $p_{wD}(t_D/C_D)$ for bottom hole pressure can be obtained by the Stehfest (1970) numerical inversion. If the production of the offset wells is 0, Eq. (3.7) is the pressure solution for an effective wellbore radius with the constant rate of a well in an infinite homogeneous reservoir. If it is a gas reservoir, the dimensionless pressure should be in the form of normalized pseudo-pressure, and the dimensionless time should be in the form of normalized pseudo-time. The form of the solution is the same as Eqs. (3.7), (3.8).

3.2.2.1.3 Pressure drawdown (PDD) type curves in the case of multiwell simultaneous production

From Eq. (3.7), it can be seen that the influence of the offset wells on the bottom hole pressure of the testing well depends mainly on the rate of the offset wells and the distance between the offset well and the testing well. Fig. 3.14 shows the PDD type curves of four wells in an infinite homogeneous reservoir when they are producing simultaneously. In case (a), the dimensionless rate of three offset wells is 1.0, 3.0, and 5.0, and the corresponding dimensionless well spacing is 10^3, 10^4, and 10^5. In case (b), the dimensionless rate of the offset wells is still 1.0, 3.0, and 5.0, but the dimensionless well spacing is 10^5, 10^4, and 10^3. In both cases, the pressure derivative curves show four horizontal radial flow lines; the closer the offset well is to the testing well, the earlier the influence on the curves will be.

(1) Different rate and different well spacing for offset wells

In case (a), the part before the first radial flow segment (line 0.5) is a manifestation of the characteristics of the testing well itself; the second radial flow segment (line 1.0) is a manifestation of the production characteristics of the testing well and the nearest offset well, similar to the characteristics of the testing well that is located near a sealed fault; the third radial flow segment (line 2.5) is a manifestation of the production characteristics of the testing well, and the two nearest offset wells; the fourth radial flow segment (line 5.0)

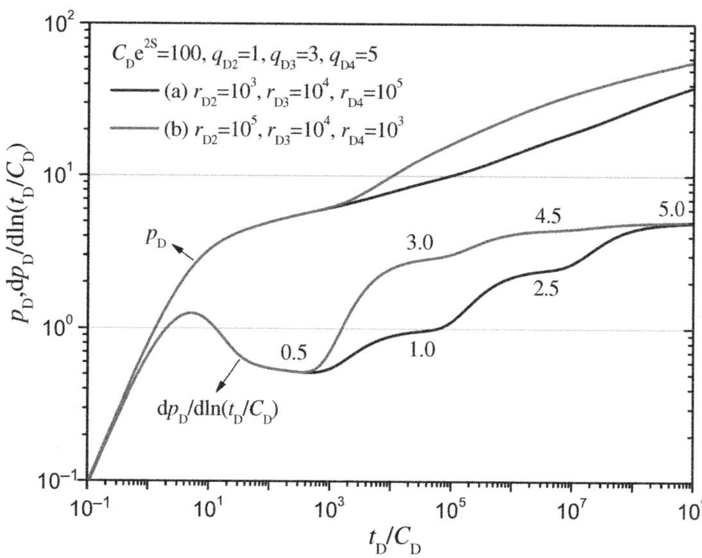

Fig. 3.14 PDD type curves in the case of multiwell simultaneous production with different rates and different well spacing for the offset well (four wells).

is a manifestation of the production characteristics of the testing well and three offset wells. The ratio of the height of the second, third, and fourth radial flow stabilization lines to the height of the first radial flow stabilization line is the algebraic sum of the dimensionless rate of the testing well and the offset well that produces an influence, that is, $\left(1+\sum_{j=2}^{X} q_{j\mathrm{D}}\right)$, in which X is the number of offset wells that have an influence on the testing wells. For example, in case (a), the ratio of the height of the fourth radial flow stabilization line to the height of the first radial flow stabilization line is $5.0/0.5=10.0$, which is equal to the algebraic sum of the dimensionless rate of 4 wells, that is, $\left(1+\sum_{j=2}^{4} q_{j\mathrm{D}}\right)=1.0+1.0+3.0+5.0=10.0$. The same is true in case (b), where $\left(1+\sum_{j=2}^{4} q_{j\mathrm{D}}\right)=1.0+5.0+3.0+1.0=10.0$.

(2) Identical rates and different well spacing for offset wells

If the offset wells have the same rate as the testing well, but different well spacing from the testing well, there are four horizontal radial flow lines (Fig. 3.15). The characteristics of the pressure derivative curves are the same as those shown in Fig. 3.14. The first radial flow segment is a manifestation of the characteristics of the testing well itself (line 0.5); the ratio of the height of the second, third and fourth radial flow stabilization lines to the height of the first radial flow stabilization line is the algebraic sum of the dimensionless rate of the testing well and the affecting offset wells.

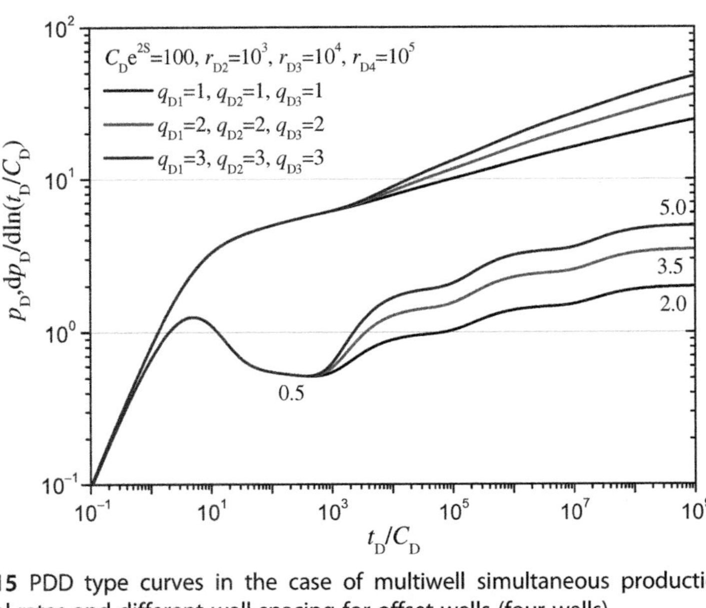

Fig. 3.15 PDD type curves in the case of multiwell simultaneous production with identical rates and different well spacing for offset wells (four wells).

(3) Different rates and similar well spacing for offset wells

If the distance between the offset well and the testing well is roughly the same, then only two radial flow stabilization lines will appear (Fig. 3.16). The first radial flow stabilization line is a manifestation of the characteristics of the testing well itself (line 0.5); and the second radial flow stabilization line is a manifestation of the production characteristics of the whole system (line 2.0, 4.0, and 5.0).

(4) An asymptotic solution in the case of multiwell simultaneous production.

When the variable σ is small enough, the Bessel function is expressed as

$$K_0(\sigma) = -\left(\ln\frac{\sigma}{2} + \gamma \right) \tag{3.9}$$

$$K_1(\sigma) = \frac{1}{\sigma} \tag{3.10}$$

Substituting Eqs. (3.9), (3.10) into Eq. (3.7) to obtain the inverse Laplace transform, we have

$$p_{\text{wD}} = \frac{1}{2}\left(1 + \sum_{j=2}^{N} q_{j\text{D}}\right)\left[\ln\left(\frac{t_\text{D}}{C_\text{D}}\right) + 0.80908 + \ln C_\text{D}\text{e}^{2S}\right] - \sum_{j=2}^{N} q_{j\text{D}} \ln r_{j\text{D}} \tag{3.11}$$

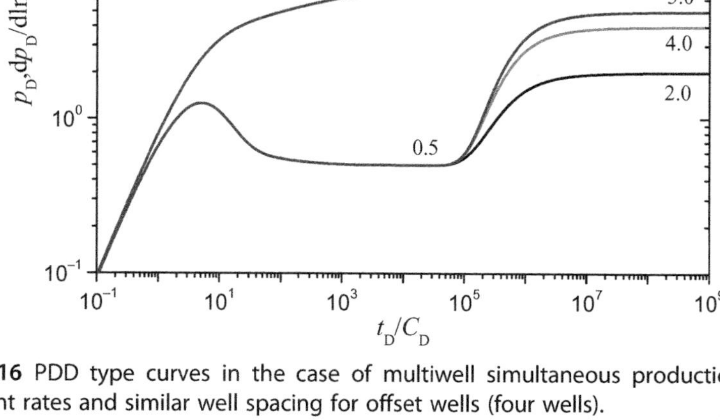

Fig. 3.16 PDD type curves in the case of multiwell simultaneous production with different rates and similar well spacing for offset wells (four wells).

Using Eq. (3.11) to take a derivative with respect to logarithmic time, we have

$$\frac{dp_{\mathrm{wD}}}{d\ \ln\left(\frac{t_{\mathrm{D}}}{C_{\mathrm{D}}}\right)}=\frac{1}{2}\left(1+\sum_{j=2}^{N}q_{j\mathrm{D}}\right) \tag{3.12}$$

Obviously, in the late stage, the ratio of the radial flow stabilization line height of the pressure drawdown derivative curve in the case of multiwell simultaneous production to the single-well radial flow stabilization line height is the algebraic sum of the dimensionless rate of the testing well and the connected well, that is, $\left(1+\sum_{j=2}^{N}q_{j\mathrm{D}}\right)$. Therefore, when the pressure derivative curve presents a "stepped rise," it may be a manifestation of characteristics of the radial composite model with an impermeable boundary or poor peripheral conditions, or it may be caused by interwell interference.

3.2.2.1.4 Pressure buildup (PBU) type curves in the case of multiwell simultaneous shut-ins

(1) Characteristics of PBU type curves

When the testing well and the offset wells are shut in simultaneously, the PBU type curve for the multiwell system is shown in Fig. 3.17. When$t_{\mathrm{pD}}\gg\Delta t_{\mathrm{D}}$, the PDD type curve and the PBU type curve are essentially overlapped. When the semi-log approximation condition is satisfied, the dimensionless shut-in buildup pressure $p_{\mathrm{BUD}}\left(\frac{\Delta t_{\mathrm{D}}}{C_{\mathrm{D}}}\right)$ takes the derivative with respect to $\left(\frac{\Delta t_{\mathrm{D}}}{C_{\mathrm{D}}}\right)$

$$\frac{dp_{\mathrm{BUD}}\left(\frac{\Delta t_{\mathrm{D}}}{C_{\mathrm{D}}}\right)}{d\left(\frac{\Delta t_{\mathrm{D}}}{C_{\mathrm{D}}}\right)}\left(\frac{\Delta t_{\mathrm{D}}}{C_{\mathrm{D}}}\right)\left(\frac{t_{\mathrm{pD}}+\Delta t_{\mathrm{D}}}{t_{\mathrm{pD}}}\right)=\frac{1}{2}\left(1+\sum_{j=2}^{N}q_{j\mathrm{D}}\right) \tag{3.13}$$

Hence, when the offset wells are shut in simultaneously and interfere with the testing well, the PBU derivative curve of the testing well will gradually "upwarp" in the middle and late stages.

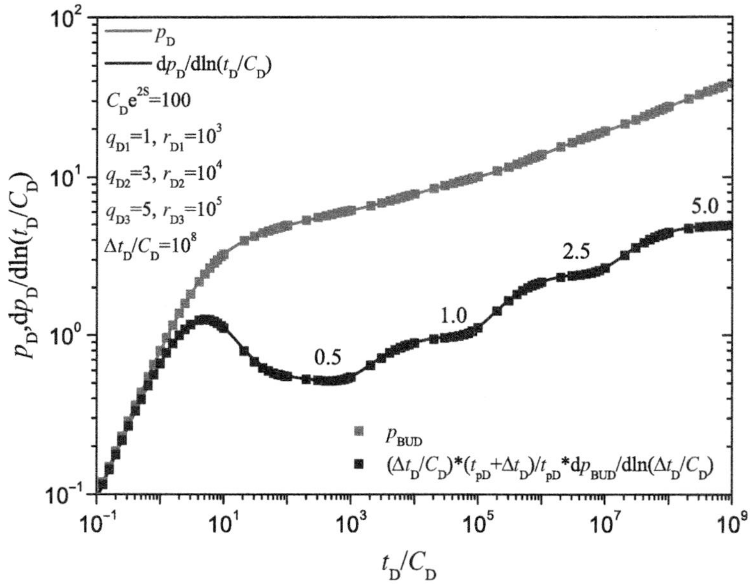

Fig. 3.17 Comparison of PDD and PBU type curves for the multiwell system in an infinite homogeneous reservoir (4 wells, PDD—multiwell simultaneous production; PBU—multiwell simultaneous shut-ins).

(2) Procedure of PBU test analysis

The procedure of PBU test analysis for the multiwell system, essentially the same as that for the single-well system, is described as follows:

(a) Calculate the pressure difference $\Delta p_{ws}(\Delta t)=p_{ws}(\Delta t)-p_{wf}(t_p)$ at each pressure measuring point and its derivative, $\frac{d\Delta p_{ws}}{d\Delta t}\times\Delta t\times\frac{t_p+\Delta t}{t_p}$. In the case of the gas well, the pseudo-pressure should be used for analysis.

(b) Match by using the PDD type curve. According to the x-coordinate matching value, y-ordinate matching value and curve matching value, such parameters as the formation capacity Kh, skin factor S and wellbore storage constant C are calculated, respectively.

(c) Determine the ratio of the height of each radial flow stabilization line to the height of the first radial flow stabilization line, namely $\left(1+\sum_{j=2}^{N}q_{jD}\right)$, and then estimate the sequence and number of interferences from the offset wells.

(d) Where necessary, conduct the history matching analysis of the numerical well test based on the log-log analysis results, to further analyze the well test results.

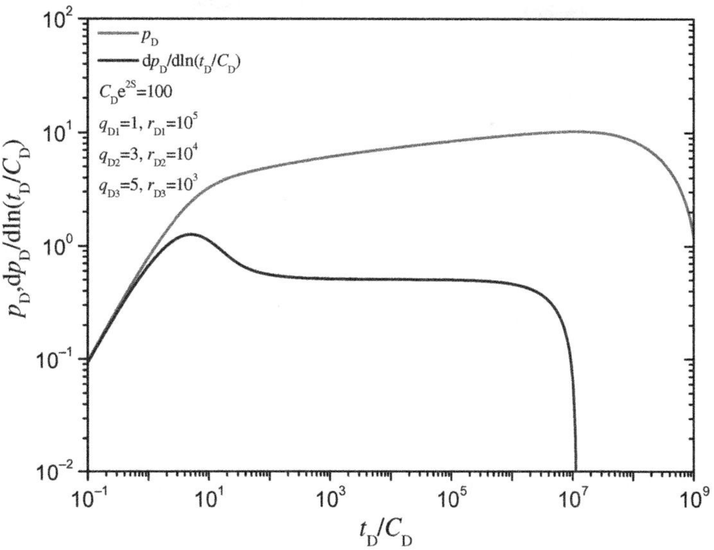

Fig. 3.18 Log-log plot in the case of offset wells in production.

3.2.2.1.5 PBU type curves in the case of offset wells in production

When the testing well is shut in and the offset wells are producing, the derivative of the asymptotic solution for the pressure buildup of the testing well is

$$\frac{dp_{\mathrm{BUD}}\left(\frac{\Delta t_{\mathrm{D}}}{C_{\mathrm{D}}}\right)}{d\left(\frac{\Delta t_{\mathrm{D}}}{C_{\mathrm{D}}}\right)}\left(\frac{\Delta t_{\mathrm{D}}}{C_{\mathrm{D}}}\right)\left(\frac{t_{pD}+\Delta t_{\mathrm{D}}}{t_{pD}}\right)=\frac{1}{2}\left[1-\sum_{j=2}^{N}q_{j\mathrm{D}}\times\left(\frac{\Delta t_{\mathrm{D}}}{t_{pD}}\right)\right] \tag{3.14}$$

Therefore, when the offset wells are producing and interfere with the testing well, the PBU derivative curve of the testing well will gradually "downwarp" in the middle and late stages, and the "downwarping" speed depends on the dimensionless rate $\sum_{j=2}^{X}q_{j\mathrm{D}}$ of the affecting offset wells and the ratio $\frac{\Delta t_{\mathrm{D}}}{t_{\mathrm{pD}}}$ of shut-in time to production time, as shown in Fig. 3.18.

3.2.2.2 Multiwell system in infinite dual-porosity reservoirs

3.2.2.2.1 Long-term asymptotic solutions in the case of multiwell simultaneous production

(1) Long-term asymptotic solutions for pressure drawdown (PDD) in the case of multiwell simultaneous production

The long-term asymptotic solution for PDD in the case of multiwell simultaneous production is

$$p_{\mathrm{wfD}} = \frac{1}{2}\left(1 + \sum_{j=2}^{N} q_{j\mathrm{D}}\right)\left[\ln\left(\frac{t_\mathrm{D}}{C_\mathrm{D}}\right) + Ei\left(-a\frac{t_\mathrm{D}}{C_\mathrm{D}}\right) - Ei\left(-b\frac{t_\mathrm{D}}{C_\mathrm{D}}\right)\right.$$

$$\left. + \ln C_\mathrm{D}e^{2S} + 0.80908 - 2\ln r_{j\mathrm{D}}\right] \tag{3.15}$$

The pressure derivative is

$$\frac{dp_{\mathrm{wfD}}}{d\ln\left(\frac{t_\mathrm{D}}{C_\mathrm{D}}\right)} = \frac{1}{2}\left(1 + \sum_{j=2}^{N} q_{j\mathrm{D}}\right)\left[1 - \exp\left(-a\frac{t_\mathrm{D}}{C_\mathrm{D}}\right) + \exp\left(-b\frac{t_\mathrm{D}}{C_\mathrm{D}}\right)\right] \tag{3.16}$$

(2) An asymptotic solution for pressure buildup (PBU) in case of offset wells simultaneous shut-in or production

Similar to the homogeneous reservoir, the PBU when both the testing well and the offset wells are shut in can be expressed as

$$p_{\mathrm{BUD}}\left(\frac{\Delta t_\mathrm{D}}{C_\mathrm{D}}\right) = p_\mathrm{D}\left(\frac{t_{\mathrm{pD}}}{C_\mathrm{D}}\right) - p_\mathrm{D}\left(\frac{t_{\mathrm{pD}} + \Delta t_\mathrm{D}}{C_\mathrm{D}}\right) + p_\mathrm{D}\left(\frac{\Delta t_\mathrm{D}}{C_\mathrm{D}}\right) \tag{3.17}$$

The pressure derivative is

$$\frac{\mathrm{d}p_{\mathrm{BUD}}\left(\frac{\Delta t_\mathrm{D}}{C_\mathrm{D}}\right)}{\mathrm{d}\left(\frac{\Delta t_\mathrm{D}}{C_\mathrm{D}}\right)}\left(\frac{\Delta t_\mathrm{D}}{C_\mathrm{D}}\right)\left(\frac{t_{\mathrm{pD}} + \Delta t_\mathrm{D}}{t_{\mathrm{pD}}}\right) = \frac{1}{2}\left(1 + \sum_{j=2}^{N} q_{j\mathrm{D}}\right)$$

$$\left[1 - \exp\left(-a\frac{t_\mathrm{D}}{C_\mathrm{D}}\right) + \exp\left(-b\frac{t_\mathrm{D}}{C_\mathrm{D}}\right)\right] \tag{3.18}$$

The pressure derivative when the testing well is shut in and the offset wells are producing is

$$\frac{1}{2}\left(1 - \sum_{j=2}^{N} q_{j\mathrm{D}}\frac{\Delta t_\mathrm{D}}{t_{\mathrm{pD}}}\right) + \frac{1}{2}\left(1 + \sum_{j=2}^{N} q_{j\mathrm{D}}\right)\left(\frac{\Delta t_\mathrm{D}}{t_{\mathrm{pD}}}\right)\left[\exp\left(-a\frac{\Delta t_\mathrm{D} + t_{\mathrm{pD}}}{C_\mathrm{D}}\right)\right.$$

$$\left. - \exp\left(-b\frac{\Delta t_\mathrm{D} + t_{\mathrm{pD}}}{C_\mathrm{D}}\right)\right] + \frac{1}{2}\left(\frac{\Delta t_\mathrm{D} + t_{\mathrm{pD}}}{t_{\mathrm{pD}}}\right)\left[\exp\left(-b\frac{\Delta t_\mathrm{D}}{C_\mathrm{D}}\right) - \exp\left(-a\frac{\Delta t_\mathrm{D}}{C_\mathrm{D}}\right)\right] \tag{3.19}$$

3.2.2.2.2 PDD type curves in the case of multiwell simultaneous production

Similar to an infinite homogeneous reservoir, in the case of simultaneous production of the multiwell system in a dual-porosity reservoir, the ratio of the radial flow stabilization line height of the PDD derivative curve in the case of multiwell simultaneous production to the single-well radial flow stabilization line height is also the algebraic sum of the dimensionless rate of the testing well and the connected wells, as shown in Fig. 3.19. The PBU case is similar to that of the homogeneous reservoir, so it will not be described here.

3.2.2.3 Simulation example

3.2.2.3.1 Basic data

Under the assumption that two gas wells in a gas reservoir produce for a long time simultaneously, the permanent downhole pressure gauge is used for performance monitoring. The basic parameters of the gas reservoir are shown in Table 3.2. The first PBU test was conducted after the testing well produced at an average rate of $40 \times 10^4\,m^3/d$ for 10,000 h, with the shut-in time of 1000 h. Subsequently, the second PBU test was conducted after the testing well produced at an average rate of $40 \times 10^4\,m^3/d$ for 5000 h, with the shut-in time of 2000 h. After the offset well produced at a rate of $120 \times 10^4\,m^3/d$ for consecutive 16,000 h, it was shut in at the same time when the second PBU test was conducted in the testing well.

3.2.2.3.2 Well test analysis

The log-log plots of the two PBU tests are shown in Fig. 3.20, illustrating the completely inverse shape of the derivative curves at the late stage. The derivative curve for the first PBU test goes down in the late stage, which may be due to: (1) the improving periphery of the reservoir; and (2) interference of the offset wells.

The derivative curve for the second PBU test "rises," which may be caused by: (1) the deteriorating periphery of the reservoir; (2) the characteristics of edge-water (because the water viscosity is much higher than the gas viscosity, the gas reservoir of the edge-water drive may show characteristics similar to the deteriorating periphery of the reservoir); (3) one or more impermeable boundaries; and (4) interference from simultaneous shut-in of offset wells.

When the composite model with deteriorating periphery is used for interpretation, the log-log plot for the second PBU test (Fig. 3.21) is good,

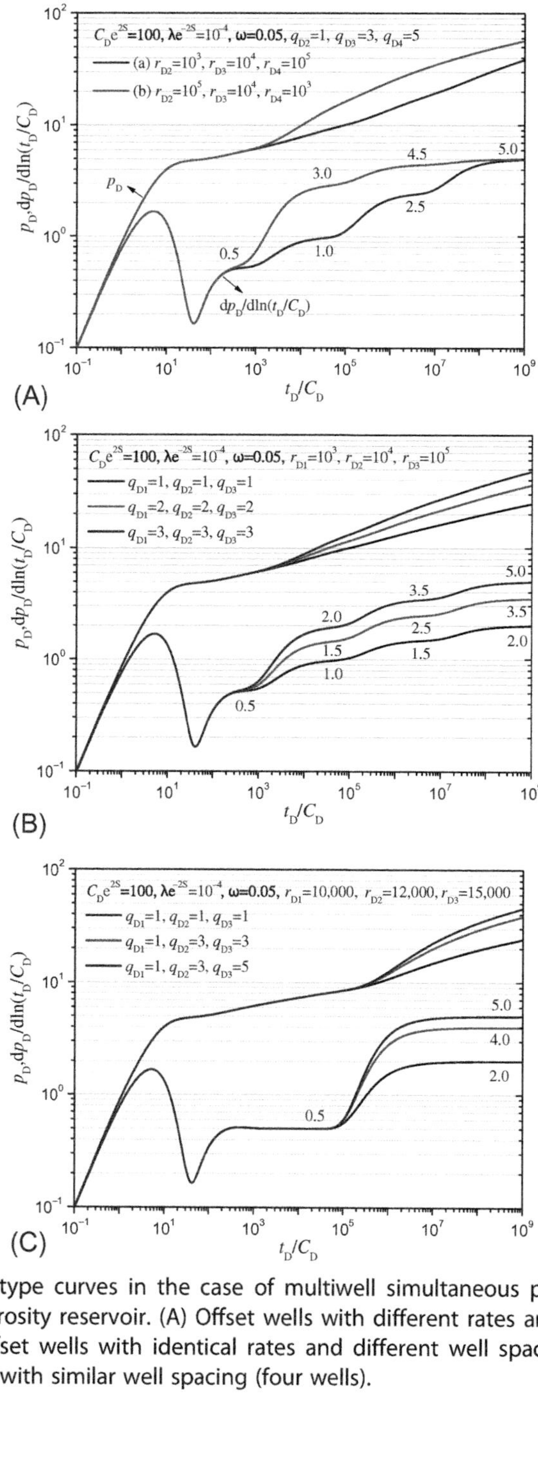

Fig. 3.19 PDD type curves in the case of multiwell simultaneous production in an infinite dual-porosity reservoir. (A) Offset wells with different rates and different well spacing. (B) Offset wells with identical rates and different well spacing (four wells). (C) Offset wells with similar well spacing (four wells).

Table 3.2 Basic parameters

Well spacing (m)	Initial pressure (MPa)	Formation temperature (°C)	Net pay thickness (m)	Porosity	Relative density of gas	Pseudo-critical temperature (K)	Pseudo-critical pressure (MPa)
3880	35.0	100	100	0.1	0.6	195.697	4.66875

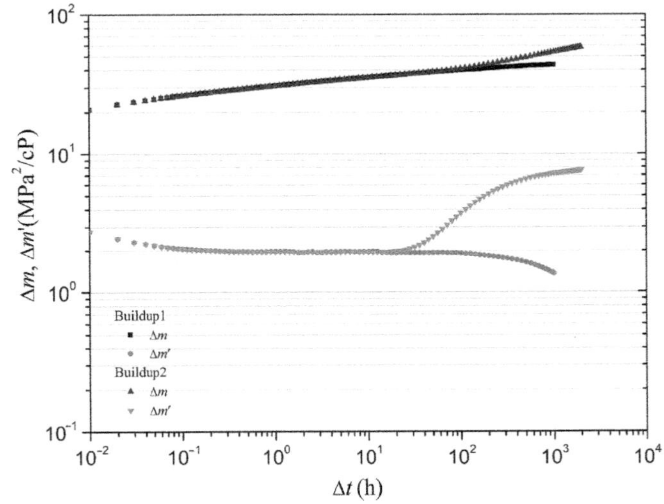

Fig. 3.20 Log-log plots for two PBU tests.

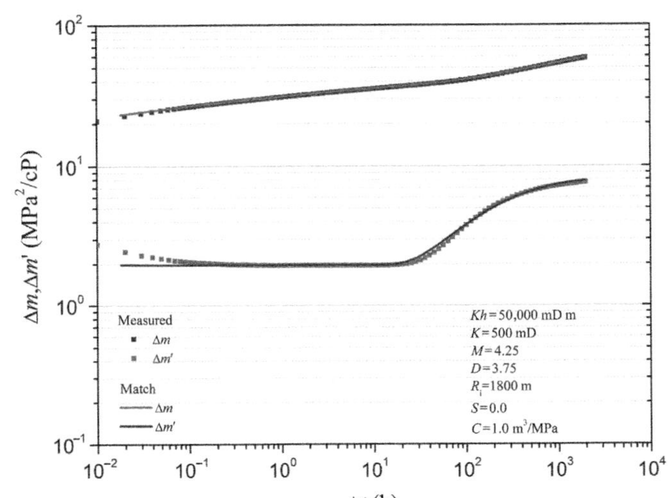

Fig. 3.21 The log-log plot for the second PBU test (composite model).

but the history matching plot (Fig. 3.23) is not ideal; furthermore, the results of the two PBU tests run in opposite directions. Therefore, the composite model is not recommended for interpretation.

The gas reservoir distributes well in the plane, without large faults or fractures, and the testing well is far away from the boundary of the gas reservoir. The "downwarping" of the pressure derivative curve in the late stage of the first PBU test might be caused by the production of offset wells, while the "upwarping" of the pressure derivative curve in the late stage of the second PBU test might result from the simultaneous shut-in of offset wells. Therefore, excluding other possibilities, the multiwell model is preferred for analysis, and the log-log matching results are shown in Fig. 3.22.

K, *S*, and *C* interpretation results are the same as those of the composite model. The ratio between the heights of the two horizontal segments of the pressure derivative curve is 8.0/2.0 = 4.0, which is exactly the ratio of the sum of the production rate of the testing well and the offset wells to the production rate of the testing well, that is, (40 + 120)/40 = 4.0. The derivative matching plot is also obviously improved compared with that in Fig. 3.21. The history matching plot is good, as shown in Fig. 3.23.

The preceding findings show that the interpretation result of the multiwell model is reasonable. Specifically, the reservoir reflects the characteristics of an infinite homogeneous reservoir, with the permeability of 500 mD, good interwell connectivity, and production interference.

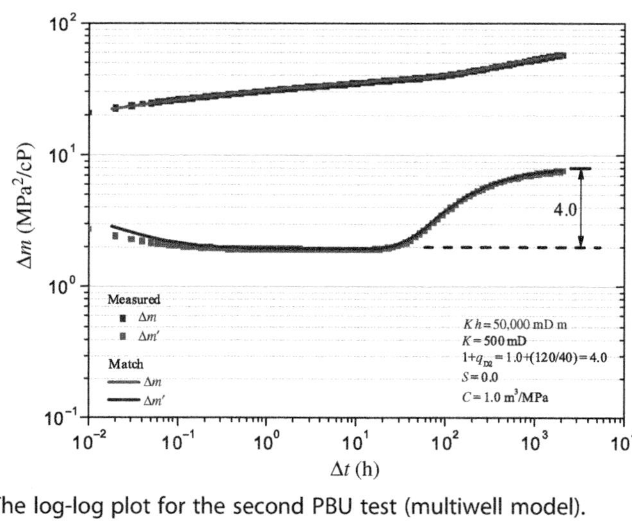

Fig. 3.22 The log-log plot for the second PBU test (multiwell model).

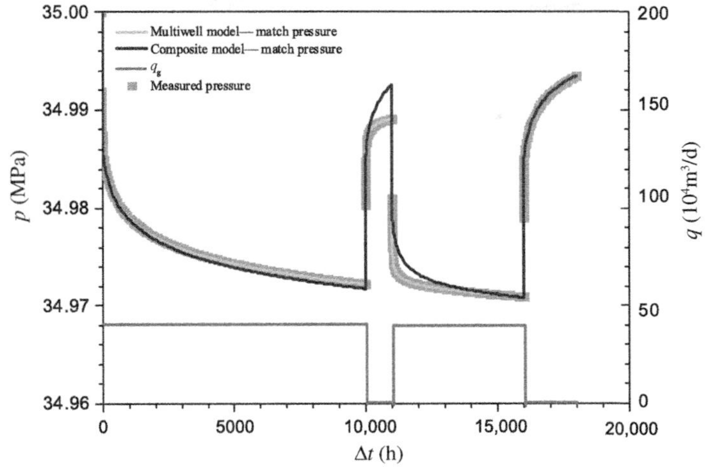

Fig. 3.23 The history matching plot of pressure (two models).

3.2.2.4 A field example

3.2.2.4.1 A gas reservoir survey

For Reservoir A, the porosity is 5.5%–8.0%, the matrix permeability is 0.01–1.0 mD, and high-angle fractures are developed. The initial formation pressure is 89.09 MPa, the formation temperature is 128°C, and the relative density of natural gas is 0.58. So far, six development wells have been drilled, as shown in Fig. 3.24.

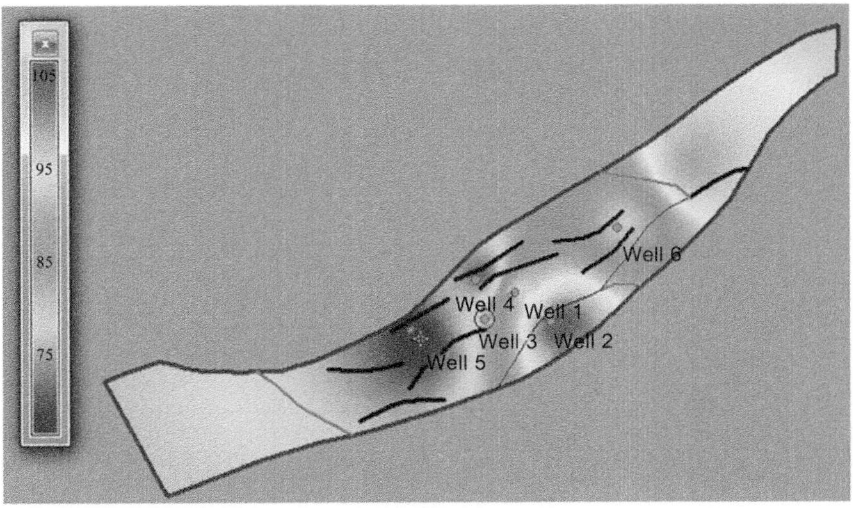

Fig. 3.24 Well location and net pay thickness in Reservoir A.

3.2.2.4.2 Production history

Six wells were put into production successively in the reservoir, with an average daily gas production rate of $170 \times 10^4 m^3/d$, cumulative gas production of $15 \times 10^8 m^3$, and daily single-well gas production rate of $(15–45) \times 10^4 m^3/d$. During production, Well 2 suffered sand flow, with significant fluctuation of flow pressure; Well 3, without fractures, revealed a low production rate and low flow pressure. In May 2016, all the wells in the reservoir were shut in, with their wellhead static pressure building up to the same value, except Well 3 (Fig. 3.25). In addition, the static pressure of

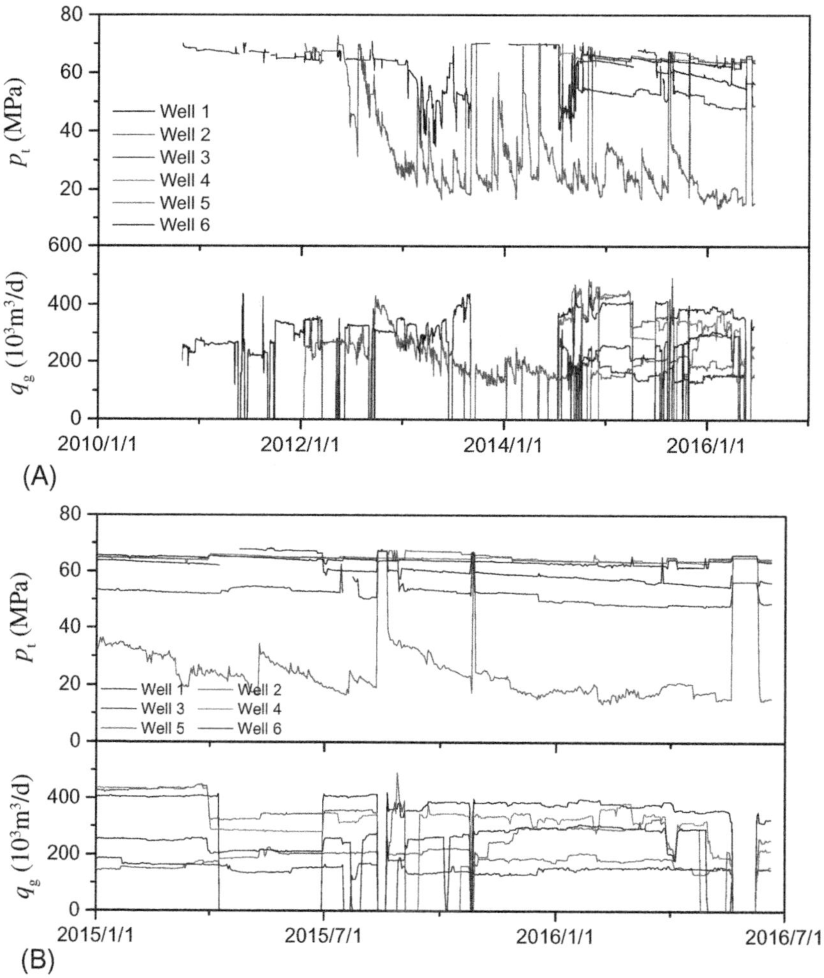

Fig. 3.25 Production curves of six wells in Reservoir A. (A) Production curves of the well cluster. (B) Partial enlargement of the production curves of the well cluster.

Table 3.3 Static pressure of producing wells in Reservoir A

Well no.	Date of initial production	Initial static pressure (MPa)	Net pay thickness (m)	Porosity (%)	Gas production rate ($10^4 m^3/d$)
Well 3	2014-7-16	79.74	87	5.6	16.5
Well 6	2014-9-29	86.06	79	6.2	25.3
Well 5	2014-9-27	84.72	65	7.1	33.0
Well 1	2010-10-30	89.09	82	6.8	33.4
Well 4	2014-7-17	85.83	69	6.9	35.3
Well 2	2012-1-11	86.74	105	7.9	20.7

the wells that were put into production later gradually decreased (Table 3.3). Therefore, it is possible that the entire reservoir is connected.

3.2.2.4.3 Well test

In August 2015, the PBU test was conducted in Well 5 after 320 days of production. The first PBU test was carried out with all wells shut in simultaneously. Then the modified isochronal test was conducted, during which Wells 1, 2, and 6 were producing. Finally, the extended PBU test was conducted, during which Wells 3 and 4 were successively opened for producing, as shown in Fig. 3.26.

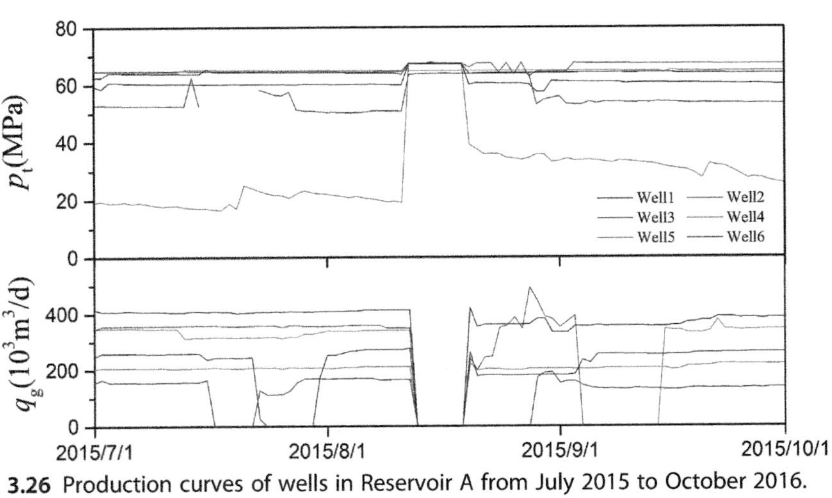

Fig. 3.26 Production curves of wells in Reservoir A from July 2015 to October 2016.

3.2.2.4.4 Data analysis

The log-log plot of Well 5 is shown in Fig. 3.27. An obvious difference is found in the pressure derivative curves of the first PBU test and the last PBU

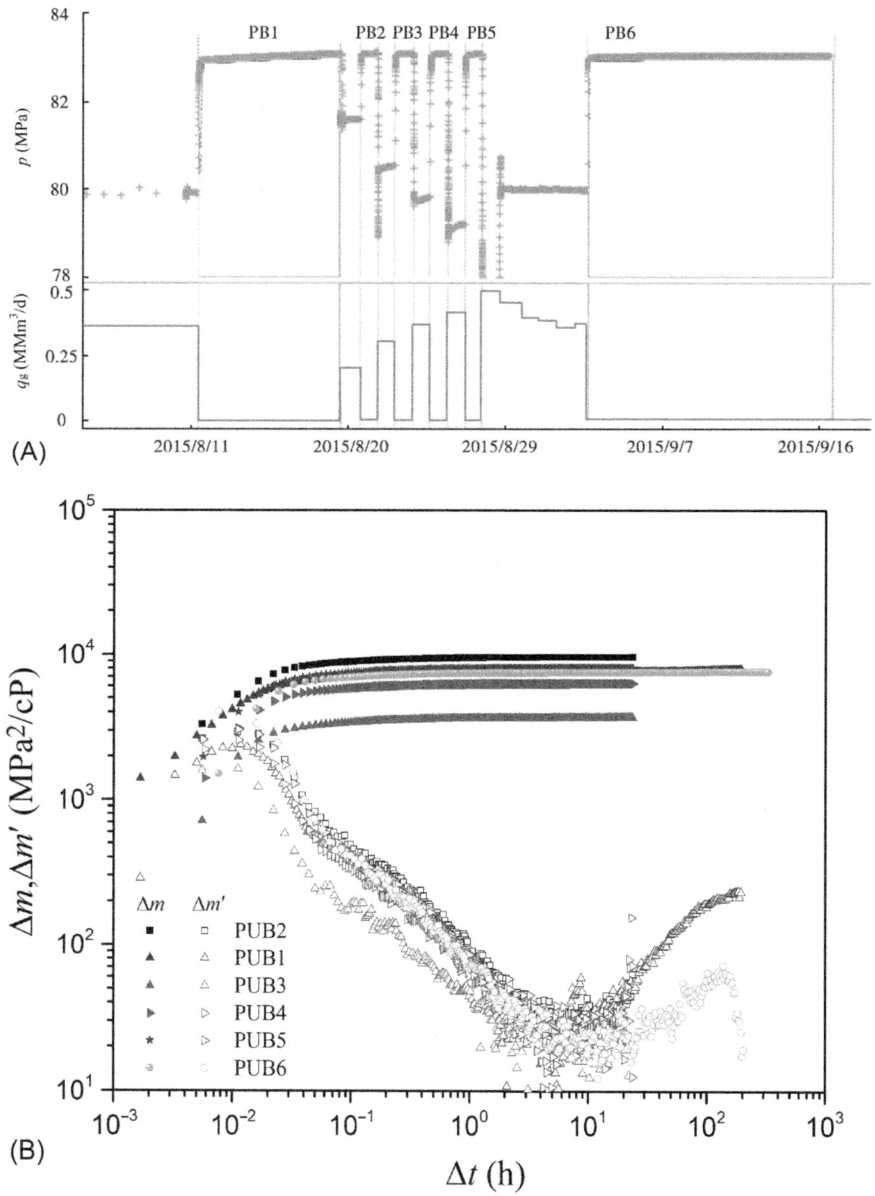

Fig. 3.27 History plot and log-log plot of Well 5 based on test data. (A) History of the test and (B) log-log curves.

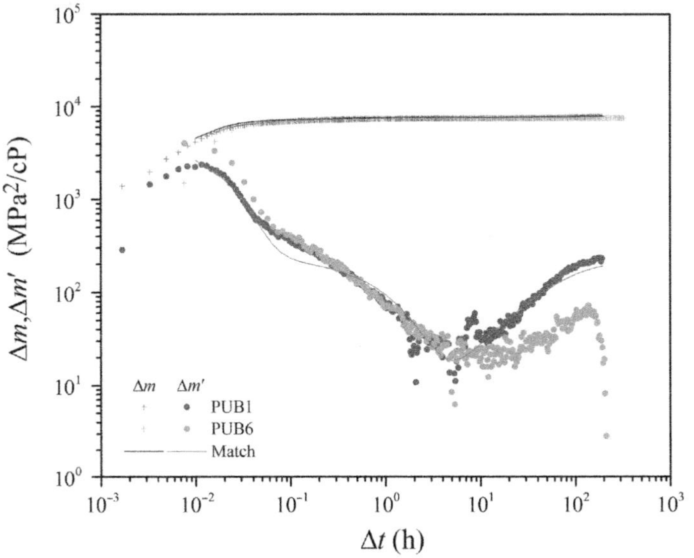

Fig. 3.28 History plot and log-log plot of Well 5 based on well test data with the dual-porosity model.

test, revealing an inverse shape. There is an obvious increase in the pressure derivatives for the modified isochronal test.

When the dual-porosity model is used for interpretation of the gas reservoir that contains fractures, the results are shown in Fig. 3.28. It is difficult to match the results of the two PBU tests and to interpret the difference in the shapes of their pressure derivative curves with a set of parameters. Based on production performance, it is advisable to adopt the multiwell model for analysis.

According to the production performance, it is indicated that the production wells in Reservoir A may be connected. The ratio of the average production rate in the block to the production rate of the testing wells is $(16.5+25.3+33.4+35.3+20.7)/33=5.0$. The ratio of the log-log stabilization line height in the first PBU test to the step line height is $(3.85\times10^7)/(20\times10^7)=5.0$. Therefore, the multiwell model for the homogeneous reservoir can be used, with the interpretation results shown in Fig. 3.29. When the well was put into production, the initial formation pressure dropped from 89 to 85 MPa, and now the formation pressure is 81.6 MPa.

3.2.2.4.5 Comparison with previous tests

Two additional tests were performed to the well in August 2016 and July 2017. The test results were consistent with those in 2015, suggesting the

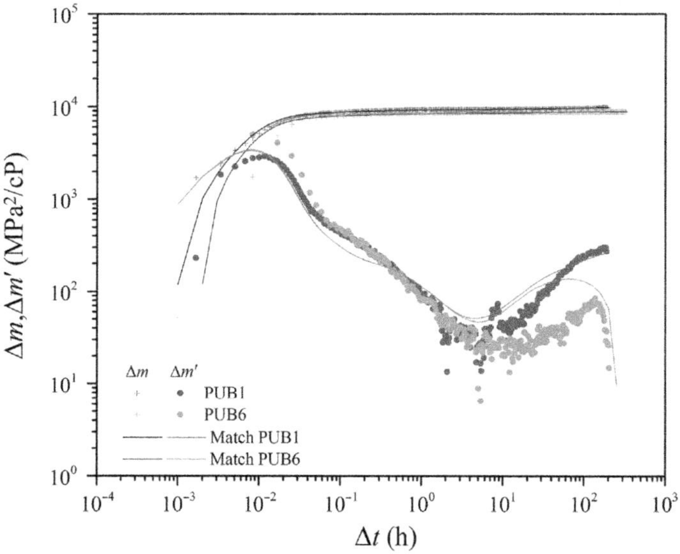

Fig. 3.29 History plot and log-log plot of Well 5 based on well test data with the multiwell model for homogeneous reservoirs.

characteristics of the homogeneous reservoir with the multiwell system (Fig. 3.30). The test results in 2017 revealed signs of water invasion.

3.2.2.5 *Discussion and conclusion*

The pressure buildup derivative curves have several rising "steps." Accordingly, the reservoir model may: (1) be a composite model with a deteriorating periphery; (2) have one or more impermeable boundaries; (3) reflect the edge-water in the case of gas wells, and (4) be subject to interwell interference.

In both cases of multiwell simultaneous production and shut-in, the PDD and PBU derivative curves present the features of the "stepped rise," and the ratio of each stabilization line-height to the first stabilization line height is the algebraic sum of the dimensionless production rate of the testing well and the affecting offset wells. In the multiwell system, if the offset wells are producing continuously, the PBU derivative curve of the testing well will gradually "downwarp" in the middle and late stages, and the "downwarping" speed depends on the algebraic sum of the dimensionless

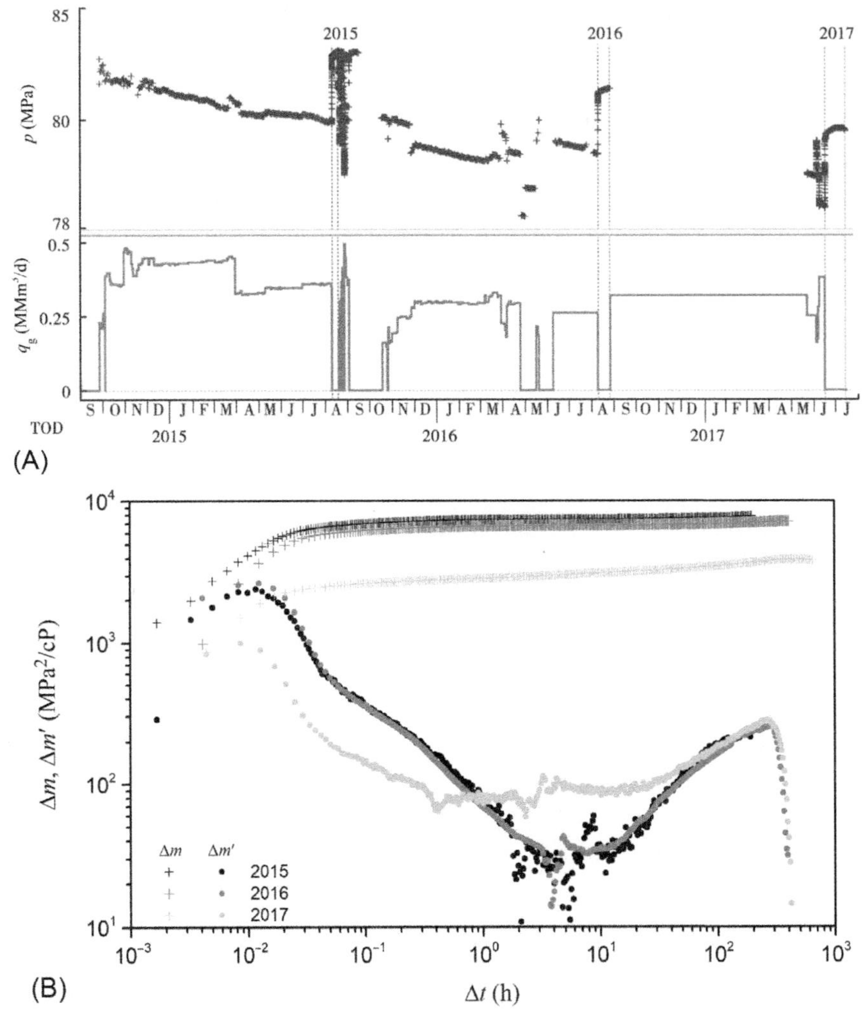

Fig. 3.30 History plot and log-log plot of Well 5 based on well test data with the multiwell model for homogeneous reservoirs. (A) History of the test and (B) log-log curves.

production rate of the affecting offset wells and the ratio of the shut-in time to the production time.

For a newly discovered gas reservoir, its boundaries and the potential interference of offset wells can be identified through two PBU tests. The first PBU test is conducted with offset wells producing, and the second PBU test is conducted with the offset wells and the testing well shut in simultaneously. The multiwell test analysis can be combined with an advanced multiwell production decline analysis to scientifically describe the performance of gas reservoirs with good connectivity, thereby providing technical support for the formulation and optimization of development strategies.

3.2.3 The three-dimensional numerical well test analysis method

3.2.3.1 Overview of the numerical well test

By the solving for the methods of the flow equation, the well test can be divided into an analytical well test and a numerical well test. An analytical well test focuses on how to use such methods as the Bessel function, Laplace transform, Fourier transform, and Green function to obtain the mathematical solution of the unsteady flow equation under new boundary conditions.

The numerical well test is used to numerically solve the flow equations describing the flow of fluids in the formation by adopting the optimized grid division technique, to realize the real-time calculation of the fluid properties and saturation changes. Specifically, based on the numerical analysis, the flow equations are discretized and linearized for a complex reservoir in a grid system to derive the linear equations, and the linear equations are solved to determine the downhole pressure at a given time; then, the calculated pressure is matched with the measured pressure to obtain the relevant formation and wellbore parameters, which are used to evaluate the complex reservoir. This method, ignoring the superposition theory followed in the traditional analytical well test, enables the theoretical simulation results to approximate the actual reservoir performance.

The numerical well test realizes the numerical solution through grid division, numerical discretization, and an equation solution. Therefore, it is slower than the analytical well, test but more capable of description. This method can quickly and reasonably interpret various problems, such as any boundary shape, faults (including partially permeable faults) in any number or any shape, multiphase flows, multiple layers (different skin factors), multiwell interference, and heterogeneity. It is more practical in description and production performance forecasting of gas reservoirs. For example, different formation parameters can be set for each specific reservoir to consider the heterogeneity; the porosity and thickness distribution can be displayed in two or three dimensions, and the pressure distribution can be expressed at any time or displayed in animation.

For the fractured-vuggy carbonate reservoirs with strong heterogeneity, conventional analytical well tests and two-dimensional numerical well tests are not sufficient for describing the reservoir performance. It is urgent to

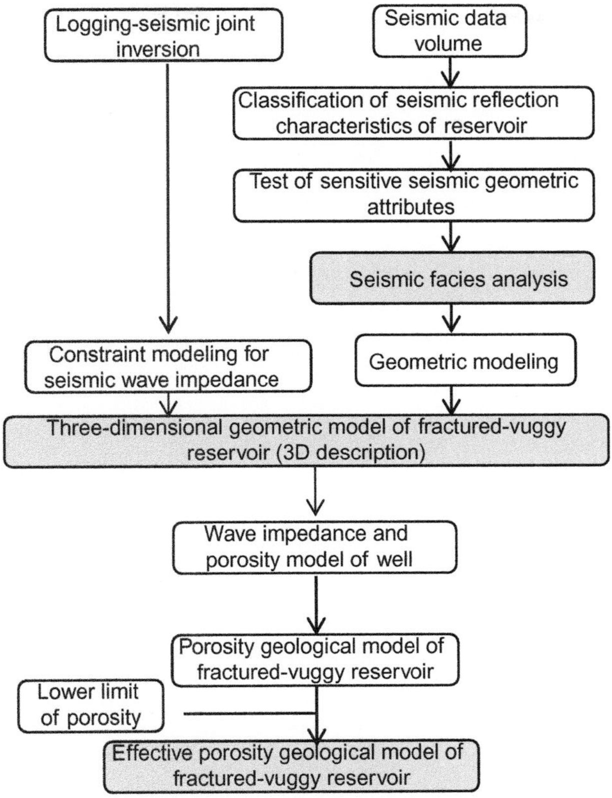

Fig. 3.31 Flow chart of quantitative description by seismic inversion.

develop a three-dimensional numerical well test analysis method in combination with seismic inversion attributes.

3.2.3.2 The three-dimensional numerical well test

3.2.3.2.1 Seismic inversion-derived attributes

The heterogeneous fractured-vuggy carbonate reservoirs can be identified and forecasted by using seismic attributes. The flow chart for a quantitative description of the fractured-vuggy reservoir in the combination of seismic attributes and seismic inversion is shown in Fig. 3.31.

Fig. 3.32 shows the well-controlled data volume obtained by seismic wave impedance inversion for Block G262 in the Tazhong gas field. Based on the correlation between wave impedance and porosity (Fig. 3.33), a seismic inversion-derived porosity model (Fig. 3.34) can be obtained, which is used to build a three-dimensional numerical model.

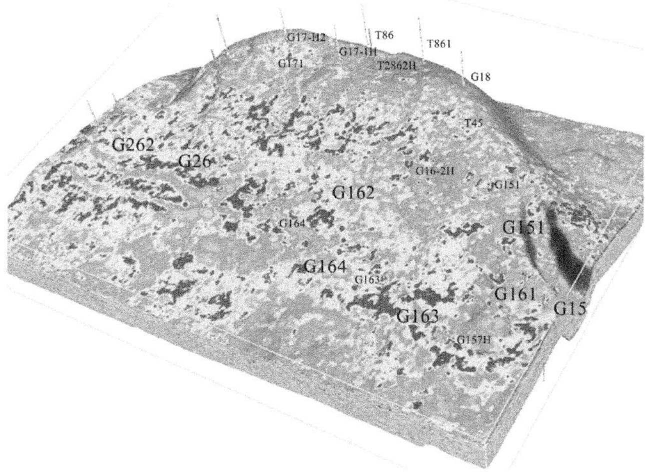

Fig. 3.32 Well-controlled data volume obtained by seismic wave impedance inversion.

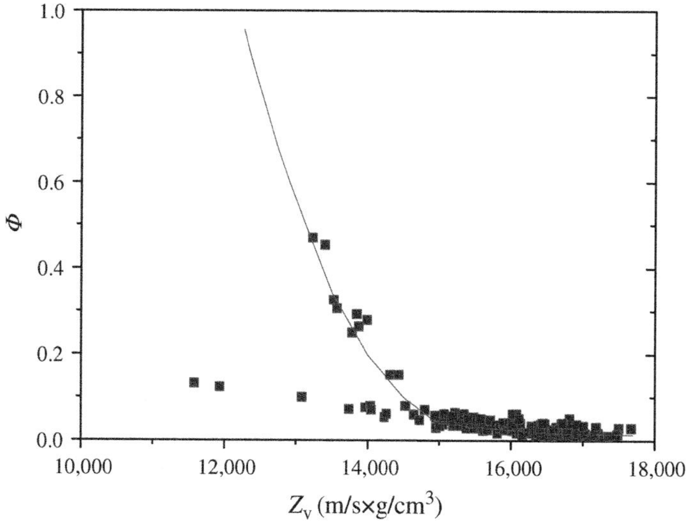

Fig. 3.33 Correlation between wave impedance and porosity.

The seismic inversion-derived porosity model is characterized by poor continuity and variable dimensions in space, and is restricted by the interpretation accuracy of seismic inversion results. Therefore, this model cannot be directly applied to numerical simulation analysis, but should be combined with well test analysis for modeling a heterogeneous reservoir.

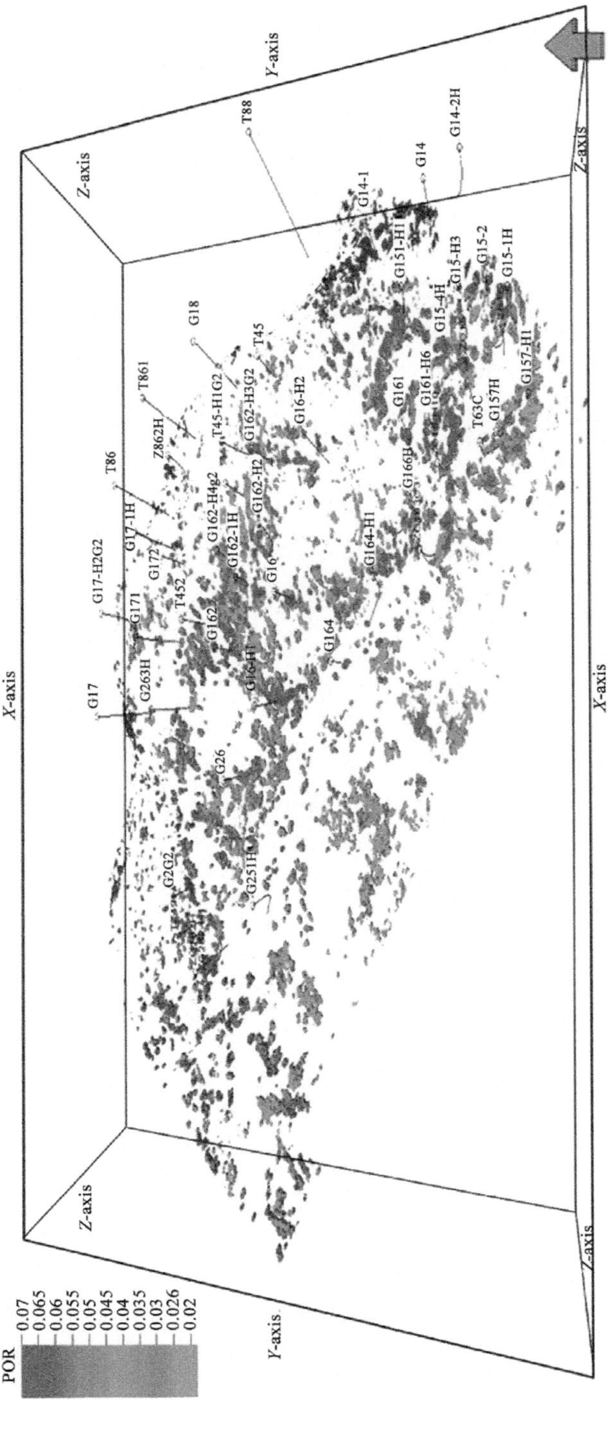

Fig. 3.34 Seismic inversion-derived porosity model of the block.

3.2.3.2.2 Building of the three-dimensional model

The reservoir is divided into zones, laterally, and layers, vertically. The lateral heterogeneity is described by way of a two-dimensional numerical well test, and the vertical heterogeneity is controlled by seismic inversion-derived attributes. With the log-log analysis and long-term production history matching as the constraints, the three-dimensional numerical well test analysis is conducted. The flow chart is shown in Fig. 3.35.

(1) Lateral zoning

The lateral porosity model (Fig. 3.36) obtained by seismic inversion is used to build a lateral zoning model according to the seismic reflection intensity and the delineation of porosity. Then, the lateral zoning is made according to the geological structures and faults, as shown in Fig. 3.37. Finally, the parameters of zones are preliminarily determined through two-dimensional numerical well test analysis.

(2) Vertical layering

Vertical layering is conducted on the vertical porosity model (Fig. 3.38) obtained by seismic inversion. Then, the average inversion-derived porosity of layers is determined (Fig. 3.39). Finally, the initial values of layer parameters are determined according to the correlation between permeability and porosity in the whole block derived from the well test.

(3) Production splitting

The data for the gas production profile test and the layered production test are analyzed to determine the initial production of each layer. In the case of less data, the production of layers can be split empirically based on the initial formation capacity.

(4) Log-log matching analysis

After the initial parameters of zones and layers are determined, the multilayer well test model is used for log-log matching analysis to identify the individual layer parameters.

(5) History matching

Based on the log-log matching analysis results, and with the long-term production history data as the constraint, the history matching analysis is conducted to further optimize the individual layer parameters (Fig. 3.40). The procedure is repeated until the geological reserves of the model are essentially consistent with the OGIP/OOIP determined by the advanced production decline analysis technique, as shown in Fig. 3.41. Finally, a three-dimensional heterogeneous equivalent reservoir model is obtained.

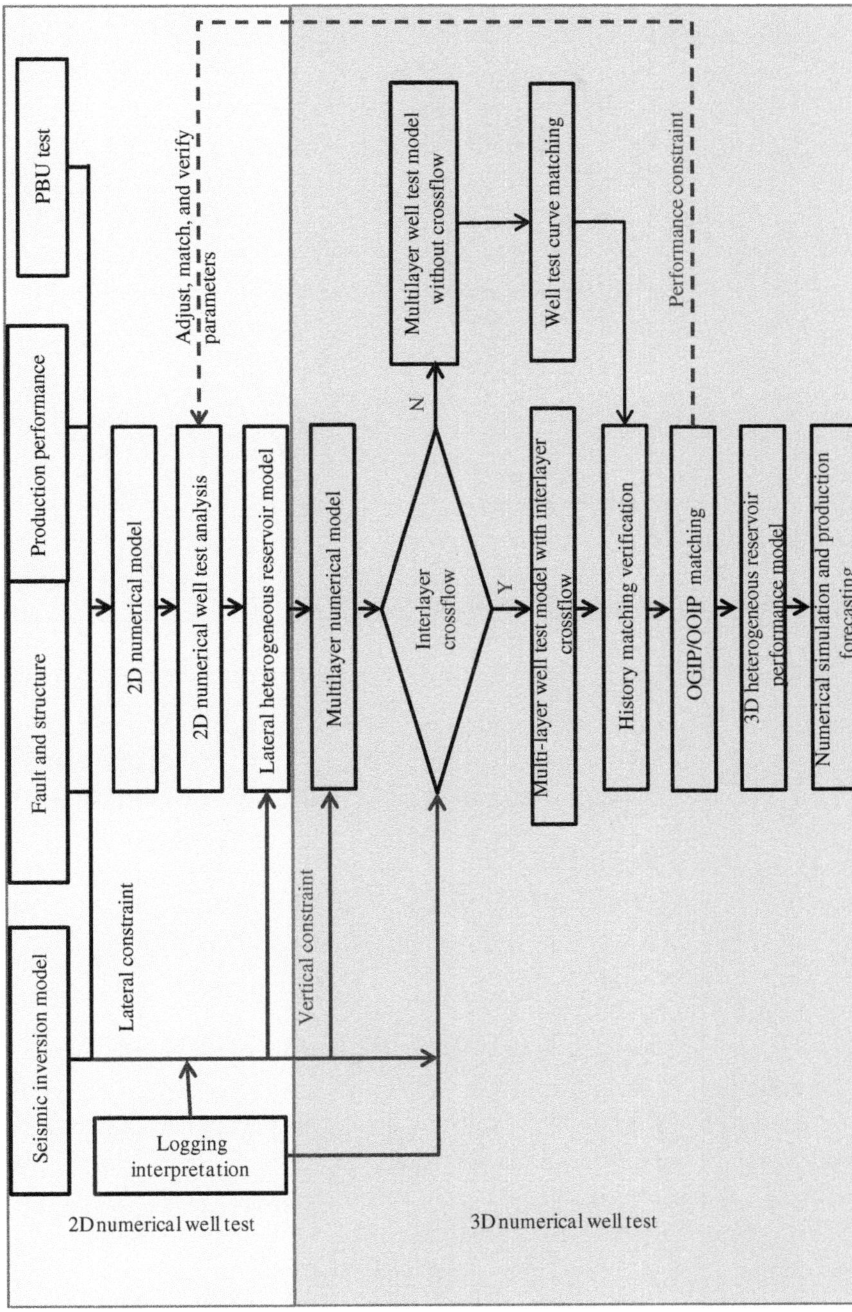

Fig. 3.35 Flow chart of three-dimensional numerical well test analysis.

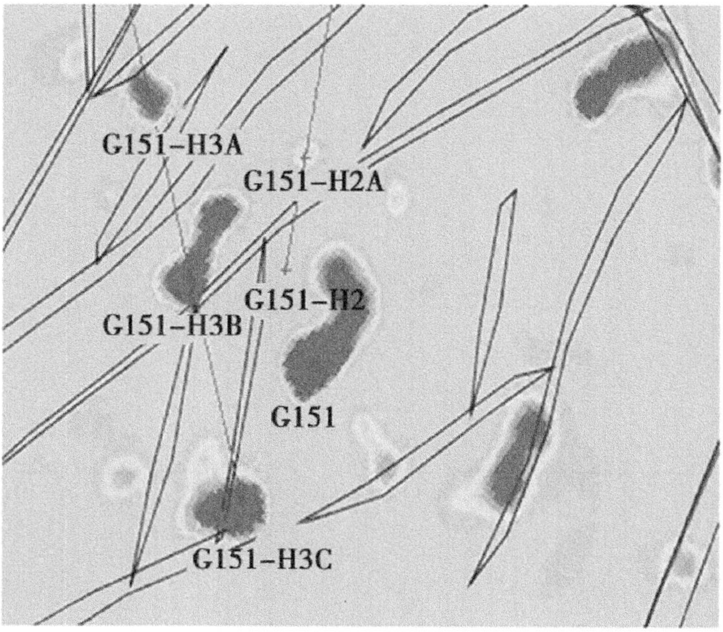

Fig. 3.36 Superposition of lateral seismic attributes.

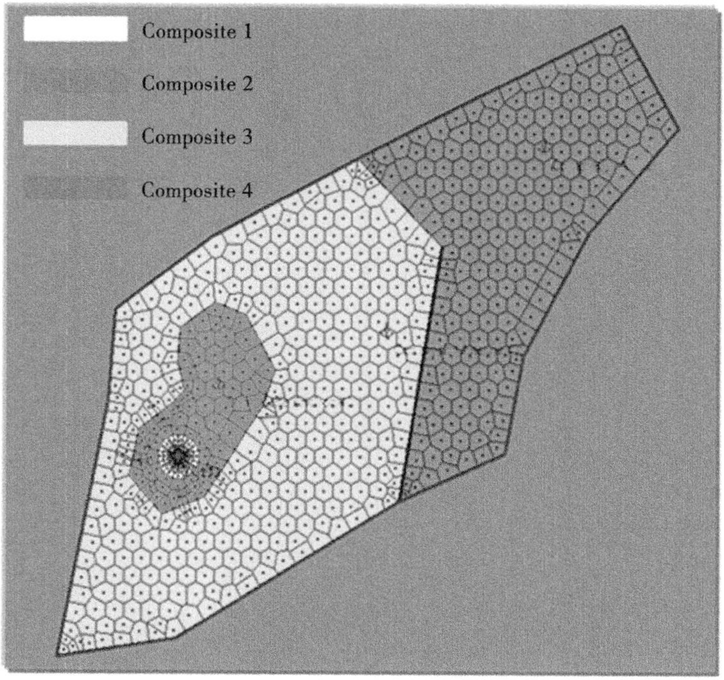

Fig. 3.37 Two-dimensional lateral zoning model.

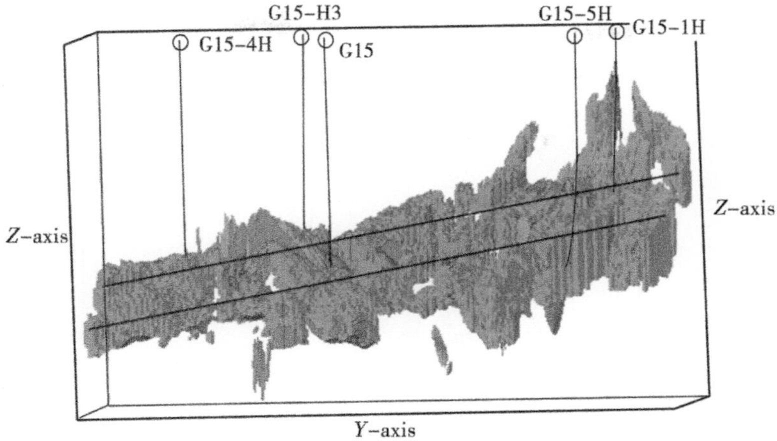

Fig. 3.38 Vertical porosity model obtained by seismic inversion.

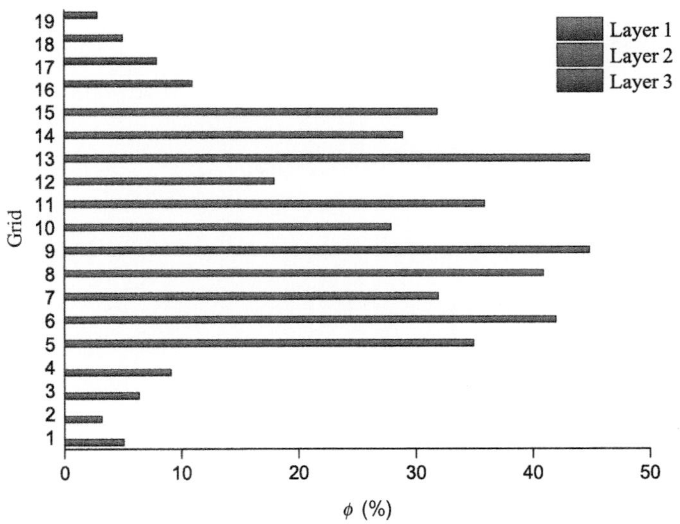

Fig. 3.39 Inversion-derived porosity.

(6) Performance forecasting

Based on the three-dimensional heterogeneous equivalent reservoir model, the production performance can be further forecasted (see Section 5.3 in Chapter 5 for details).

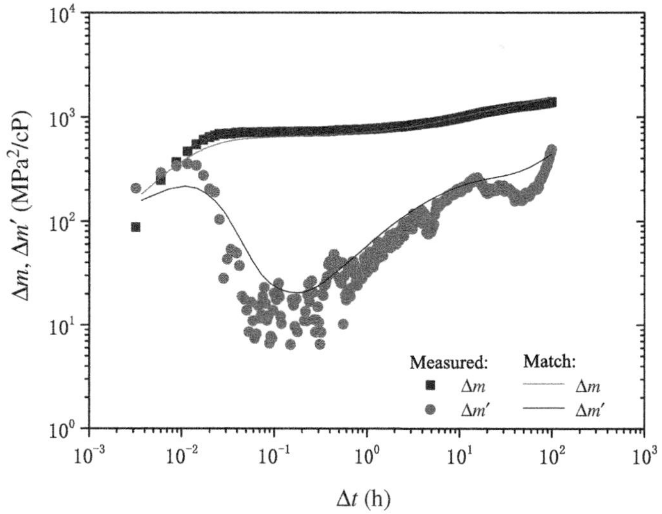

Fig. 3.40 Log-log matching plot of Well G151.

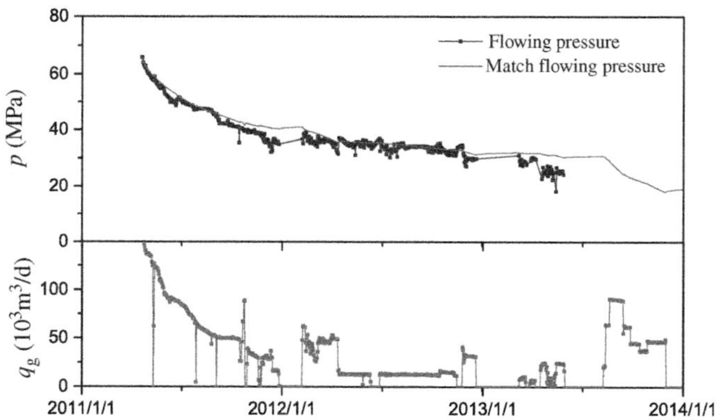

Fig. 3.41 History matching plot of Well G151.

3.2.3.2.3 Simulation example

(1) Basic data

The basic parameters used in the simulation are taken from the Tazhong I gas field. The three-dimensional heterogeneous reservoir model is categorized into four zones laterally, with the parameters shown in Table 3.4. The reservoir permeability is 0.006–50 mD or 5% on average, and the net pay thickness is 20 m.

It is assumed that the wells produce at a constant rate for 100 days, followed by the PBU test for 20 days after shut-in, and then they are opened for

Table 3.4 Parameters of the synthetic model

Layer	Zone	Net pay thickness (m)	M	D	Permeability (mD)	Porosity (%)	Wellbore storage factor (m^3/MPa)
Layer1	#1	3	/	/	3	2	0.231
	#2		5	5	0.6		
	#3		50	50	0.06		
	#4		500	500	0.006		
Layer2	#1	12	/	/	50	7	
	#2		5	5	10		
	#3		50	50	1		
	#4		500	500	0.1		
Layer3	#1	5	/	/	11.5	3	
	#2		5	5	2.3		
	#3		50	50	0.23		
	#4		500	500	0.023		
Total/average		20	/	/	33	5	/

production, with a monthly production decline rate of 5%. The simulation results of different analysis methods are shown in Fig. 3.42.

(2) Comparison with previous results

The pressure history matching results based on the theoretical forward modeling data of the gas reservoir with multiple layers vertically and heterogeneity laterally suggest that the analytical well test model provides the worst matching results, and the multilayer three-dimensional numerical well test model provides the best matching results, while the conventional well test method provides the results between the two methods, as shown in Fig. 3.43. The results of three-dimensional numerical well test method are essentially better than that of the conventional well test method. To sum up, the multilayer, three-dimensional numerical well test analysis method based on seismic inversion-derived attributes can effectively describe lateral and vertical heterogeneity, laying a foundation for well and cluster productivity evaluation and index forecasting, and thereby helping greatly improve the reliability of production allocation and performance forecasting.

3.2.3.3 Discussion and conclusion

The two-dimensional numerical well test method is used to describe the lateral heterogeneity of the reservoir, and with the seismic inversion-derived

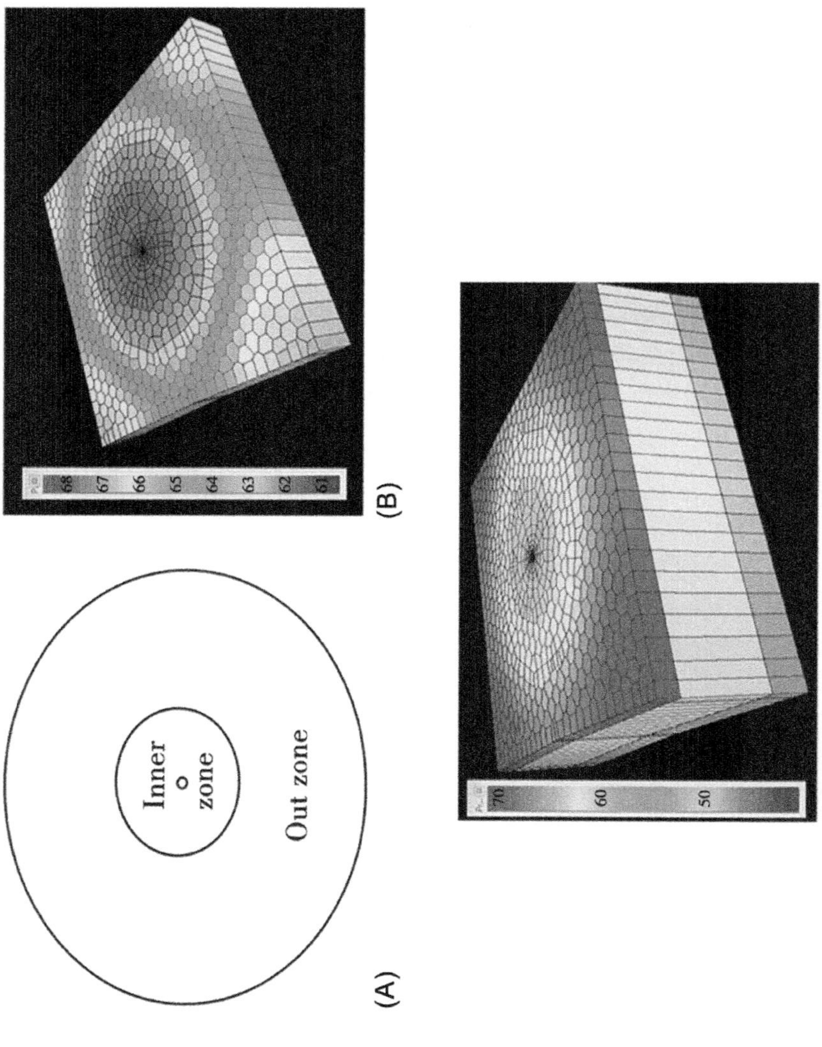

Fig. 3.42 Comparison of different well test analysis methods. (A) Analytical model: simple. (B) 2D numerical model: four zones laterally. (C) 3D numerical model: four zones laterally, three layers vertically.

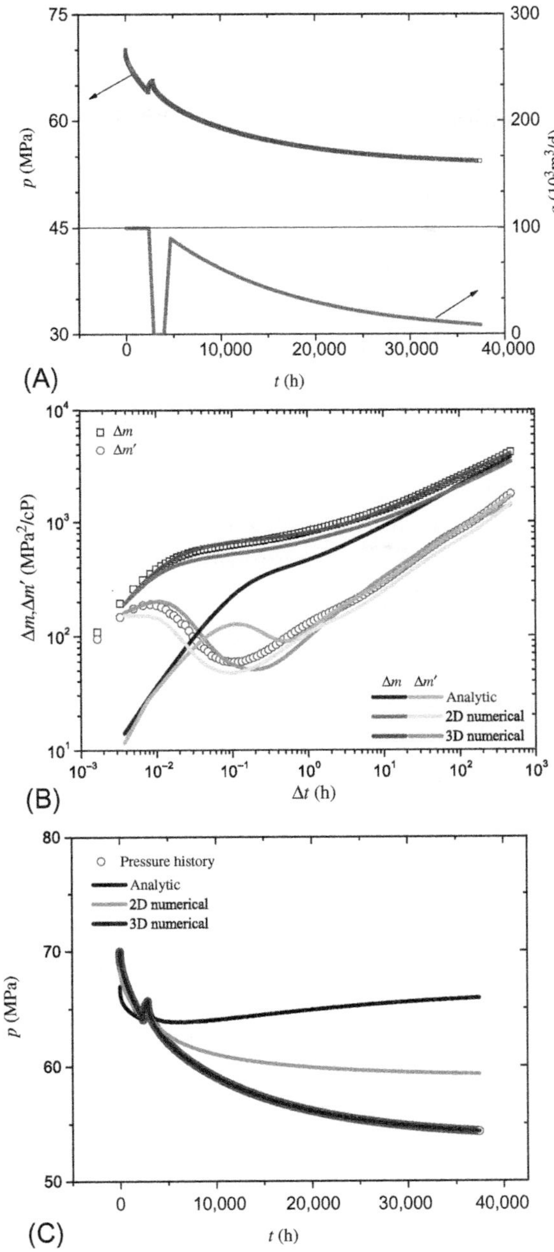

Fig. 3.43 Comparison of analytical well test, 2D and 3D numerical well test methods. (A) Pressure history plot, (b) log-log plot, and (C) pressure history matching plot.

attributes as the constraint, and considering the vertical heterogeneity of the reservoir, the multilayer three-dimensional numerical well test model is built. Then, taking the log-log plot and long-term production history matching as the constraints, the three-dimensional numerical well test analysis method for multilayer gas reservoirs is established. Thus, the numerical well test is transited from the two-dimension [single-layer M zone] to three-dimension [multilayer ΣM (N) zone, where N is the number of layers]. The analytical well test model is simple with regular boundaries; the two-dimensional numerical well test model takes into account the lateral heterogeneity; the multilayer three-dimensional numerical well test model considers both lateral heterogeneity and vertical heterogeneity, and provides a mathematical description more approximate to the actual reservoir performance.

The combination of the conventional analytical well test, the (2D and 3D) numerical well test, and advanced production decline analysis can greatly improve the reliability of interpretation results and the forecasting accuracy of key development indices. The well test, which is used to determine layer parameters, and advanced production decline analysis, which is used to determine well-controlled reserves, can be integrated to effectively describe the reservoir characteristics and forecast the performance of the well or cluster.

3.2.4 Lifecycle well test analysis method

3.2.4.1 Analysis process

Considering that the well test interpretation results are diverse, ambiguous, and complex; a well test analysis technique with the constraint of whole history matching is proposed. Based on the data of a single-well well test and production test, the history of single-well performance varying by pressure and flow rate is established to conduct well test analysis; and by reservoir types, the production performance is forecasted, as shown in Fig. 3.44.

3.2.4.1.1 Test data evaluation

The test data are sorted and evaluated to identify why the data are unsatisfactory for interpretation. Usually, six aspects are involved. First, because the test time is short, the pressure derivative curve has no stabilization line segment, making the obtained parameters unreliable. Second, the pressure derivative is in disorder due to the influence of the gas-water phase transition. Third, the pressure test data are fluctuate too much for interpretation. Fourth, the tested production is very low (only a few hundred or even tens of

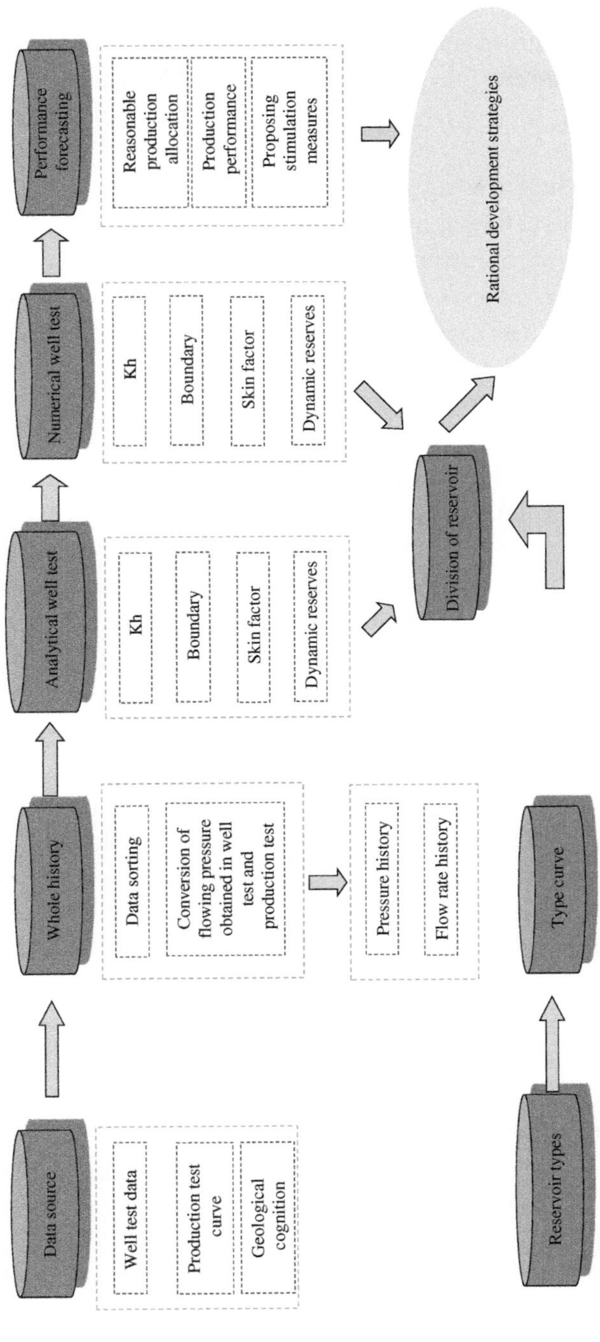

Fig. 3.44 Flow chart of lifecycle well test analysis.

square meters), but the production pressure difference is large. Fifth, channeling occurs during the test due to problems in cementing quality, leading to uncertainty in the interpretation results. Sixth, the pressure drops during the test, and the pressure derivative cannot be calculated in the late stage of buildup.

3.2.4.1.2 Bottom hole flowing pressure conversion

Based on multiphase flow in the wellbore and pressure gradient test results, the tubing pressure, or casing pressure data, are converted to the depth at which the pressure gauge is set during the well test, and incorporated with the short-term well test data, to establish the whole production history data.

3.2.4.1.3 Analytical well test analysis

For the purpose of log-log analysis, a "theoretical dynamical model" for the reservoir near the wellbore is preliminarily built based on the well test data analysis using interpretation software, together with the geological conditions. For the purpose of verifying the long-term production pressure history, a complete dynamical model is created for the reservoir where the gas well is located after thorough observation (including connectivity and boundary distribution) of the reservoirs within the well drainage area is completed.

3.2.4.1.4 Numerical well test analysis

When the analytical well test analysis method is ineffective for reservoir description, the numerical well test analysis method is adopted for the reservoir in a proper grid system according to the results of geological studies.

3.2.4.1.5 Performance forecasting

Based on the reservoir dynamical model built after the numerical well test or production performance analysis, the change of pressure or production rate of the well is forecasted at a constant rate or constant pressure.

3.2.4.2 Field example

Well T622 is a production testing well in the Tazhong gas field, with the coexistence of oil, water, and gas. Since 2005, 20 flowing pressure gradient tests have been conducted on this well.

At first, the flowing pressure gradients of 20 tests are calculated, and the flowing pressure is converted to the depth of 4800 m, where the pressure gauge is set during the PBU test. Then, the flowing pressure at 4800 m is

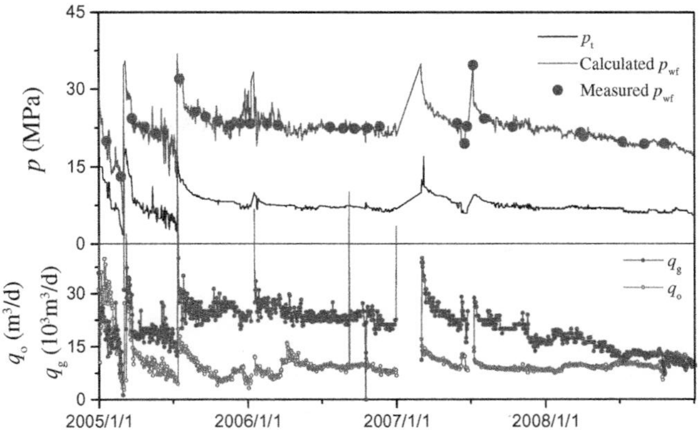

Fig. 3.45 Conversion of bottom flowing pressure for Well T622.

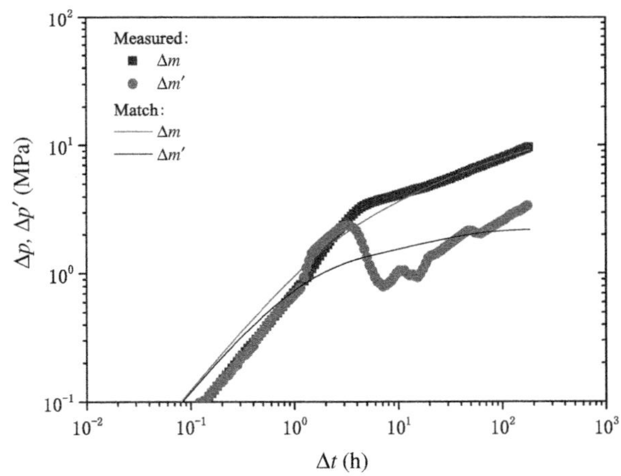

Fig. 3.46 Log-log plot for Well T622.

converted according to the multiphase flow theory and the long-term production testing data (Fig. 3.45). It is found that the converted data are well matched with the measured data.

Well test analysis is conducted using the long-term production data and the short-term well test. The well test matching plots are shown in Figs. 3.46 and 3.47. The theoretical model can well match the long-term production performance, thus achieving the dynamical description of the well.

3.2.4.3 Superiority of lifecycle well test analysis

3.2.4.3.1 Understanding the reservoir characteristics

The combination of long-term production data and short-term well testing for well test analysis has facilitated the high accuracy of well test analysis and

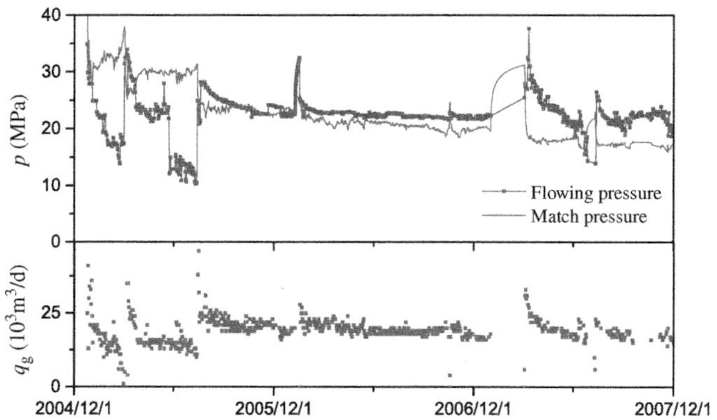

Fig. 3.47 History matching plot for Well T622.

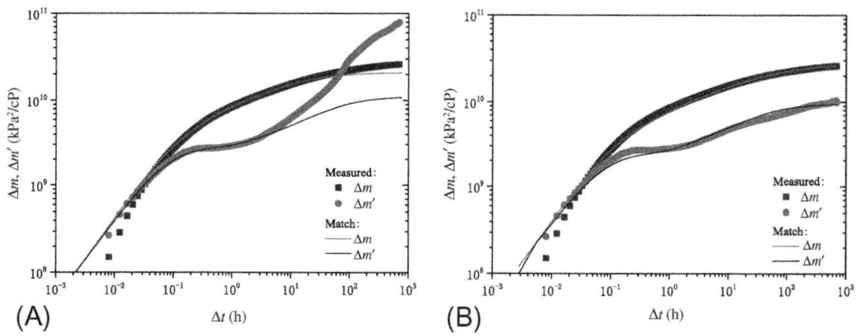

Fig. 3.48 Influence of the production history on the pressure derivative. (A) Not considering production history and (B) considering production history.

performance analysis. As shown in Fig. 3.48, the former does not fully consider the production history, resulting in the abnormalities of the log-log plot and the intersection of the pressure derivative curve with the pressure curve; when the production history is considered, the log-log plot is in a normal state, so the well presents the typical characteristics of a vertically fractured well with limited connectivity.

3.2.4.3.2 Identifying wellbore and formation reflections

As shown in Fig. 3.49, the log-log plot suggests the reflection of the boundary, which actually should be the reflection of the wellbore. The first-order derivative curve of pressure versus time can be used to judge the reflections of the wellbore and formation. The test data are carefully analyzed to sort out the initial measured pressure data, to provide authentic and reliable data for identifying the wellbore and formation reflections.

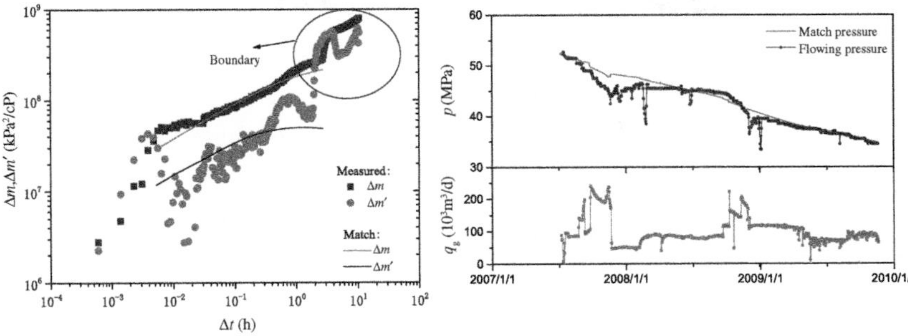

Fig. 3.49 Data sorting. (A) Influence of boundary and (B) history matching results.

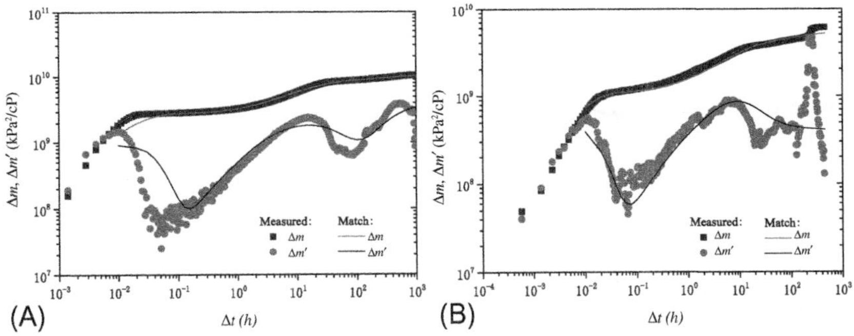

Fig. 3.50 Well test analysis for Well T62-2. (A) Log-log plot of PBU1 and (B) log-log plot of PUB2.

3.2.4.3.3 Reduce the ambiguity of reservoir evaluation

The long-term production data and the short-term well test are combined for well interpretation for Well T62-2 (Fig. 3.50). First, analytical well test analysis is conducted. The analysis results suggest that the data can be matched, but the permeability is low (only 0.348 mD), and the storability ratio is only 0.003, which is inconsistent with the large volume of fluid loss and geologic data during the drilling process. Therefore, this method is not applicable.

The dual-porosity multizone composite model is selected for the numerical well test. There is a dual-porosity reservoir near the wellbore, with the permeability of 5.2 mD and the storability ratio of 0.223, which is consistent with the drilling and geologic data. Clearly, this method can be used to accurately identify reservoir parameters, with much less ambiguity in well test interpretation.

3.2.4.3.4 Scientific performance forecasting

The dynamical model built on the basis of well testing can forecast the single-well production performance (see Sections 5.2 and 5.3 in Chapter 5 for details). It promotes the deliverability evaluation from the traditional empirical method based on absolute open flow to the new level of low-risk production allocation based on material balance, providing a scientific basis for making development plans.

3.3 Application of the well test analysis

3.3.1 Reservoir performance model

The lifecycle well test matching plots of some typical wells in the Tazhong gas field are shown in Fig. 3.51. It can be seen that the log-log plot and the long-term history plot are well matched, revealing multiple characteristics such as deteriorating periphery, boundary, and macrofractures.

On the basis of the circular bounded dual-porosity model and the numerical well test method, 13 numerical well test models are built for 6 types of reservoirs, including the radial composite reservoir, the high-permeability strip reservoir, and the reservoir with multiple (2–3) accumulation bodies. Fig. 3.52 shows these reservoirs and the corresponding log-log plots. The well location and size of the accumulation may affect the shape of the well test curves, making the curves play a guiding role in the classification of reservoir types.

The preceding results are just derived from the analytical well test analysis. If static geological cognition is combined with the numerical well test method, together with the geophysical exploration technology and data, it is believed that the reservoir can be more thoroughly described (Fig. 3.53).

The well test analysis is essentially an inverse problem, which is ambiguous, especially for complex carbonate reservoirs in the Tazhong I gas field. From the perspective of the log-log plot alone, there may be several performance models based on well tests for such carbonate reservoirs. The combination of the well test and geophysical fracture-vug delineation can achieve a more accurate description, minimize the ambiguity of well test parameter evaluation, and enhance the accuracy of performance forecasting for wells.

3.3.2 Reservoir performance evaluation

Compared with conventional sandstone gas reservoirs, the fractured-vuggy carbonate reservoir is more outstanding because of the substantial difference

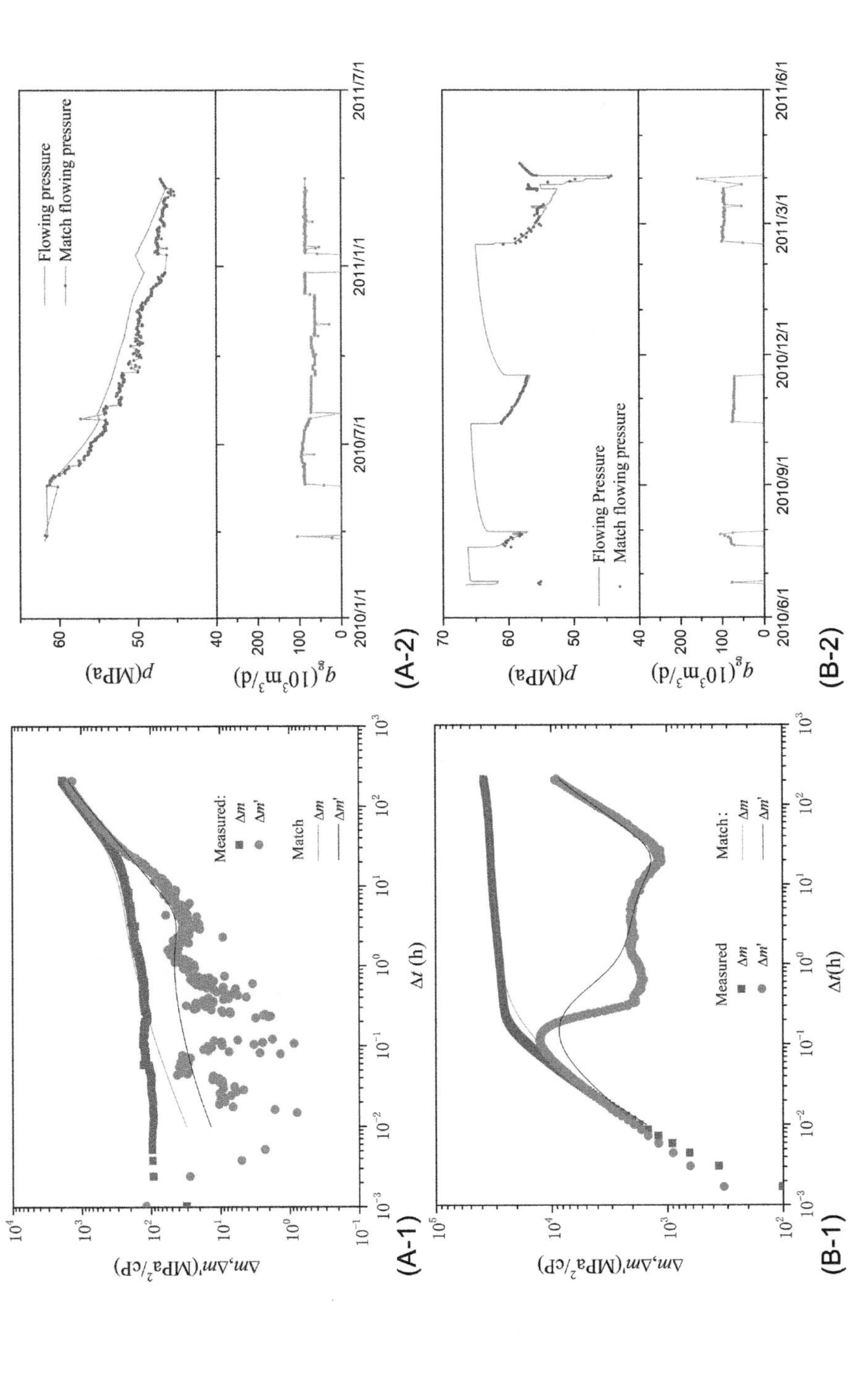

Fig. 3.51 Well test matching plots for typical wells in the Tazhong I gas field. (A-1) G43 log-log, (A-2) G43 history match. (B-1) T201C log-log, (B-2) T201C history match.

(Continued)

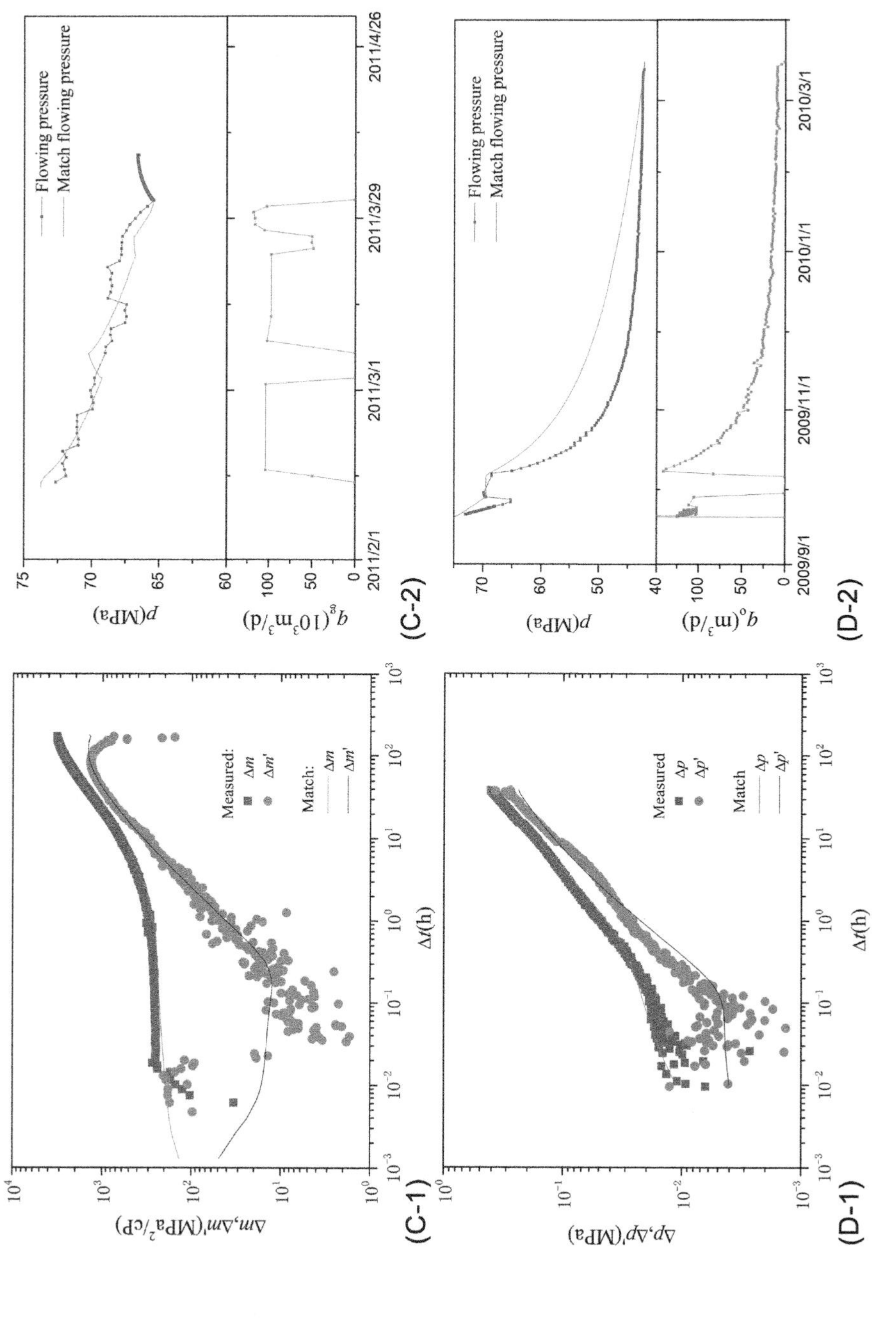

Fig. 3.51, cont'd (C-1) G14-1C log-log, (C-2) G14-1C history match. (D-1) G13 log-log, (D-2) G13 history match.

in the size of media (pores, vugs, and fractures), complex fluid properties, and other elements. Fractured-vuggy carbonate reservoirs are usually categorized into three types: cave reservoirs, fractured-vuggy reservoirs, and vuggy reservoirs, which present different characteristics in well logging interpretation, seismic reflection, drilling/completion engineering, well testing, production testing, acid fracturing, and well testing. The comprehensive identification and performance evaluation of reservoirs is shown in Table 3.5.

No.	Model			Log-log plot for different well locations		
	Base	Transform	Pattern	A	B	C
1	Circular dual-porosity reservoir	Basic	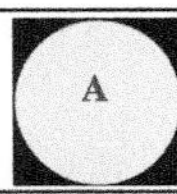	(1A)		
2		Composite	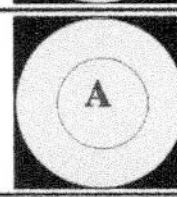	(2A)		
3		1 Circle +1 high-permeability stripe		(3A)	(3B)	
4		1 Circle +inner high-permeability circle	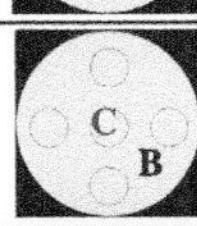	(4A)	(4B)	(4C)
5		2 Circles +1 high-permeability stripe	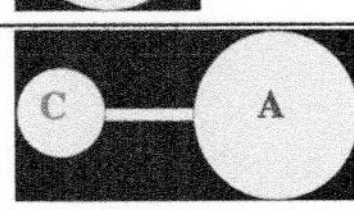	(5A)	(5B)	(5C)
6		3 Circles +2 high-permeability stripes	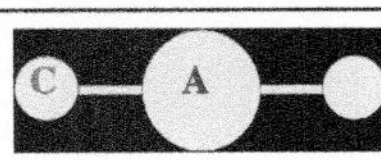	(6A)	(6B)	(6C)

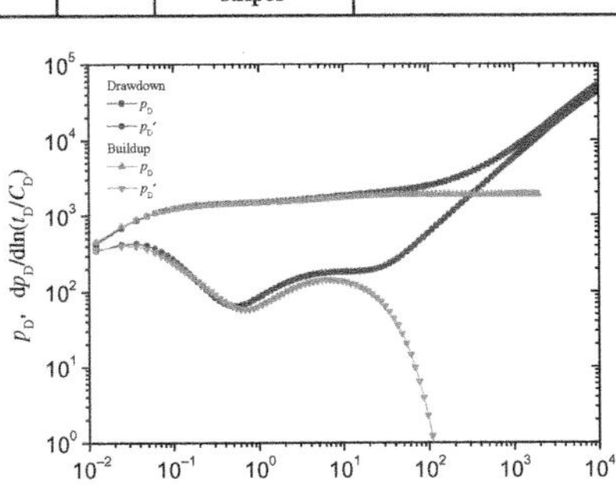

(1A)

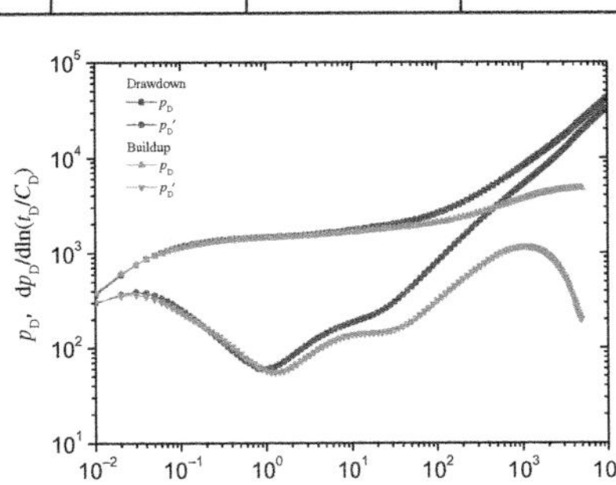

(2A)

Fig. 3.52 Typical log-log plots for six types of reservoirs.

(Continued)

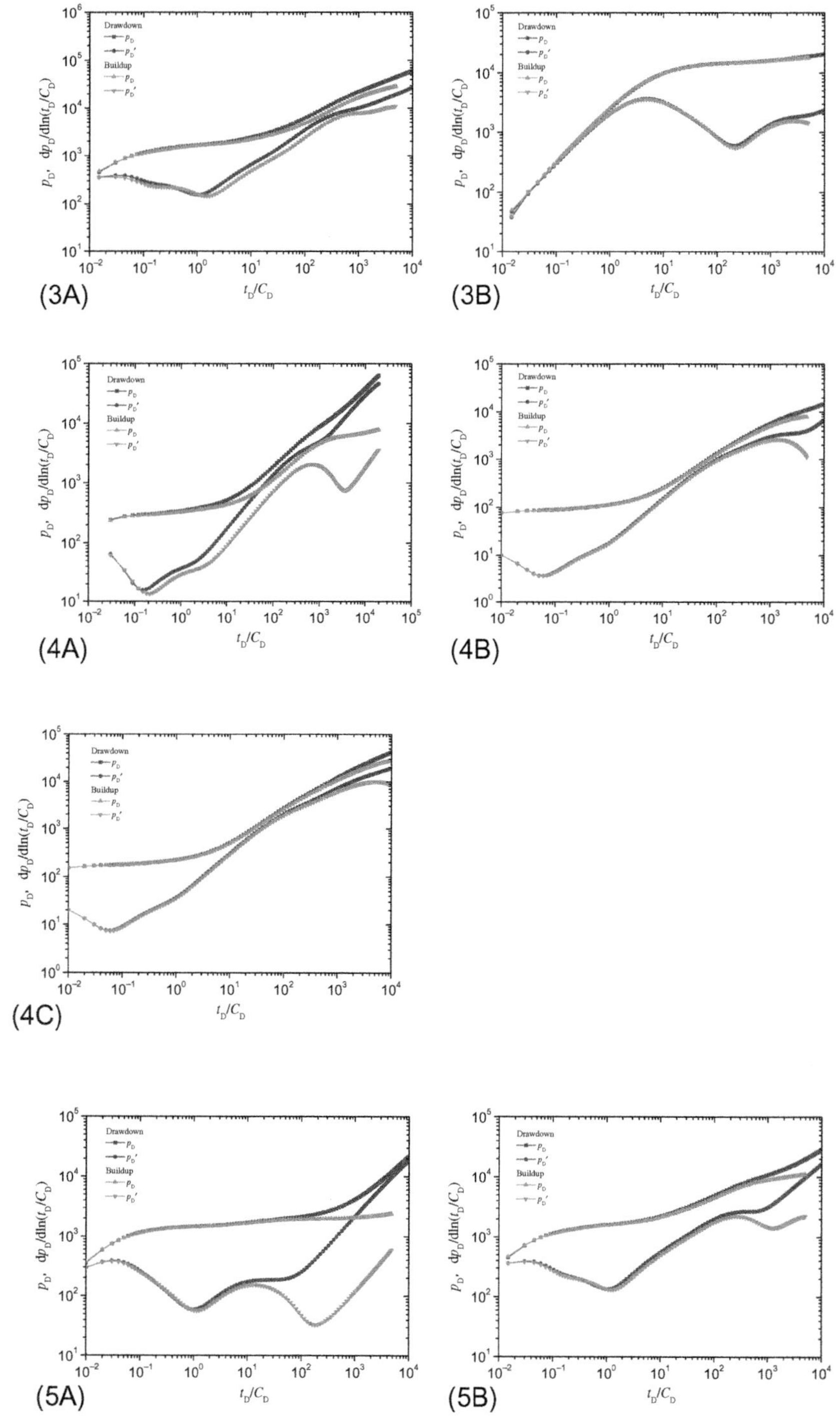

Fig. 3.52, cont'd

(Continued)

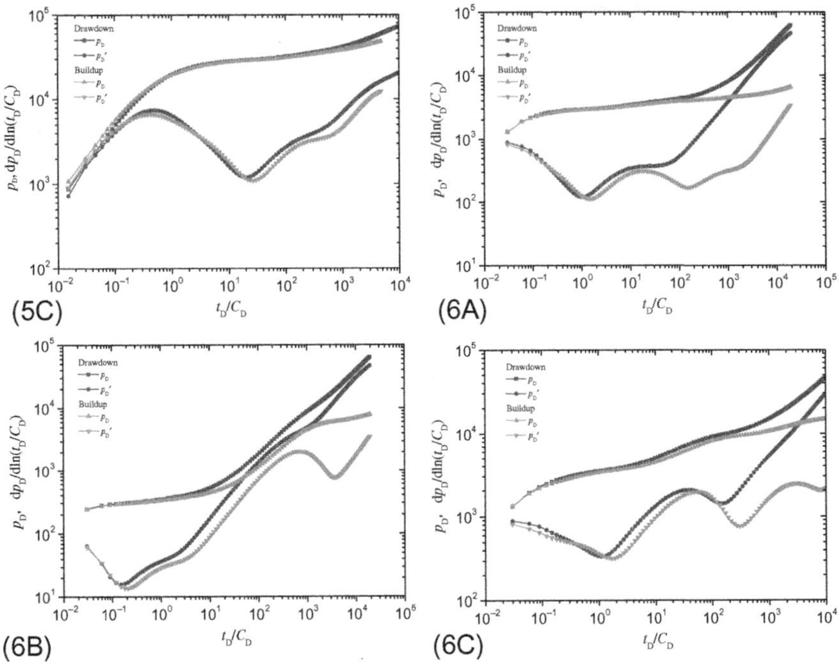

Fig. 3.52, cont'd

Typical well	Well test curve	Possible dynamical model			Delineation of fractures and vugs
G14-1	3-51-c	Well point A 	1 Circle + inner high permeability circle Well point 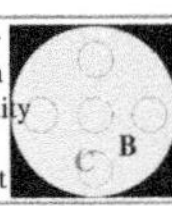	 3 Circles +2 high-permeability strips, Well point B	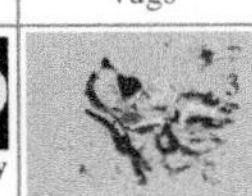
G43	3-51-a	1 Circle + high-permeability stripe Well point A 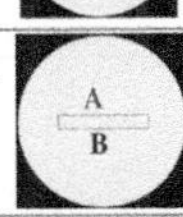			
G13	3-51-d	1 Circle + high-permeability stripe Well point A 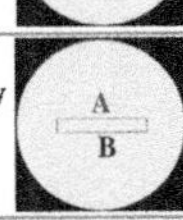	1 Circle + inner high permeability circle Well point 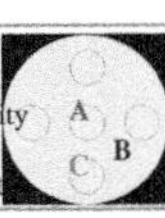		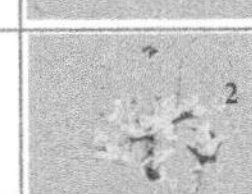
T201C	3-51-b	Well point B 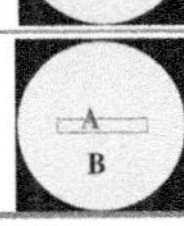		Well point C 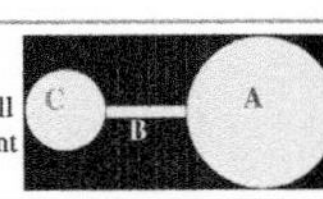	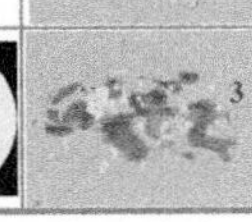

Fig. 3.53 Well test interpretation and fracture-vug delineation for typical wells in the Tazhong I gas field.

Table 3.5 Comprehensive identification and performance evaluation of fractured-vuggy carbonate reservoirs

Identification of reservoir type												
	Geological cognition			Performance analysis						Reservoir division		
No.	Drilling break and mud loss	Seismic reflections	Well log interpretation	Well test analysis	Production test	Production performance	Acid fracturing	Type of reservoir	Reservoir model	Classification	Description	Comprehensive description of reservoir
1	Yes	Bead-like	Cave	Large wellbore storage coefficient, fast pressure buildup, favorable physical properties of the inner zone of the composite model, missing or unobvious wellbore storage	High production	Large OGIP/OOIP	The pump-off PDD curve is almost straight and close to the formation pressure	Cave	Cave reservoir encountered	I	Cave reservoir + locally fractured-vuggy reservoir	(1) Favorable physical properties, strong heterogeneity, and high production according to well testing and production testing. (2) Large OGIP/OOIP and large drainage radius. (3) Strong bead-like reflection in dominance
2	No		Fractured	Small wellbore storage, fast pressure buildup, and obvious wellbore storage					Fractured reservoir encountered, connecting with the vuggy reservoir			

3	Small loss	Bead-like	Fractured	Small wellbore storage, slow pressure buildup, small difference in the physical properties of the inner zone of the composite model, obvious wellbore storage, partially with characteristics of the dual-porosity reservoir	Moderate production	Moderate OGIP/OOIP	The pump-off PDD curve slowly falls and is obviously higher than the formation pressure	Fractured-vuggy	Fractured-vuggy reservoir encountered	II	Fractured-vuggy reservoir	(1) Moderate physical properties, and relatively developed fractures (2) Moderate OGIP/OOIP (3) Weak bead-like or strong sheet-like reflection in dominance
4	No	Nonbead-like	No									
5	No	Nonbead-like	Vuggy	Small wellbore storage, slow pressure buildup, obvious wellbore storage, partially with the characteristics of the pseudo-homogenous reservoir	Low production	Small OGIP/OOIP	The pump-off PDD curve is horizontal and much higher than the formation pressure	Vuggy	Vuggy reservoir encountered	III	Vuggy reservoir	(1) Poor physical properties and undeveloped fractures. (2) Small OGIP/OOIP and limited well drainage area. (3) Nonbead-like reflection in dominance.

3.3.2.1 *Cave reservoir*

The cave reservoir is the most popular type of fractured-vuggy carbonate reservoir, with caves and dissolved vugs with a diameter of >20 mm serving as the accumulation space. It presents bead-like reflections on the seismic profile, and often meets with drilling breaks and lost circulation during drilling. If no cave reservoir is directly encountered, the acid fracturing curve commonly suggests low pump-off pressure (<10 MPa), the pump-off PDD curve presents as a horizontal stabilization line, and the backflow rate of the acid liquid is low. Distinct large dark or black patches can be seen on FMI and EMI images. The conventional logging curve reveals relatively low resistivity, the deep and shallow lateral resistivity logging curves are significantly different, and the tri-porosity logging curve reflects great changes, indicating that the reservoir has favorable porosity and permeability. The formation capacity obtained by well test interpretation is usually >100 mD m; at the initial stage of production, the tubing pressure is high (>20 MPa), the production is high and stable, and the tubing pressure drops linearly or exponentially; the performance in oil replacement by water injection and shut-in pressure coning is good, as shown in Table 3.6.

3.3.2.2 *Fractured-vuggy reservoir*

The fractured-vuggy reservoir contains secondary vugs as the main accumulation space, and fractures with both seepage and storage capacity to communicate with the vugs. The reservoir is encountered without drilling breaks, with a loss of <200 m^3. If the reservoir is communicated through acid fracturing, the acid fracturing curve usually shows a high pump-off pressure, the pump-off PDD curve presents a fast decline or a slow decline, and the backflow rate of residual acid is 10%–50%. The conventional logging curves are characterized by low resistivity, low GR, and jumping of acoustic, neutron, and density; and the imaging logs demonstrate clearly that the vugs are connected by fractures or growing along fractures. The formation capacity obtained from well test interpretation is usually <100 mD m. At the initial stage of production, the tubing pressure is low (<20 MPa) and declines rapidly, and the production is low; the performance in oil replacement by water injection and shut-in pressure coning is poor, as shown in Table 3.7.

3.3.2.3 *The vuggy reservoir*

When a gas well is drilled into a vuggy reservoir, there are usually no obvious engineering anomalies, such as drilling breaks and mud loss, according to the

Table 3.6 Comprehensive identification criteria for cave reservoirs (gas wells)

Type	Loss (m^3)	Drilling break	Acid fracturing	Formation capacity (mD·m)	Production test performance	Logging
Cave	Large loss >500	No	Low pump-off pressure (<10 MPa), the horizontal stabilization line for the pump-off PDD curve, and low backflow rate of acid liquid	>100	High tubing pressure (>20 MPa), high and stable production, linear or exponential drop of tubing pressure, suggesting good performance in oil replacement by water injection and shut-in pressure coning	Caves, vugs and fractures-vugs by electrical logging interpretation
		Yes	Direct production			Commonly no logging

Table 3.7 Comprehensive identification criteria for the fractured-vuggy reservoir (gas well)

Type	Drilling break	Loss (m^3)	Acid fracturing	Formation capacity (mD·m)	Production test performance	Logging
Fractured-vuggy	No	Small loss <200 No	High pump-off pressure, fast decline or slow decline of the pump-off PDD curve, and the residual acid backflow rate of 10%–50% Direct production	<100	Low tubing pressure (<20 MPa), fast tubing pressure decline, low production, poor performance in oil replacement by water injection and shut-in pressure coning	Fractures-vugs; imaging logs present meso-microfractures with certain apertures and density

drilling and completion data. Logging interpretation suggests the dominance of low-porosity and low-permeability vuggy reservoirs, which present "chaotic weak reflections" on the seismic profile.

3.3.2.4 Comprehensive evaluation

Based on the analysis of the static and dynamic data for 257 producing wells in typical blocks in the Tazhong I gas field, the reservoir types are identified (Table 3.8). The proportion of cave reservoirs is >82% and increases gradually from the east to the west.

3.3.3 Evaluation of flow patterns

The well test interpretation and the statistics of dynamic and static data for typical wells in the Tazhong I gas field are shown in Table 3.9. In spite of the bead-like reflections on the seismic profile, the reservoirs are mainly cave and fractured-vuggy types. According to the well test interpretation, the average formation capacity is 415 mD m and the average permeability is 13.1 mD, suggesting medium-low-permeability reservoirs, for which the classical flow theory is applicable for performance analysis. In addition, the average production rate and production pressure difference obtained in the production test and the production index obtained in well testing also suggest the characteristics of medium-low-permeability reservoirs.

Table 3.8 Reservoir types in typical blocks, the Tazhong I gas field

Block	Reservoir type	Subtype	Number of wells	Proportion
I	Cave	Single reservoir body Multiple reservoir bodies	54	70%
	Fractured-vuggy		23	35%
II	Cave	Single reservoir body Multiple reservoir bodies	115	85%
	Fractured-vuggy		21	15%
III	Cave	Single reservoir body Multiple reservoir bodies	42	95%
	Fractured-vuggy		2	5%

Table 3.9 Well test interpretation and drilling, well testing, and production test results for typical wells in the Tazhong I gas field

			Well test interpretation			Well testing performance				Production test for 1 year	
Well no.	Drilling break	Reservoir type	Model	*Kh*	Skin factor	Nozzle	Tubing pressure	Daily gas production	Production index	Daily gas production	producing pressure drop
	m			mD·m		mm	MPa	$10^4 m^3$	$10^4 m^3/MPa$	$10^4 m^3$	MPa
G14-1		Cave	Composite	1620	1.8	5	49.1	15	3.2	10.8	6.2
G12	2.3	Small cave	Homogeneous	100.3	11.3	6	43.4	17	1.5	5.2	4.8
G11C	4.7	Small cave	Dual-porosity boundary	44	30	5	30	9.5	0.4	5.6	8.6
G462		Cave	Dual-porosity boundary	194	−0.9	4	39.9	7.4	18.5	6.9	4.0
G10		Cave	Composite	3540	−2.3	6	49.6	21.9	166.7	13.4	5.5
G43		Cave	Composite	583	−6.4	6	39.2	14	81.9	7.7	10.0
T201C		Cave	Composite	27.8	1.5	4	35	7.6	0.7	8.8	4.5

YM2 is a typical cave reservoir. Its well test characteristics are shown in Fig. 3.54, and its interpretation results are shown in Table 3.10. The results are obviously different from those in the Tazhong I gas field. For example, the wells that encounter large caves generally demonstrate a large wellbore storage coefficient, high permeability, and the dominance of pipe flow or cavity flow.

3.3.4 Reserves estimation

The well test analysis technique with the constraint of whole history matching, which is designed to cope with such challenges as the diversity, ambiguity, and complexity of well test interpretation, can greatly improve the utilization of data, minimize the ambiguity of well test interpretation results, and estimate the OGIP/OOIP of a single well. This technique is essentially based on material balance, that is, the short-term well test is combined with the long-term production performance to estimate the drainage area of a gas well and calculate the single-well performance-based reserves with the volumetric method.

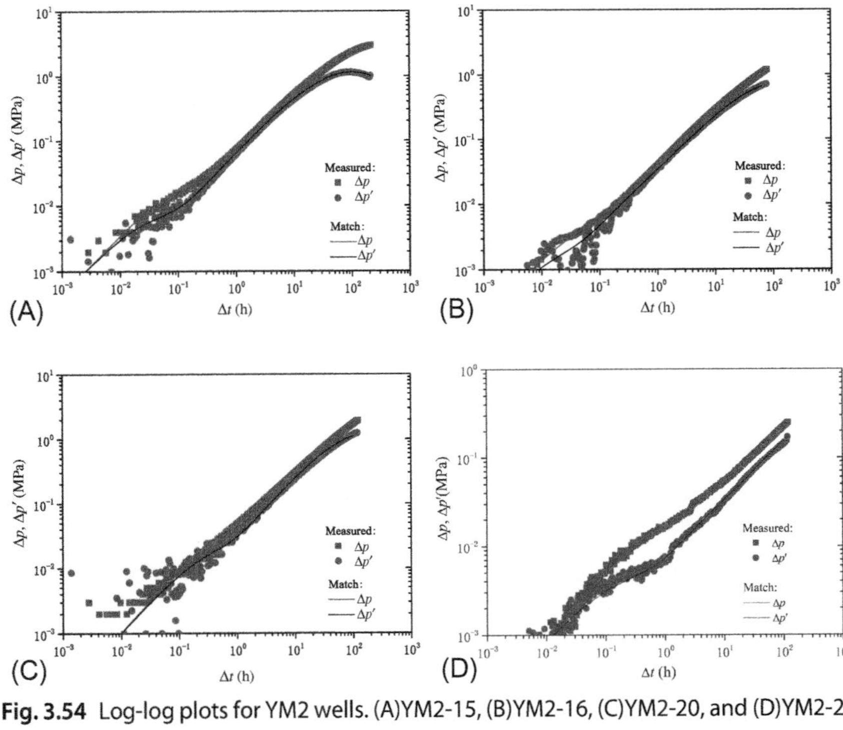

Fig. 3.54 Log-log plots for YM2 wells. (A)YM2-15, (B)YM2-16, (C)YM2-20, and (D)YM2-28.

Table 3.10 Characteristic parameters of typical YM2 wells in the well test

Well no.	Type curve	Wellbore storage coefficient (m^3/MPa)	Skin factor	Permeability (inner/outer) (mD)	Description
YM2-15	3-54-a	5.65	−4.64	1800/12.5	(1) The well test curve presents the "inverted bow shape," after radial flow, the curve keeps rising and gradually flattens in the late stage. (2) The wellbore storage coefficient is large (usually 1–20 m^3/MPa), and the permeability in the inner zone is in Darcy order of magnitude. (3) Characteristics of large caves are observed, and physical properties of the reservoir outside the caves are poor.
YM2-16	3-54-b	14.2	−4.44	1950/2.5	
YM2-20	3-54-c	21.85	−5.0	560/12.5	
YM2-28	3-54-d	32.9	−6.4	3100/10.5	

The single-well performance-based reserves can be determined by three methods. First, for the wells with a boundary reflection on the log-log plot, the volumetric method can be directly used to calculate the reserves. Second, for the wells with no boundary reflection on the log-log plot, the infinite approximation method is adapted to match the long-term production history to determine the boundary area, and then the performance-based reserves within the equivalent range of the current pressure wave propagation are obtained, as shown in Fig. 3.55 (where the current drainage radius of the well is 350 m). Third, for the wells with a long production time and the

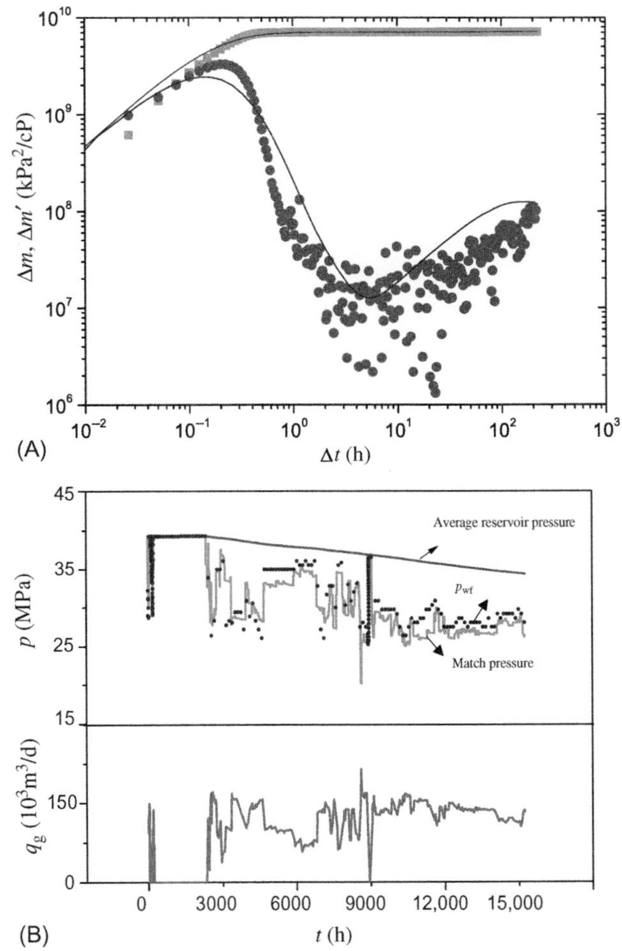

Fig. 3.55 Well test analysis matching plot with the infinite approximation method for Well A (boundary 350m). (A) Log-log plot and (B) pressure history matching plot.

PBU test performed, the deconvolution method is used to determine the boundary, as shown in Fig. 3.56 (where the well presents the obvious characteristics of the closed boundary, which is interpreted to be 500m).

When the analytical well test analysis method is ineffective for reservoir description, the numerical well test analysis method can be selected to describe the reservoir by properly adjusting the distribution characteristics of the reservoir section, as shown in Fig. 3.57.

Well C presents as a dual-porosity multizone composite model. In the area around the well, there is a dual-porosity reservoir with a radius of

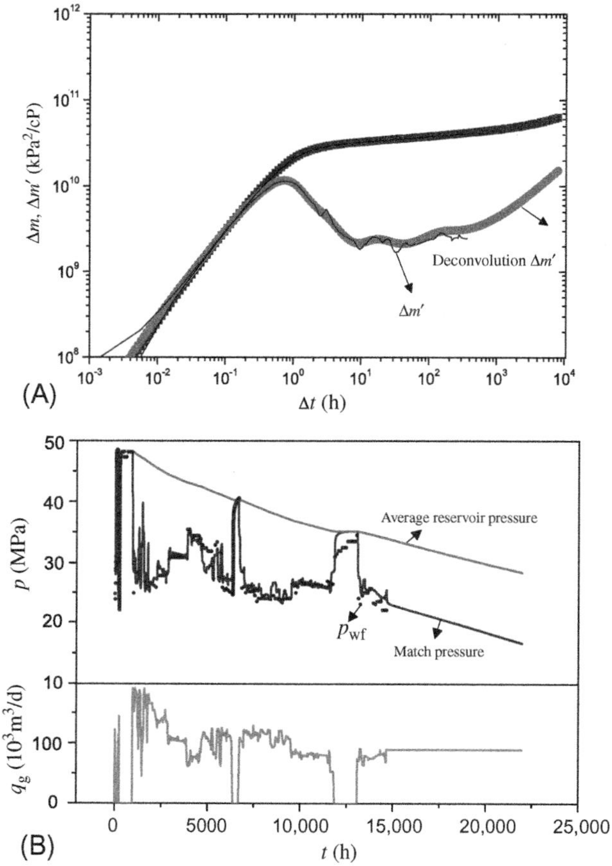

Fig. 3.56 Well test analysis matching plot with the deconvolution method for Well B. (A) Log-log plot and (B) pressure history matching plot.

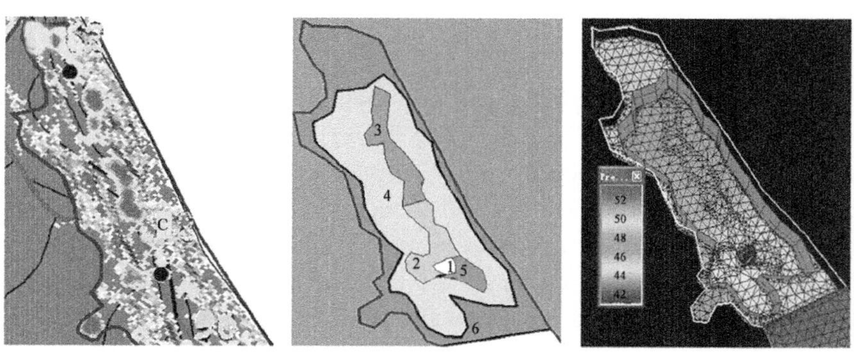

Fig. 3.57 Reservoir forecasting, classification of numerical well test heterogeneity, and pressure field distribution for Well C.

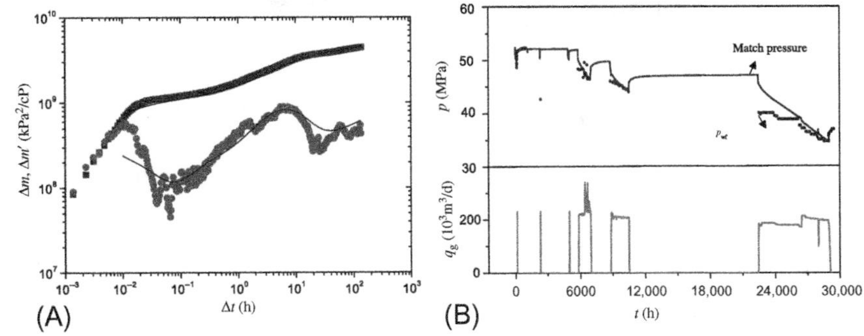

Fig. 3.58 Numerical well test analysis matching plot for Well C. (A) Log-log plot and (b) pressure history matching plot.

66 m, permeability of 5.2 mD, and storability ratio ω of 0.223, which are consistent with the drilling and geological results. The matching results are shown in Fig. 3.58.

Based on the single-well performance description through the combination of the short-term well test and long-term production performance, the volumetric method is used to calculate the single-well performance-based reserves. The single-well control area is 2.95 km^2, and the single-well performance-based reserves calculated by the well test performance description method are 6.0×10^8 m^3 (Table 3.11).

The technical process greatly enhances the function of well test analysis, which can be used not only in evaluating the reservoir stimulation performance and identifying reservoir parameters, but also in estimating performance-based reserves, forecasting production performance, allocating reasonable production, and analyzing multiwell connectivity.

Table 3.11 Numerical well test interpretation results for Well C

Area	*C* (m³/MPa)	*Kh* (mD·m)	S	ω	λ	M	D	*A* (km²)	OGIP (10⁸ m³)
1	0.18	221	−3.36	0.22	1.42E−05			0.014	0.040
2				0.01	8.20E−07	40	40	0.06	0.004
3				0.05	2.00E−06	2	0.2	0.17	2.476
4				0.10	2.00E−06	0.25	1	0.19	0.547
5						1.75	1.55	1.41	2.619
6						10	10	1.10	0.317
Total								2.95	6.000

3.3.5 Deliverability evaluation

The statistical results suggest that there is a clear relationship between well test characteristics and deliverability, as shown in Table 3.1.

(1) Pseudo-homogeneous model

The derivative curve has an obvious radial flow segment, which shows that the reservoir has good properties. It also demonstrates the best reservoir stimulation performance, with the highest success rate, favorable conditions for stable production, and high cumulative production of the single well.

(2) The dual-porosity model

The fractured-vuggy unit contains rich dissolved pores and microfractures, thus providing a certain accumulation space.

(3) The fractured model

The matrix porosity and permeability are poor, and the energy supply is insufficient. Accordingly, the deliverability is low and declines fast.

(4) The fractured-vuggy model

The matrix porosity and permeability are poor, and the deliverability is low. However, the energy supply is available, contributing to a certain period of stable production.

(5) Composite model

It is usually a model with poor peripheral conditions. The inner zone is rather limited, the flowing pressure drops rapidly, and there is essentially no stable production period.

3.4 Summary

In order to address such challenges as the diversity of well test curves, the ambiguity of well test interpretation, and the complexity of analytical models for fractured-vuggy carbonate gas reservoirs, the well test analysis technique, with the constraint of whole history matching, is proposed. It can effectively characterize the physical properties and heterogeneity of the reservoir and forecast the reasonable production allocation and production performance of gas wells under different conditions. The single-well performance-based reserves estimation method based on well test analysis can help to verify the reserves calculated by the advanced production decline analysis technique, thus minimizing the uncertainty of reserves estimation. The single-well performance forecasting (for reasonable production allocation) based on well test analysis facilitates the deliverability evaluation to upgrade from the traditional absolute open flow method to the material balance method, thus greatly reducing the uncertainty in production allocation.

The pressure buildup derivative curves have several rising "steps." Accordingly, the reservoir model may: (1) be a composite model with deteriorating periphery; (2) have one or more impermeable boundaries; (3) reflect the edge-water in case of gas wells, and (4) be subject to interwell interference. In both cases of multiwell simultaneous production and shut-in, the PDD and PBU derivative curves present the features of "stepped rise," and the ratio of each stabilization line-height to the first stabilization line height is the algebraic sum of the dimensionless production rate of the testing well and the affecting offset wells. In the multiwell system, if the offset wells are producing continuously, the PBU derivative curve of the testing well will gradually "downwarp" in the middle and late stages, and the "downwarping" speed depends on the algebraic sum of the dimensionless production rate of the affecting offset wells and the ratio of the shut-in time to the production time.

The analytical well test model is simple with regular boundaries; the two-dimensional numerical well test model takes into account the lateral heterogeneity; the multilayer three-dimensional numerical well test model considers both lateral heterogeneity and vertical heterogeneity and provides a mathematical description more approximate to the actual reservoir performance. The combination of the conventional analytical well test, the (2D and 3D) numerical well test, and advanced production decline analysis can greatly improve the reliability of interpretation results and the forecasting accuracy of key development indices.

Accurate formation pressure data acquired are fundamental for well test analysis. In conjunction with geological studies, the high-precision pressure gauge should be landed at a due time in the well test process into the middle part of the gas reservoir to acquire pressure data in a timely manner.

CHAPTER 4

Reserves estimation methods of fractured vuggy carbonate gas reservoirs

Contents

4.1 Performance-based reserves estimation

4.1.1 Definition of performance-based reserves

As described by Sun (2015), "performance-based reserves or original gas in place (OGIP)" refers to cumulative gas production obtained by gas reservoir engineering methods, provided that the flow rate of the gas well drops to zero, or the reservoir pressure within the drainage area drops to 1 atm. It is based on the production performance data (e.g., production, and pressure) of the well or reservoir, and under the condition that the current production technology and well pattern remain unchanged. The performance-based reserves: (1) can represent a gas reservoir or gas well; (2) are movable theoretically, and usually less than the reserves estimated by the volumetric method; (3) are obtained using dynamic data; (4) contain both recoverable

Dynamic Description Technology of Fractured Vuggy Carbonate Gas Reservoirs
https://doi.org/10.1016/B978-0-12-818324-3.00004-8

and nonrecoverable reserves, and reflect a quantity between the measured hydrocarbon initially in place and the estimated ultimate recovery, being dependent on the current well pattern and production technology; and (5) are time-sensitive.

4.1.2 Estimation methods of performance-based reserves

The performance-based reserves can be estimated by the material balance method, rate transient analysis method, well test method, and others (Table 4.1). In oilfield applications, these methods can be combined, depending on the specific characteristics of the gas reservoirs.

4.1.2.1 Material balance method

The material balance method is used to estimate the original gas in place (OGIP) based on the shut-in pressure test data of a gas reservoir (or condensate gas reservoir) at different production stages. Its theoretical basis is the law of conservation of mass.

The material balance method is often used for estimating the performance-based reserves with consideration of three aspects. First, proper calculation methods should be selected, depending on the type of gas reservoir (e.g., constant-volume gas reservoir, water-drive gas reservoir, ultra-high-pressure gas reservoir, and condensate gas reservoir). Second, a certain quantity of cumulative gas production should be available for the gas reservoir, and the recovery percent should be greater than 10% for conventional gas reservoirs, or even greater than 20% for the gas reservoirs in complex lithologic traps or with the multifracture system, low-permeability, or strong heterogeneity. Third, there should be adequate shut-in pressure data for the whole reservoir acquired through at least 3–5 tests. Thus, a reliable estimate can be ensured.

4.1.2.1.1 Material balance analysis of constant-volume gas reservoir

When the compressibility of rocks/fluids is neglected, and there is no water drive, the material balance equation for a constant-volume closed gas reservoir is expressed as:

$$G_p B_g = G\left(B_g - B_{gi}\right) \tag{4.1}$$

The preceding equation can be converted to a pressure drawdown (PDD) equation:

$$\frac{p}{Z} = \frac{p_i}{Z_i}\left(1 - \frac{G_p}{G}\right) \tag{4.2}$$

Table 4.1 Performance-based reserves estimation methods and range of applications

Type	Application conditions	Calculation method	Application range
Material balance method	(1) Gas reservoirs or gas wells have been produced at a certain recovery ratio. (2) Gas reservoirs or gas wells with shut-in pressure data.	Constant-volume gas reservoir	Constant-volume closed gas reservoir
		Water-drive gas reservoir	Water-drive index >0.1
		High and ultra-high-pressure gas reservoirs	Pressure coefficient >1.3
		Condensate gas reservoir	Condensate oil content $>50\,g/m^3$
Advanced production decline analysis	(1) Some flow rate data are available and shut-in pressure data are not necessary. (2) Gas flow in the wells reaches the pseudo-steady flow state.	Fetkovich method	Various gas reservoirs producing at constant pressure and a declining rate
		Blasingame method	Variable rate and variable pressure, boundary-dominated flow state
		Agarwal-Gardner method	
		NPI method	
		Flowing material balance method	
Elastic two-phase method	(1) Small-scale constant-volume gas reservoirs (2) Gas wells with transient well test data	Elastic two-phase method	Small-scale constant-volume closed gas reservoir, pressure drawdown test data under stable production conditions
Well test method	(1) Pressure buildup test (2) Production test	Volumetric method	Reach boundary

In a rectangular coordinate system, there is a linear relationship between p/Z and G_p, and the y-intercept of the straight line is the initial reservoir pseudo-pressure. When the straight line is extrapolated to $p/Z=0$, the intersection between the straight line and the x-axis represents the performance-based reserves of the gas reservoir.

Example 1

For a constant-volume gas reservoir (with the flow rate data shown in Table 4.2), the initial reservoir pressure is 12.397 MPa, the reservoir temperature is 330 K, the relative density of the gas is 0.6, the critical pressure of the gas is 4.71 MPa, and the critical temperature is 193 K.

Table 4.2 Average reservoir pressure and cumulative gas production of the constant-volume gas reservoir

Time t (a)	Average reservoir pressure p (MPa)	Cumulative gas production G_p (10^8 m^3)
0.0	12.397	0
0.5	11.584	0.2717
1.0	10.618	0.6009
1.5	9.846	0.9084
2.0	9.205	1.1094

The performance-based reserves of the reservoir are calculated in the following procedure:

(1) Determine the deviation factor Z based on the gas properties (Table 4.3).

(2) Determine p/Z (Table 4.3).

(3) Plot the p/Z vs. G_p relationship curve (Fig. 4.1)

(4) Derive the $p/Z-G_p$ relationship equation by the linear regression method:

$$p/Z=13.972-3.39399G_p$$

Table 4.3 Production calculation for the constant-volume gas reservoir

Time, t (a)	Reservoir pressure, p (MPa)	Cumulative gas production, G_p (10^8 m^3)	Deviation factor, Z	Apparent pressure, p/Z (MPa)
0.0	12.397	0	0.8854	14.00
0.5	11.584	0.2717	0.8890	13.03
1.0	10.618	0.6009	0.8941	11.88
1.5	9.846	0.9084	0.8989	10.95
2.0	9.205	1.1094	0.9032	10.19

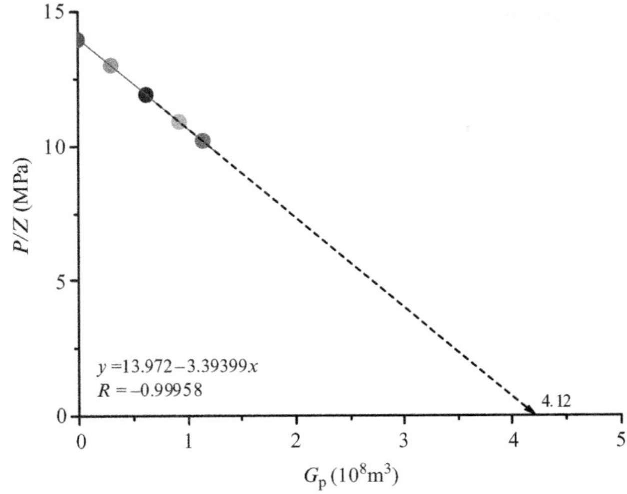

Fig. 4.1 Relationship between *p/Z* and G_p of the constant-volume gas reservoir.

According to the preceding equations, when $p/Z=0$, G_p represents the performance-based reserves:

$$G = 13.972/3.39399 = 4.12 \times 10^8 \text{ m}^3$$

Alternatively, the performance-based reserves are estimated by the linear extrapolation method to be 4.12×10^8 m^3.

4.1.2.1.2 Material balance analysis of water-drive gas reservoirs

The material balance equation for water-drive gas reservoirs is:

$$G = \frac{G_p B_g - 10^{-4}\left(W_e - W_p B_w\right)}{B_g - B_{gi}} \tag{4.3}$$

The PDD equation for the water-drive gas reservoirs is:

$$\frac{p}{Z} = \frac{p_i}{Z_i}\left[\frac{G - G_p}{G - 10^{-4}\left(W_e - W_p B_w\right)\dfrac{p_i T_{sc}}{p_{sc} Z_i T}}\right] \tag{4.4}$$

The water influx is calculated by different methods, depending on the geological conditions. If the natural water is considered to be a transient flow, then:

$$W_e = B \Delta p Q(t_D)$$

Eq. (4.3) can be transformed to:

$$\frac{G_p B_g + 10^{-4} W_p B_w}{B_g - B_{gi}} = G + 10^{-4} B \frac{\sum_0^t \Delta p Q(t_D)}{B_g - B_{gi}} \tag{4.5}$$

Let

$$y = \frac{G_p B_g + 10^{-4} W_p B_w}{B_g - B_{gi}} \tag{4.6}$$

$$x = 10^{-4} \frac{\sum_0^t \Delta p Q(t_D)}{B_g - B_{gi}} \tag{4.7}$$

Then

$$y = G + Bx \tag{4.8}$$

Obviously, the material balance equation for a water-drive gas reservoir can also be simplified to a linear relationship. The intercept of the straight line is the original gas in place (OGIP) of the gas reservoir, and the slope of the straight line is the water invasion coefficient.

In Eq. (4.8), the left side is a known quantity, while the right side includes unknown quantities, so it is difficult to calculate the x value by strictly using the flow model. In practice, two empirical graphic methods have been developed for material balance analysis of water-drive gas reservoirs, based on the Schilthuis steady-state influx model. The product of production time and reservoir pressure variation is directly proportional to water influx. In the early and middle stages of gas reservoir development with a lower gas flow rate, the time variation is much larger than the reservoir pressure variation, and water influx is approximately proportional to time. Thus, it is inferred that, under the preceding conditions, if the gas reservoir remains stable, water influx and cumulative gas production are also in a proportional relationship. Limited by the application scope of the Schilthuis steady-state influx model and some assumptions in subsequent derivations, the preceding two methods are not applicable in all cases.

The relationship between y and the production time t, or between y and cumulative gas production G_p can be plotted. If there is a good linear relationship, it indicates that the empirical methods are applicable to the

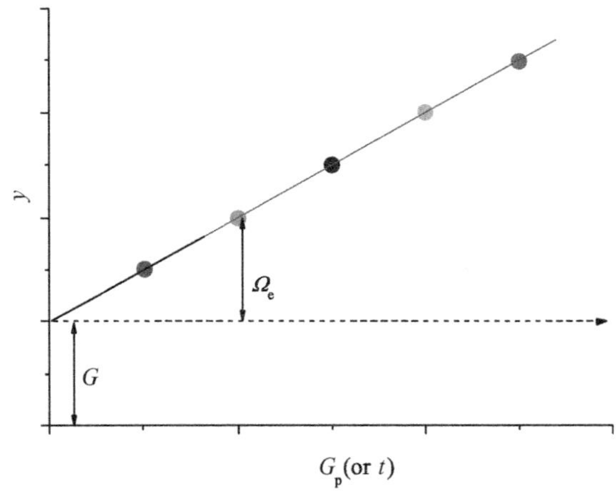

Fig. 4.2 Relationship between y and G_p (or t) of the water-drive gas reservoir.

example reservoir. The intercept of the straight line obtained by the regression method is the original gas in place G, the difference between the cumulative gas production G_p or production time t on the straight line, and the stabilization line of OGIP is the corresponding cumulative pseudo-influx (Fig. 4.2), and the cumulative water influx can be obtained by multiplying the cumulative pseudo-influx with $(B_g - B_{gi})$.

When the unsteady-state water invasion is considered, theoretical research shows that, for the radial water influx model of an infinite aquifer, the dimensionless water influx $Q(t_D) = 2\sqrt{\frac{t_D}{\pi}}$ when $0 < t_D < 0.01$, that is, in the case of a large edge-water area, the water influx in the early production stage is positively proportional to $\sum_0^t \Delta p\sqrt{t}$; for the linear water influx model of an infinite aquifer, $Q(t_D) = 2\sqrt{\frac{t_D}{\pi}}$ at any time. Reserves and water influx can be calculated by plotting the relationship between y and $\frac{\sum_0^t \Delta p\sqrt{t}}{B_g - B_{gi}}$.

Example 2

For a belt-shaped water-drive gas reservoir (with the flow rate data for the first three years of production shown in Table 4.4), the initial reservoir pressure is 19.7 MPa, the original volume factor is 5.3601×10^{-3}, and the OGIP obtained by the volumetric method is $(339.85 - 489.94) \times 10^8\ m^3$.

Continued

Example 2—cont'd

Table 4.4 Flow rate data of the water-drive gas reservoir

No.	Time, t (mon)	Cumulative gas production, G_p (10^8 m^3)	Cumulative water production, W_p (10^4 m^3)	Reservoir pressure, p (MPa)
0	0	0	0	19.7
1	12	6.88	0	19.4
2	20	17.0	0	19.0
3	28	27.5	0	18.7
4	36	37.3	0	18.2

(1) $y \sim t$ analysis

(a) Determine B_g under different pressures according to the gas properties (Table 4.5).

(b) Determine y according to Eq. (4.6) (Table 4.5).

(c) Plot the relationship between y and t (Fig. 4.3), and perform the linear regression to obtain the performance-based reserves (432.21×10^8 m^3).

(2) $y \sim G_p$ analysis

The relationship between y and G_p (Fig. 4.4) is plotted using the data in Table 4.3, and the linear regression is performed to obtain the performance-based reserves (438.73×10^8 m^3).

(3) $y \sim \sum_0^t \Delta p \sqrt{t}$ analysis

(a) Determine B_g under different pressures according to the gas properties (Table 4.6).

(b) Determine y according to Eq. (4.6) (Table 4.6).

(c) Determine $\sum_0^t \Delta p \sqrt{t}$.

(d) Determine x according to the formula in Table 4.6.

Table 4.5 Basic data of the water-drive gas reservoir obtained by the empirical graphic method

No.	t (mon)	G_p (10^8 m^3)	W_p (10^4 m^3)	p (MPa)	B_g	y (10^8 m^3)
0	0	0	0	19.7	0.005360	0
1	12	6.88	0	19.4	0.005444	445.9
2	20	17.0	0	19.0	0.005569	453.0
3	28	27.5	0	18.7	0.005701	459.8
4	36	37.3	0	18.2	0.005820	471.9

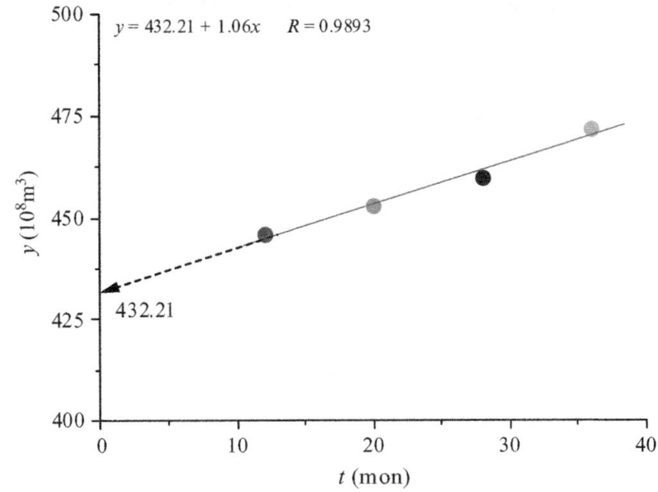

Fig. 4.3 Relationship between y and t of the water-drive gas reservoir.

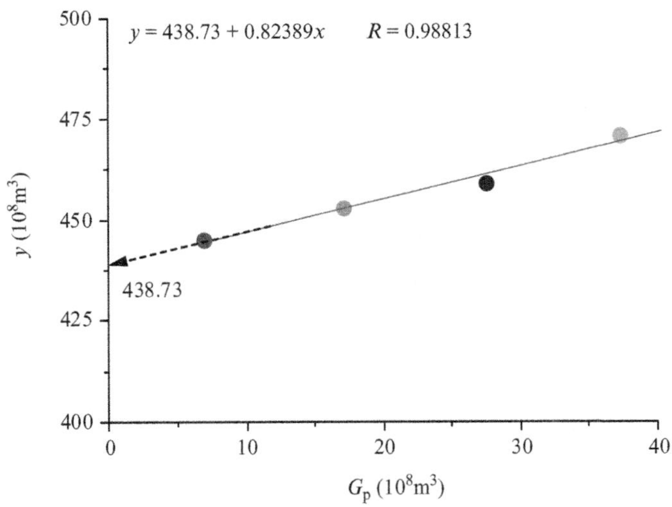

Fig. 4.4 Relationship between y and G_p of the water-drive gas reservoir.

(e) Plot the relationship between y and x (Fig. 4.5), and perform the linear regression to obtain the performance-based reserves (413.1×10^8 m^3).

4.1.2.1.3 Material balance analysis of ultra-high pressure gas reservoirs

The material balance equation for ultra-high pressure gas reservoirs is:

$$G_pB_g = G(B_g - B_{gi}) + GB_{gi}\left(\frac{C_wS_{wi} + C_f}{1 - S_{wi}}\right)\Delta p \tag{4.9}$$

Table 4.6 Basic data of the water-drive gas reservoir obtained by the approximate solution method

No.	t (mon)	G_p (10^8 m^3)	W_p (10^4 m^3)	p (MPa)	B_g	y (10^8 m^3)	$\sum_0^t \Delta p \sqrt{t}$ (MPa $\sqrt{\text{mon}}$)	$x = \frac{\sum_0^t \Delta p \sqrt{t}}{B_g - B_{gi}}$ (MPa $\sqrt{\text{mon}}$)
0	0	0	0	19.7	0.005360	0.0	0.0	
1	12	6.88	0	19.4	0.005444	445.9	0.52	6190.48
2	20	17.0	0	19.0	0.005569	453.0	1.66	7952.15
3	28	27.5	0	18.7	0.005701	459.8	3.17	9310.85
4	36	37.3	0	18.2	0.005820	471.9	5.18	11,260.87

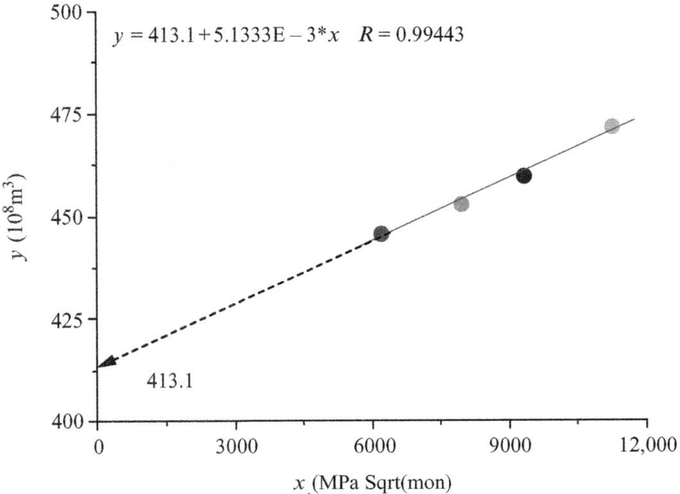

Fig. 4.5 Relationship between y and $\frac{\sum_0^t \Delta p\sqrt{t}}{B_g - B_{gi}}$ of the water-drive gas reservoir.

The PDD equation is:

$$\frac{p}{Z} = \frac{p_i}{Z_i}\left(\frac{1 - G_p/G}{1 - C_e\Delta p}\right) \tag{4.10}$$

where, $\Delta p = p_i - p$.

Apparent reservoir pressure correction method Eq. (4.10) can be transformed to:

$$\frac{p}{Z}(1 - C_e\Delta p) = \frac{p_i}{Z_i} - \frac{p_i}{Z_i}\frac{G_p}{G} \tag{4.11}$$

According to Eq. (4.11), on the $\frac{p}{Z}(1 - C_e\Delta p)$ versus the G_p relationship curve, the intercept of the straight line is $a = p_i/Z_i$ and the slope of the straight line is $b = -\frac{p_i}{Z_i}\frac{1}{G}$. When the straight line is extrapolated to $\frac{p}{Z}(1 - C_e\Delta p) = 0$, the intersection between the straight line and the x-axis is the original gas in place G of the gas reservoir.

Example 3

For an ultra-high pressure gas reservoir in offshore Louisiana, in the United States (with the flow rate data shown in Table 4.7), the burial depth is 4055 m, the initial reservoir pressure is 78.904 MPa, the reservoir temperature is 128.4°C, the relative density of gas is 0.6, the initial water saturation is 0.22, and the compressibility of formation water is 4.41×10^{-4} MPa^{-1}.

Continued

Example 3—cont'd

Table 4.7 Flow rate data of the ultra-high pressure gas reservoir in offshore Louisiana (Billy and Fred, 1981, Permission to publish by the SPE, Copyright SPE)

No.	Date	G_p (10^8 m^3)	p (MPa)
0	Jan. 25, 1966	0	78.904
1	Feb. 1, 1967	2.809	73.595
2	Feb. 1, 1968	8.105	69.851
3	June 1, 1969	15.180	63.797
4	June 1, 1970	21.997	59.116
5	June 1, 1971	28.723	54.510
6	June 1, 1972	34.087	50.883
7	Sep. 1, 1973	41.067	47.208
8	Aug. 1, 1974	45.491	44.044
9	Aug. 1, 1975	51.640	40.176
10	Aug. 1, 1976	55.998	37.294
11	Aug. 1, 1977	61.076	34.474
12	Aug. 1, 1978	66.763	31.026
13	Aug. 1, 1979	69.640	28.751

The reserves of the ultra-high pressure gas reservoir are calculated in the following procedure:

(1) Determine Z under different pressures according to the gas properties (Table 4.8).

(2) Determine the rock compressibility of the reservoir using the empirical formula depending on the burial depth:

$$c_f = \left(8.82 \times 10^{-3} H - 2.51\right) \times 10^{-4}$$
$$c_f = \left(8.82 \times 10^{-3} \times 4055 - 2.51\right) \times 10^{-4} = 3.325 \times 10^{-3}\ \text{MPa}^{-1}$$

(3) Determine c_e:

$$c_e = \left(4.41 \times 10^{-4} \times 0.22 + 33.25 \times 10^{-4}\right)/(1 - 0.22)$$
$$= 4.387 \times 10^{-3}\ \text{MPa}^{-1}$$

(4) Determine $p(1 - c_e\Delta p)/Z$, as shown in Table 4.8.

(5) Plot the relationship between $p(1 - c_e\Delta p)/Z$ and G_p (Fig. 4.6), and obtain the performance-based reserves (126.16×10^8 m^3).

Table 4.8 Reservoir pseudo-pressure of the ultra-high pressure gas reservoir

No.	Date	G_p (10^8 m^3)	p (MPa)	Z	$p(1-c_e\Delta p)/Z$ (MPa)
0	Jan. 25, 1966	0	78.904	1.496	52.743
1	Feb. 1, 1967	2.809	73.595	1.438	49.987
2	Feb. 1, 1968	8.105	69.851	1.397	48.015
3	June 1, 1969	15.180	63.797	1.330	44.789
4	June 1, 1970	21.997	59.116	1.280	42.175
5	June 1, 1971	28.723	54.510	1.230	39.574
6	June 1, 1972	34.087	50.883	1.192	37.440
7	Sep. 1, 1973	41.067	47.208	1.154	35.220
8	Aug. 1, 1974	45.491	44.044	1.122	33.252
9	Aug. 1, 1975	51.640	40.176	1.084	30.766
10	Aug. 1, 1976	55.998	37.294	1.057	28.842
11	Aug. 1, 1977	61.076	34.474	1.033	26.868
12	Aug. 1, 1978	66.763	31.026	1.005	24.387
13	Aug. 1, 1979	69.640	28.751	0.998	22.470

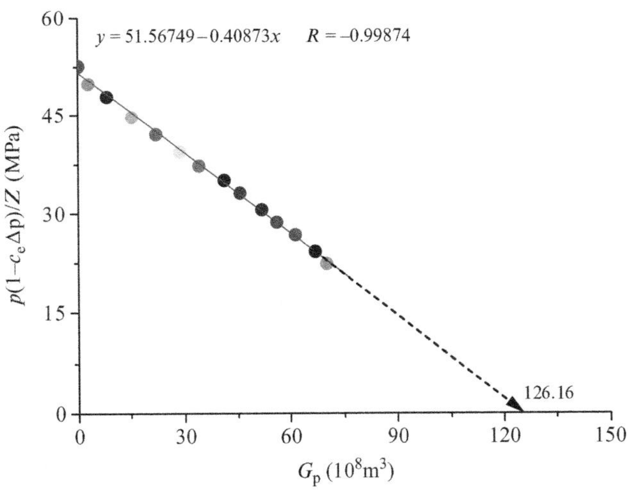

Fig. 4.6 Apparent reservoir pressure of the ultra-high-pressure gas reservoir.

Linear regression method Eq. (4.10) is modified to:

$$\frac{1-\psi}{\psi\Delta p} = -c_e + \frac{1}{G}\left(\frac{G_p}{\psi\Delta p}\right) \tag{4.12}$$

where,

$$\psi = \frac{p/Z}{p_i/Z_i} \tag{4.13}$$

Let

$$y = (1 - \Psi)/\Psi \Delta p \quad (4.14)$$

$$x = G_p/\psi \Delta p \quad (4.15)$$

$$a = -c_e \quad (4.16)$$

$$b = 1/G \quad (4.17)$$

Then

$$y = a + bx \quad (4.18)$$

According to Eqs. (4.16), (4.17), the absolute value of the intercept a in Eq. (4.18) is equal to the effective compressibility of the gas reservoir, and the reciprocal of the slope is equal to the OGIP of the gas reservoir.

The simple regression analysis is conducted as follows:

(1) Determine Z under different pressures according to the gas properties (Table 4.9).

(2) Determine ψ according to Eq. (4.13) (Table 4.9).

(3) Determine x and y according to Eqs. (4.14), (4.15) (Table 4.9).

(4) Plot the relationship between x and y (Fig. 4.7), and calculate the performance-based reserves, that is, $G = 1/0.00763 = 131.06 \times 10^8\ \text{m}^3$. The intercept obtained by the graphic method, that is, the effective compressibility, is $0.00332\,\text{MPa}^{-1}$.

Analytical method For a constant-volume closed ultra-high-pressure gas reservoir, the p/Z versus G_p relationship curve is usually manifested in a fold line. The first straight line segment represents the influence of the recompaction of the ultra-high-pressure gas reservoir, and the original gas in place derived by extrapolating this segment to $p/Z = 0$ is the pseudo-OGIP G_{pseudo}. The second straight line segment indicates the normal pressure change after the recompaction of the reservoir disappears, and the original gas in place derived by extrapolating this segment to $p/Z = 0$ is the real-OGIP G_{real}. G_{real} can be obtained according to G_{pseudo} with the following equation:

$$G_{real} = \frac{G_{pseudo}}{1 + \dfrac{c_e(p_i - p_{ws})}{\dfrac{p_i/Z_i}{p_{ws}/Z_{ws}} - 1}} \quad (4.19)$$

The real-OGIP can be obtained according to the intercept a_1 and slope b_1 of the first straight line segment using the following equation:

Table 4.9 Data of the ultra-high gas reservoir obtained by the linear regression method

No.	Date	G_p (10^8 m^3)	p (MPa)	Z	Ψ	y (1/MPa)	x(10^8 m^3/MPa)
0	Jan. 25, 1966	0	78.904	1.496	1.000		
1	Feb. 1, 1967	2.809	73.595	1.438	0.970	0.0058	0.5453
2	Feb. 1, 1968	8.105	69.851	1.397	0.948	0.0061	0.9444
3	June 1, 1969	15.180	63.797	1.330	0.909	0.0066	1.1049
4	June 1, 1970	21.997	59.116	1.280	0.876	0.0072	1.2695
5	June 1, 1971	28.723	54.510	1.230	0.840	0.0078	1.4013
6	June 1, 1972	34.087	50.883	1.192	0.809	0.0084	1.5031
7	Sep. 1, 1973	41.067	47.208	1.154	0.776	0.0091	1.6705
8	Aug. 1, 1974	45.491	44.044	1.122	0.744	0.0099	1.7534
9	Aug. 1, 1975	51.640	40.176	1.084	0.703	0.0109	1.8975
10	Aug. 1, 1976	55.998	37.294	1.057	0.669	0.0119	2.0118
11	Aug. 1, 1977	61.076	34.474	1.033	0.633	0.0131	2.1726
12	Aug. 1, 1978	66.763	31.026	1.005	0.585	0.0148	2.3824
13	Aug. 1, 1979	69.640	28.751	0.998	0.546	0.0166	2.5422

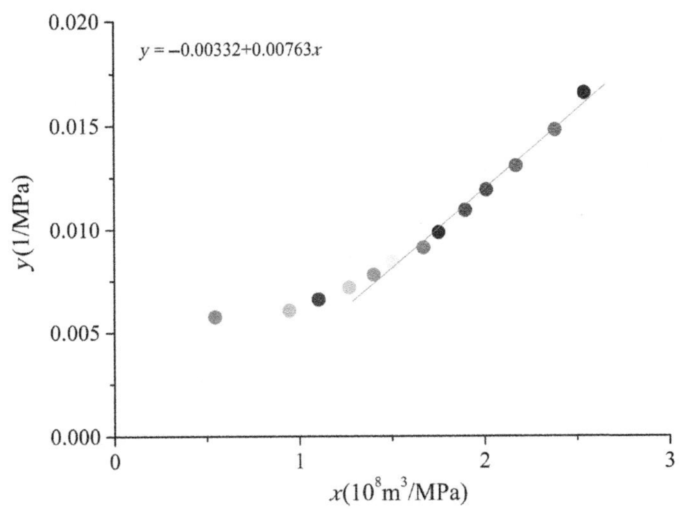

Fig. 4.7 Linear regression analysis for the ultra-high pressure gas reservoir.

$$G_{\text{real}} = \frac{a_1/b_1}{1 + \dfrac{c_e(p_i - p_{ws})}{\dfrac{p_i/Z_i}{p_{ws}/Z_{ws}} - 1}} \tag{4.20}$$

The analytical analysis is completed in the following procedure:

(1) Determine Z under different pressures according to the gas properties (Table 4.10).

(2) Determine p/Z (Table 4.10).

(3) Plot the $p/Z - G_p$ relationship curve, and perform regression analysis for the first straight line segment (Fig. 4.8).

(4) Calculate the performance-based reserves.

At the burial depth of 4055 m, the hydrostatic pressure is 40.55 MPa, the deviation factor is 1.09, and the hydrostatic apparent-pressure is 37.202 MPa. After regression, the intercept and slope of the first straight line segment are 52.3448 and −0.28255, respectively. The pseudo-OGIP of the gas reservoir is:

$$G_{\text{pseudo}} = 52.3448/0.28255 = 185.26 \times 10^8 \text{ m}^3$$

Substituting the preceding parameters into Eq. (4.20), the performance-based reserves are obtained:

Table 4.10 Data of the ultra-high pressure gas reservoir obtained by the analytical method

No.	Date	G_p (10^8 m^3)	p (MPa)	Z	p/Z
0	Jan. 25, 1966	0	78.904	1.496	52.743
1	Feb. 1, 1967	2.809	73.595	1.438	51.179
2	Feb. 1, 1968	8.105	69.851	1.397	50.001
3	June 1, 1969	15.180	63.797	1.330	47.968
4	June 1, 1970	21.997	59.116	1.280	46.184
5	June 1, 1971	28.723	54.510	1.230	44.317
6	June 1, 1972	34.087	50.883	1.192	42.687
7	Sep. 1, 1973	41.067	47.208	1.154	40.908
8	Aug. 1, 1974	45.491	44.044	1.122	39.255
9	Aug. 1, 1975	51.640	40.176	1.084	37.063
10	Aug. 1, 1976	55.998	37.294	1.057	35.283
11	Aug. 1, 1977	61.076	34.474	1.033	33.373
12	Aug. 1, 1978	66.763	31.026	1.005	30.872
13	Aug. 1, 1979	69.640	28.751	0.998	28.809

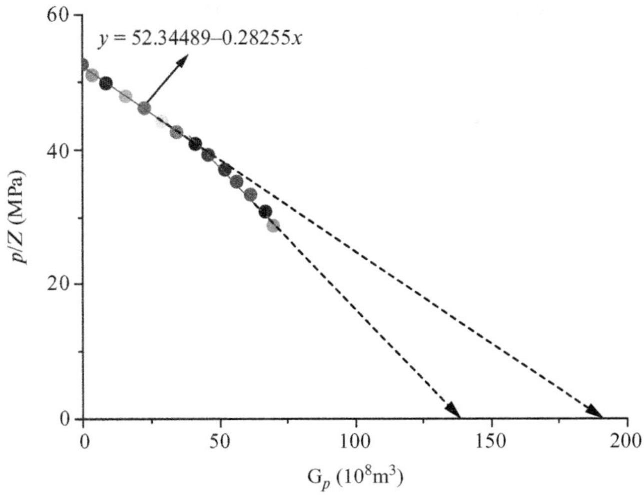

Fig. 4.8 Analysis of the first straight line segment of the ultra-high pressure gas reservoir.

$$G_{\text{real}} = \frac{a_1/b_1}{1 + \frac{c_e(p_i - p_{ws})}{\frac{p_i/Z_i}{p_{ws}/Z_{ws}} - 1}} = \frac{185.26}{1 + \frac{4.387 \times 10^{-3}(78.904 - 40.55)}{\frac{52.743}{37.202} - 1}}$$
$$= 132.07\left(10^8\ \text{m}^3\right)$$

4.1.2.1.4 Material balance analysis of condensate gas reservoirs

The PDD equation used for estimating the reserves of the dry gas reservoir is adopted. Assuming that the initial reservoir pressure is higher than the dew point pressure, the rock compressibility is ignored, there is no water vapor, and the adsorption of the porous medium is not considered; then the material balance equation is:

$$G_i B_{gi} = \left(G_t - G_{pt}\right) B_g + \frac{S_o\left(G_t - G_{pt}\right) B_g}{1 - S_o} \tag{4.21}$$

The condensate oil saturation in the core after the occurrence of retrograde condensation can be determined through multiple tests.

The PDD equation is:

$$(1 - S_o)\frac{p}{Z} = \frac{p_i}{Z_i}\left(1 - \frac{G_{pt}}{G_t}\right) \tag{4.22}$$

There is a linear relationship between $(1-S_o)p/Z$ and G_{pt}. Then, the OGIP can be calculated by extrapolating the PDD curve and linear regression.

$$G_{pt} = G_{sg} + 10^{-4} N_o \overline{E}_{og} \tag{4.23}$$

Where,

$$\overline{E}_{og} = 24{,}056 \times \frac{\gamma_o}{M_o} \tag{4.24}$$

The average molecular weight of the condensate can be calculated by the following equation:

$$M_o = \frac{44.29\gamma_o}{1.03-\gamma_o} \tag{4.25}$$

If the reservoir pressure is lower than the dew point pressure, or the condensate gas reservoir contains an oil rim, then it is two-phase in the original state, and the corresponding material balance equation can be expressed as:

$$G_t B_{2gi} = \left(G_t - G_{pt}\right) B_{2g} \tag{4.26}$$

Converting the volume factor, then:

$$\frac{p}{Z_{2g}} = \frac{p_i}{Z_{2gi}}\left(1-\frac{G_{pt}}{G_t}\right) \tag{4.27}$$

At present, Z_{2g} and Z_{2gi} can only be determined through laboratory experiments.

It can be seen that p/Z_{g2} is in a linear relationship with G_{pt}, through which the OGIP can be determined.

4.1.2.2 Elastic two-phase method

For small-scale constant-volume elastic gas-drive closed gas reservoirs, single-well fracture systems, small faulted gas reservoirs, or a single well with a relatively small drainage area, the fluid flow can more easily reach a pseudo-steady state. Based on the test data of pressure drawdown (straight line segment in Fig. 4.9) under stable production conditions, the elastic two-phase method can be adapted to estimate the OGIP.

Through pseudo-pressure analysis, we have:

$$m(p_{wf}) = a - bt \tag{4.28}$$

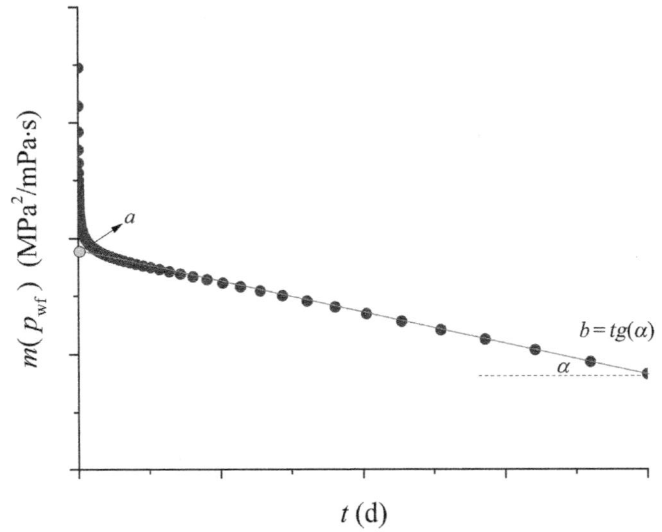

Fig. 4.9 PDD curve obtained by the elastic two-phase method.

Where,

$$a = m(p_R) - \frac{3.684 \times 10^4 q p_{sc} T}{Kh \quad T_{sc}} \left[\ln\left(\frac{r_e}{r_{wa}}\right) - \frac{3}{4} \right] \quad (4.29)$$

$$b = \frac{2 q p_R (1 - S_w)}{G c_t \mu Z} \quad (4.30)$$

Therefore, the OGIP in the well drainage area can be calculated based on the slope b of the straight line:

$$G = \frac{2 q p_R (1 - S_w)}{b c_t \mu Z} \quad (4.31)$$

The solution of Eq. (4.31) is the remaining reserves in the well drainage area under the reservoir pressure at the initial stage of the stable production. Z, μ, and c_t in the equation are parameters corresponding to this reservoir pressure.

Within the lower pressure range, there is an approximately linear relationship between pressure-squared and pseudo-pressure; the pressure-squared equation can help realize simplified and reliable calculation. Within the higher pressure range, there is an approximately linear relationship between pressure and pseudo-pressure; the pressure equation can be directly adopted for analysis. See Table 4.11 for details.

Table 4.11 Pressure, pressure-squared, and pseudo-pressure equations for the elastic two-phase method

Pressure	Pressure-squared	Pseudo-pressure
$p_{wf} \sim t$ $G = \frac{q(1-S_w)}{bc_t}$	$p_{wf}^2 \sim t$ $G = \frac{2qp_R(1-S_w)}{bc_t}$	$\psi(p_{wf}) \sim t$ $G = \frac{2qp_R(1-S_w)}{bc_t \mu Z}$
$c_t = S_g c_g + S_w c_w + c_f$		

The equations in the table take into account the water saturation of the reservoir, but they are not applicable to the gas-water two-phase flow with variable water saturation.

4.1.2.3 Lifecycle well test method

The well test analysis with the constraint of "whole history matching" can greatly improve the data utilization, minimize the ambiguity of well test interpretation results, and estimate the performance-based reserves of single-wells. This technique is essentially based on material balance; that is, the short-term well test is combined with the long-term production performance to estimate the drainage area of a gas well and calculate the single-well performance-based reserves with the volumetric method.

The single-well performance-based reserves can be determined by three methods. First, for the wells with a boundary response on the log-log plot, the volumetric method can be directly used to calculate the reserves. Second, for the wells with no boundary response on the log-log plot, the infinite approximation method is adapted to match the long-term production history to determine the boundary area, and then the performance-based reserves within the equivalent range of the current pressure wave propagation are obtained. Third, for the wells with a long production time and the PBU test performed, the deconvolution method is used to determine the boundary. See Section 3.3.4 in Chapter 3 for details.

4.1.2.4 Advanced production decline analysis method

This series of methods uses the gas well production performance data and type-curve matching to quantitatively analyze the single-well performance-based reserves and relevant formation parameters. The gas well production performance data must cover transient flow at an early stage and boundary-dominated flow at a late stage. The methods mainly include the traditional Fetkovich type curve and the modern Blasingame, Agarwal-Gardner, and NPI type curves (Sun, 2015).

The Fetkovich type-curve method only requires the flow rate data under constant pressure in the boundary-dominated flow production period, while the Blasingame, Agarwal-Gardner, and NPI type-curve methods use the performance data (e.g., bottom hole pressure and flow rate) during the production period for analysis, and also require the boundary-dominated flow production period (or the pseudo-steady flow in the case of constant-rates). All these methods are based on the theory of single-phase gas flow. If there is water invasion or retrograde condensation in the formation, these methods should be selected, depending on the actual situations. Because the bottom hole flowing pressure (BHFP) is not monitored continuously for most gas wells during production, it is usually converted from the wellhead pressure (WHP). If there is a two-phase flow in the wellbore, or the calculation by the flow gas column method in case of a high flow rate may cause a significant conversion error, the test data of the flow pressure gradient are used for the constraint.

4.1.2.4.1 Fetkovich method

Fetkovich introduced the transient flow equations applied in well test analysis into the decline analysis and established a log-log type-curve matching method that is very similar to the well test analysis. The dimensionless time for the decline curve is:

$$t_{\mathrm{Dd}} = \frac{3.6Kt}{\phi\mu c_t r_w^2} \cdot \frac{1}{\frac{1}{2}\left[\left(\frac{r_e}{r_w}\right)^2 - 1\right]\left[\ln\left(\frac{r_e}{r_w}\right) - \frac{1}{2}\right]} \tag{4.32}$$

The dimensionless decline flow rate function q_{Dd} is

$$q_{\mathrm{Dd}} = \frac{q(t)}{q_i} = q_D\left[\ln\left(\frac{r_e}{r_w}\right) - \frac{1}{2}\right] \tag{4.33}$$

In the boundary-dominated flow period, the Fetkovich-Arps type curve consists of two parts (Fig. 4.10). The first part represents the transient flow period, with the curve shape depending on the value of r_{eD}. The second part follows the Arps decline equation, with the curve shape depending on the value of the decline exponent b. When the actual production data are matched with the Fetkovich-Arps type curve, the second part is used to determine the Arps decline parameters q_i, D_i, and b, and the first part is used to determine r_{eD}; then, such parameters as the permeability K, skin factor S, drainage radius r_e, single-well

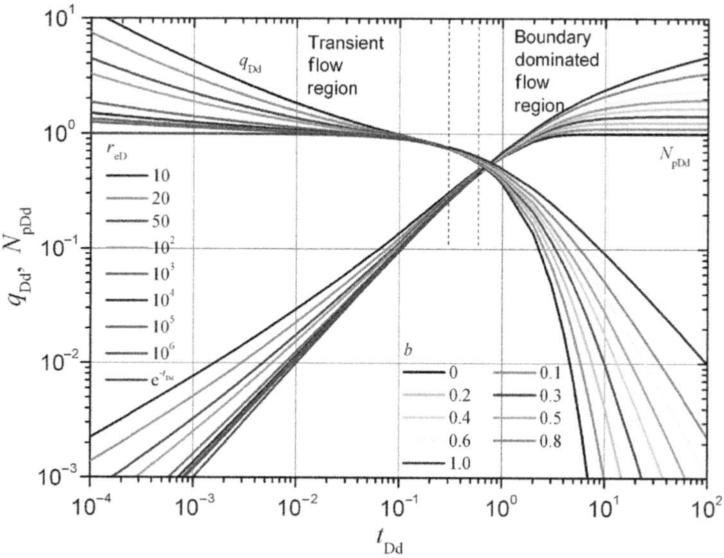

Fig. 4.10 Fetkovich type curve (Fetkovich, 1980, Permission to publish by the SPE. Copyright SPE).

performance-based reserves, and cumulative production upon abandonment are determined. Fetkovich-Arps type curve matching is performed as follows:

(1) Plot the $q \sim t$ log-log type curve.

(2) Match the $q \sim t$ type curve with theoretical type curve, similar to the log-log well test analysis.

(3) Record the values of the dimensionless drainage radius r_{eD} and the decline exponent b according to the matching result.

(4) Select any match point and record the actual match point $(t, q)_M$ and the corresponding theoretical match point $(t_{Dd}, q_{Dd})_M$.

(5) Determine the initial flow rate $q_i = \left(\frac{q}{q_{Dd}}\right)_M$ based on the rate match point.

(6) Determine the initial decline rate D_i based on the time match point $D_i = \left(\frac{t_{Dd}}{t}\right)_M$.

(7) Determine the permeability K based on r_{eD} and q_i obtained in Step (3).

$$K = \frac{\mu B(\ln r_{eD} - 0.5)}{2\pi h(p_i - p_{wf})} \left(\frac{q}{q_{Dd}}\right)_M$$

(8) Determine the effective wellbore radius r_{wa} based on the time match point and r_{eD} obtained in Step (3).

$$r_{wa} = \sqrt{\frac{2K/\phi\mu c_t}{\left(r_{eD}^2 - 1\right)\left(\ln r_{eD} - \frac{1}{2}\right)}\left(\frac{t}{t_{Dd}}\right)_M}$$

then get the skin factor S by $S = \ln \frac{r_w}{r_{wa}}$.

(9) Determine the estimated ultimate recovery (EUR).

For exponential decline, the EUR is calculated by, $EUR = N_p + Q_f = N_p + \frac{q_f - q_{ab}}{D_i}$

And the production time upon abandonment is

$$\Delta t = t_{ab} - t_f = \frac{1}{D_i} \ln\left(\frac{q_f}{q_{ab}}\right)$$

For hyperbolic decline, the EUR is calculated by

$$EUR = N_p + Q_f = N_p + \frac{q_i^{\,b}}{(1-b)D_i}\left(q_f^{\,1-b} - q_{ab}^{\,1-b}\right)$$

And the production time upon abandonment is

$$\Delta t = t_{ab} - t_f = \frac{\left(\frac{q_i}{q_{ab}}\right)^b - \left(\frac{q_i}{q_f}\right)^b}{bD_i}$$

For harmonic decline, the EUR is calculated by

$$EUR = N_p + Q_f = N_p + \frac{q_i}{D_i} \ln\left(\frac{q_f}{q_{ab}}\right)$$

And the production time upon abandonment is

$$\Delta t = t_{ab} - t_f = \frac{q_i}{D_i}\left(\frac{1}{q_{ab}} - \frac{1}{q_f}\right)$$

(10) Determine the well drainage volume or reservoir pore volume.

The reservoir pore volume V_p can be determined based on time and rate match points.

$$V_p = \frac{B}{c_t(p_i - p_{wf})}\left(\frac{t}{t_{Dd}}\right)_{MP}\left(\frac{q}{q_{Dd}}\right)_{MP}$$

(11) Determine the effective drainage radius.
The effective drainage radius is calculated by, $r_e = \sqrt{\frac{V_p}{\pi h \phi}}$
(12) Determine the reserves within the drainage area.
The reserves within the drainage area are determined by converting the subsurface volume to the surface standard state, that is, $N = \frac{V_p(1-S_w)}{B_i}$

Or, based on the Arps matching results, that is,

$$N = \int_0^{t_\infty} \frac{q_i}{(1 + bD_i t)^{1/b}} dt = \frac{q_i}{(1-b)D_i}\left[1 - \frac{1}{(1 + bD_i t_\infty)^{1/b-1}}\right] \approx \frac{q_i}{(1-b)D_i}$$

4.1.2.4.2 Blasingame method

Both the Arps and Fetkovich methods assume that the BHFP is constant to mainly analyze the production data without considering the change of gas PVT with pressure. The Blasingame method introduces the pseudo-pressure normalized rate ($q/\Delta p_p$) and the material balance pseudo-time t_{ca} to generate type curves, which considers the production at variable BHFP and the PVT variation with the change of reservoir pressure. As shown in Fig. 4.11, three types of curves in relation to the flow rate and material balance time can be plotted, that is, normalized rate curve, normalized rate integral curve, and normalized rate integral derivative curve. The three curves can be used collectively or individually for type curve analysis of actual production data. They are a set of curves with variable r_e/r_{wa} in the

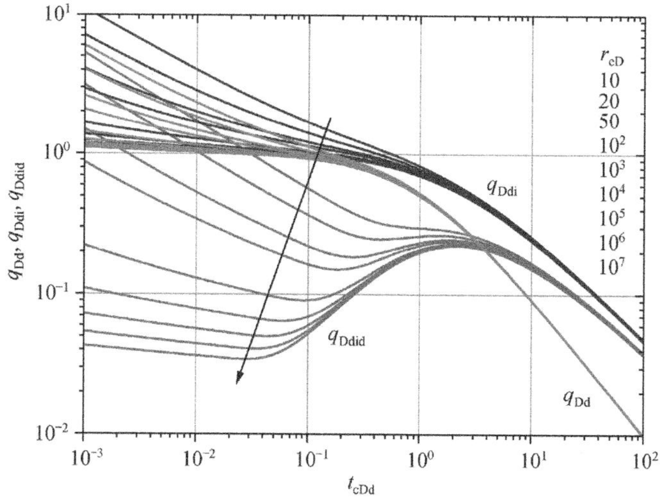

Fig. 4.11 Blasingame type curve (Sun, 2015). (*This article was published in Sun Hedong, Advanced production Decline analysis and application [M], Copyright © 2015 Petroleum Industry Press. Published by Elsevier Inc. All rights reserved*).

early transient flow period, and converge to one harmonic decline curve in the boundary-dominated flow period. By way of type curve matching, such parameters as permeability, skin factor, drainage radius, and original gas in place (OGIP) can be determined. The Blasingame method is superior for the application of rate integral for derivation, which allows for a smooth and identifiable derivative curve of the measured data point. However, the rate integral is very sensitive to the error at the early data point. A minor error at the early data point may lead to significant cumulative errors in the rate integral or the rate integral derivative curve. Matching analysis is performed as follows:

(1) Determine the material balance pseudo-time.

Assuming a value of G, the material balance pseudo-time is calculated for each data point:

$$t_{ca} = \frac{(\mu c_t)_i}{q} \int_0^t \frac{q}{\mu(\overline{p}) c_t(\overline{p})} dt = \frac{G c_{ti}}{q} \left(p_{p_i} - p_p \right)$$

(2) Determine the normalized rate.

$$\frac{q}{\Delta p_p} = \frac{q}{p_{p_i} - p_{p_{wf}}}$$

(3) Determine the normalized rate integral.

$$\left(\frac{q}{\Delta p_p} \right)_i = \frac{1}{t_{ca}} \int_0^{t_{ca}} \frac{q}{\Delta p_p} d\tau$$

where, the subscript i represents the integral.

(4) Determine the normalized rate integral derivative.

$$\left(\frac{q}{\Delta p_p} \right)_{id} = -\frac{d\left(\frac{q}{\Delta p_p} \right)_i}{d \ln t_{ca}} = -t_{ca} \frac{d\left(\frac{q}{\Delta p_p} \right)_i}{dt_{ca}}$$

where, the subscript i represents the integral and *d* represents the derivative.

(5) Plot the $\frac{\Delta p_p}{q} \sim t_{ca}$ curve in a rectangular coordinate system, and perform linear regression. Determine G according to the slope of the regression straight line. Repeat Steps (1)–(5) for iterative calculation until convergence, to satisfy the allowable error for G.

(6) Plot the log-log type curves between the normalized rate, normalized rate integral, and normalized rate integral derivative and the material balance time in the condition of G, that is, $\frac{q}{\Delta p_p} \sim t_{ca}$, $\left(\frac{q}{\Delta p_p}\right)_i \sim t_{ca}$, and $\left(\frac{q}{\Delta p_p}\right)_{id} \sim t_{ca}$.

(7) Match the preceding three curves, or any combination thereof, with the theoretical type curve to ensure each match realizes a favorable result.

(8) Record the dimensionless drainage radius r_{eD} according to the matching results.

(9) Select any match point, and record the actual match point $(t_{ca}, q/\Delta p_p)_M$ and the corresponding theoretical match point $(t_{caDd}, q_{Dd})_M$. Given the net pay thickness, total compressibility and effective wellbore radius, the parameters such as reservoir permeability, skin factor, drainage area, and OGIP can be determined.

(10) According to the rate match point, the permeability is determined.

$$K = \frac{(q/\Delta p_p)_M}{(q_{Dd})_M} \frac{\mu B}{2\pi h}\left(\ln r_{eD} - \frac{1}{2}\right)$$

(11) Based on the time match point and r_{eD} determined in Step (8), the effective wellbore radius r_{wa} is determined.

$$r_{wa} = \sqrt{\frac{2K/\phi\mu c_t}{(r_{eD}^2 - 1)\left(\ln r_{eD} - \frac{1}{2}\right)}\left(\frac{t_{ca}}{t_{caDd}}\right)_M}$$

(12) Determine the skin factor S, $S = \ln \frac{r_w}{r_{wa}}$

(13) Determine the drainage radius and drainage area, that is, $r_e = r_{wa} r_{eD}$ and $A = \pi r_e^2$, and then calculate OGIP.

$$G = \frac{1}{c_t}\left(\frac{t_{ca}}{t_{caDd}}\right)_M \left(\frac{q/\Delta p_p}{q_{Dd}}\right)_M (1 - S_w)$$

4.1.2.4.3 Agarwal-Gardner method

Agarwal et al. established the Agarwal-Gardner decline type curve based on the relationship between the pseudo-pressure normalized rate (q/p_p) and material balance pseudo-time t_{ca} and the dimensionless parameters in transient well test analysis. Due to a different definition of the dimensionless

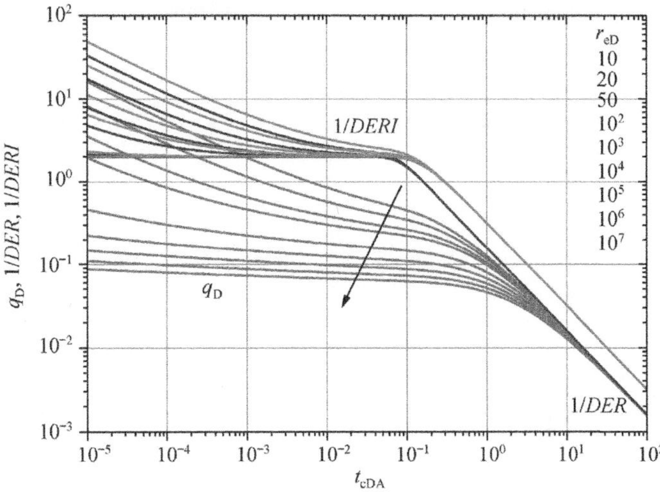

Fig. 4.12 Agarwal-Gardner type curve (Sun, 2015). (*This article was published in Sun Hedong, Advanced production Decline analysis and application [M], Copyright © 2015 Petroleum Industry Press. Published by Elsevier Inc. All rights reserved*).

quantity, the early part of the type curve is relatively scattered compared with the Blasingame type curve, which is helpful in reducing the ambiguity of the matching analysis (Fig. 4.12).

(1) Determine the material balance pseudo-time.

Assuming a value of G, the material balance pseudo-time is calculated for each data point:

$$t_{ca} = \frac{(\mu c_t)_i}{q} \int_0^t \frac{q}{\mu(\overline{p})c_t(\overline{p})} dt = \frac{Gc_{ti}}{q}(p_{p_i} - p_p)$$

(2) Determine the normalized rate, $\frac{q}{\Delta p_p} = \frac{q}{p_{p_i} - p_{p_{wf}}}$

(3) Determine the normalized rate integral, $\left(\frac{\Delta p_p}{q}\right)_i = \frac{1}{t_{ca}} \int_0^{t_{ca}} \frac{\Delta p_p}{q} d\tau$

(4) Determine the normalized rate integral derivative,

$$\left(\frac{\Delta p_p}{q}\right)_{id} = \frac{d\left(\frac{\Delta p_p}{q}\right)_i}{d \ln t_{ca}} = t_{ca} \frac{d\left(\frac{\Delta p_p}{q}\right)_i}{dt_{ca}}$$

(5) Determine the reciprocal of the normalized pressure derivative,

$$\frac{1}{DER} = \frac{1}{\frac{\partial\left(\frac{\Delta p_p}{q}\right)}{\partial \ln t_{ca}}} = \frac{1}{t_{ca} \frac{\partial\left(\frac{\Delta p_p}{q}\right)}{\partial t_{ca}}}$$

(6) Determine the reciprocal of the normalized pressure integral derivative,

$$\frac{1}{DERI}=\frac{1}{\dfrac{\partial\left(\dfrac{\Delta p_{\mathrm{p}}}{q}\right)}{\partial \ln t_{\mathrm{ca}}}}=\frac{1}{t_{\mathrm{ca}}\dfrac{d\left(\dfrac{\Delta p_{\mathrm{p}}}{q}\right)_{\mathrm{i}}}{dt_{\mathrm{ca}}}}$$

(7) Plot the $\frac{\Delta p_{\mathrm{p}}}{q} \sim t_{\mathrm{ca}}$ curve in a rectangular coordinate system, and perform linear regression. Determine G according to the slope of the regression straight line. Repeat Steps (1)–(7) for iterative calculation until convergence, to satisfy the allowable error for G.

(8) Plot the log-log type curves between the normalized rate, reciprocal of the normalized pressure derivative, and reciprocal of the normalized pressure integral derivative and the material balance time in the condition of G, that is, $\frac{q}{\Delta p_{\mathrm{p}}} \sim t_{\mathrm{ca}}$, $\frac{1}{DER} \sim t_{\mathrm{ca}}$, and $\frac{1}{DERI} \sim t_{\mathrm{ca}}$.

(9) Match the preceding three curves, or any combination thereof, with the theoretical type curve to ensure each matching realizes a favorable result.

(10) Record the dimensionless drainage radius r_{eD} according to the matching results.

(11) Select any match point, and record the actual match point $(t_{\mathrm{ca}}, q/\Delta p_{\mathrm{p}})_{\mathrm{M}}$ and the corresponding theoretical match point $(t_{\mathrm{caDd}}, q_{\mathrm{Dd}})_{\mathrm{M}}$. Given the net pay thickness, total compressibility, and effective wellbore radius, the parameters such as reservoir permeability, skin factor, drainage area, and OGIP can be determined.

(12) According to the rate match point, the permeability is determined, $K=\frac{(q/\Delta p_{\mathrm{p}})_{\mathrm{M}}}{(q_{\mathrm{D}})_{\mathrm{M}}}\frac{\mu B}{2\pi h}$.

(13) Based on the time match point and K determined in Step (12), the drainage radius r_{e} is determined.

$$r_{\mathrm{e}}=\sqrt{\frac{K}{\pi\phi\mu c_{\mathrm{t}}}\left(\frac{t_{ca}}{t_{\mathrm{caDA}}}\right)_{\mathrm{M}}}$$

(14) Based on the matching result of the dimensionless radius r_{eD} determined in Step (10), the effective wellbore radius r_{wa} is determined, $r_{\mathrm{wa}}=\frac{r_{\mathrm{e}}}{r_{\mathrm{eD}}}$.

(15) Determine the skin factor S, $S=\ln\frac{r_{\mathrm{w}}}{r_{\mathrm{wa}}}$.

(16) Determine the OGIP by the volumetric method, $G=\frac{\pi r_{\mathrm{e}}^2\phi h S_{\mathrm{g}}}{B_{\mathrm{gi}}}$.

4.1.2.4.4 NPI method

Both Blasingame and Agarwal-Gardner methods use the pseudo-pressure normalized rate ($q/\Delta p_{\mathrm{p}}$) and material balance pseudo-time function t_{ca} to plot type curves, whereas the NPI (normalized pressure integral) method uses rate normalized pressure integrals to develop an analysis method that is relatively reliable and not affected by the scatters of data (Fig. 4.13).

The x-coordinate of an NPI type curve is the material balance pseudo-time t_{ca}, identical to the Blasingame and Agarwal-Gardner type curves; the three y-coordinates are:

Rate normalized pseudo-pressure, $\frac{\Delta p_{\mathrm{p}}}{q}$

Rate normalized pseudo-pressure integral, $\left(\frac{\Delta p_{\mathrm{p}}}{q}\right)_{\mathrm{i}} = \frac{1}{t_{ca}} \int_0^{t_{ca}} \frac{\Delta p_{\mathrm{p}}}{q} dt$

And the derivative of rate normalized pseudo-pressure integral, $\left(\frac{\Delta p_{\mathrm{p}}}{q}\right)_{\mathrm{id}} = \frac{t_{\mathrm{ca}} d\left(\frac{\Delta p_{\mathrm{p}}}{q}\right)_{\mathrm{i}}}{dt_{\mathrm{ca}}}$.

Through the NPI type-curve matching based on actual production data, K, S, r_{e}, and OGIP can be obtained in the same procedure as the Agarwal-Gardner method.

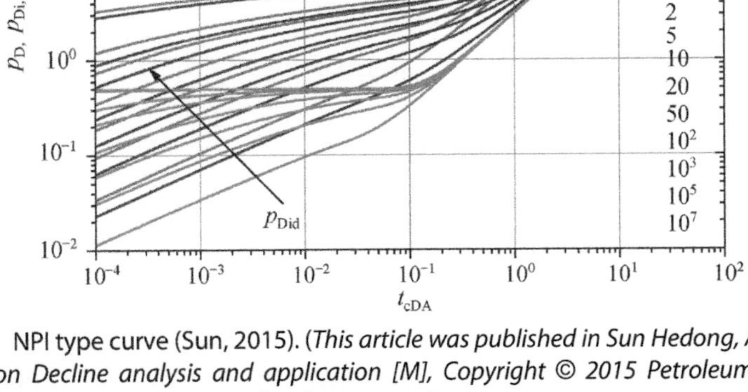

Fig. 4.13 NPI type curve (Sun, 2015). (*This article was published in Sun Hedong, Advanced production Decline analysis and application [M], Copyright © 2015 Petroleum Industry Press. Published by Elsevier Inc. All rights reserved*).

4.1.2.5 Flowing material balance (FMB) method

The FMB method, essentially based on the material balance theory, only analyzes the production data after the gas well production reaches the boundary-dominated flow state (or the pseudo-steady state in case of a constant rate).

4.1.2.5.1 Mattar method

Under the conditions that the flow rate of the gas well is constant and the BHFP is consistent with the reservoir pressure decline rate, the material balance curve based on the BHFP p_{wf} should be parallel to the material balance curve based on the average static reservoir pressure p_R (Fig. 4.14). Therefore, the BHFP data can be used to plot the relationship between p_{wf}/Z and the cumulative gas production, and thus to obtain the slope of the straight line. Then, a straight line passing the point $(0,p_i/Z_i)$ is plotted with this slope. The intersection between the straight line and the x-axis is the original gas in place (OGIP), as shown in Fig. 4.15. As a precondition, the flow should have reached the pseudo-steady flow state.

In the case of a variable rate, we have

$$\frac{p_{p_i}-p_{p_{wf}}}{q}=\frac{t_{ca}}{Gc_{ti}}+\frac{(\mu B)_i}{2\pi Kh}\left[\frac{1}{2}\ln\frac{4A}{C_A e^{\gamma} r_w^2}\right] \tag{4.34}$$

that is,

$$\frac{\Delta p_p}{q}=m_a t_{ca}+b_{a,pss} \tag{4.35}$$

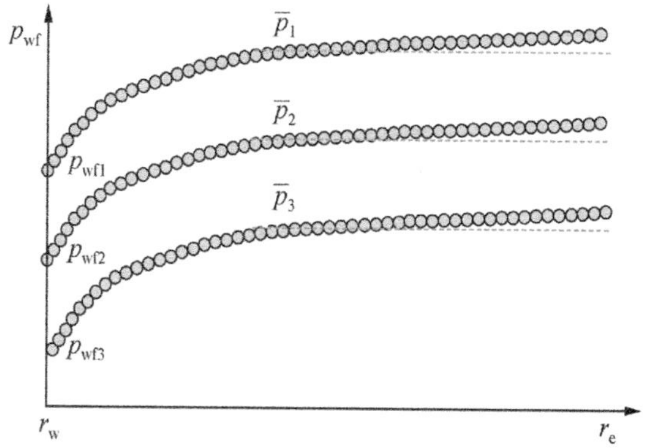

Fig. 4.14 FMB type curve at constant rate (Mattar and McNeil, 1998, Permission to publish by the SPE, Copyright SPE).

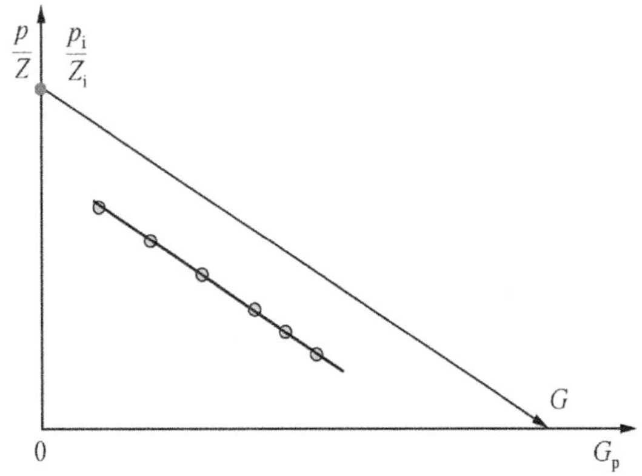

Fig. 4.15 FMB type curve for gas well at a constant rate.

where,

$$\Delta p_p = p_{p_i} - p_{p_{wf}} \tag{4.36}$$

$$m_a = \frac{1}{G c_{ti}} \tag{4.37}$$

$$b_{a,pss} = \frac{(\mu B)_i}{2\pi Kh}\left[\frac{1}{2}\ln\frac{4A}{C_A e^{\gamma} r_w^2}\right] \tag{4.38}$$

According to the preceding equation, after the gas well reaches the pseudo-steady flow state, the total pressure drawdown (from the initial reservoir pressure to the BHFP) consists of two parts. One part is the pressure drawdown formed by depletion development, that is, $\frac{t_{ca}}{GC_{ti}}$, and the other part is the pressure loss when the gas flows from the reservoir to the bottom hole, that is, $b_{a,pss} = \frac{(\mu B)_i}{2\pi Kh}\left[\frac{1}{2}\ln\frac{4A}{C_A e^{\gamma} r_w^2}\right]$.

According to Eq. (4.35), $\frac{\Delta p_p}{q} \sim t_{ca}$ presents a linear relationship in the rectangular coordinate system. The FMB curve of the gas well at a variable rate can be obtained by the following steps:

(1) Convert the pressure data into the normalized pseudo-pressure Δp_p
(2) Assuming the value of G, determine the normalized pseudo-time t_{ca}
(3) Plot the $\frac{\Delta p_p}{q} \sim t_{ca}$ curve, with the intercept of the straight line being $b_{a,\,pss}$.
(4) Calculate the average normalized pseudo-pressure of the reservoir.
(5) Convert the average normalized pseudo-pressure of the reservoir into the average reservoir pressure p_R.
(6) Determine the value of p_R/Z and plot the relationship between p_R/Z and the cumulative gas production G_p, just like the conventional

material balance curve of the gas reservoir, with the intercept on the x-axis being the original gas in place G.

(7) Repeat Steps (2)–(6) based on the new G value until convergence.

4.1.2.5.2 Agarwal-Gardner (A-G) method

Assuming a vertical well is in the center of a columnar gas reservoir with the radius r_e, in case of variable-rate or variable-pressure production, by normalizing the rate and material balance pseudo-time based on the normalized pseudo-pressure, we can obtain the following relationship.

For a gas well,

$$\frac{q}{\Delta p_p} = \frac{1}{b_{a,pss}} - \frac{q}{\Delta p_p c_{ti}} \frac{1}{G b_{a,pss}} t_{ca} \tag{4.39}$$

If the $\frac{q}{\Delta p_p} \sim \frac{q}{\Delta p_p c_{ti}} t_{ca}$ curve is plotted, the intercept of the straight line on the x-axis is the original gas in place G. The procedure of the A-G method is similar to that of the Mattar method, as follows (Fig. 4.16):

(1) Estimate the OGIP.

(2) Determine the normalized pseudo-time, normalized pseudo-pressure, and cumulative gas production,. that is, $\frac{q}{\Delta p_p}, t_{ca}, \frac{q}{\Delta p_p c_{ti}} t_{ca}$.

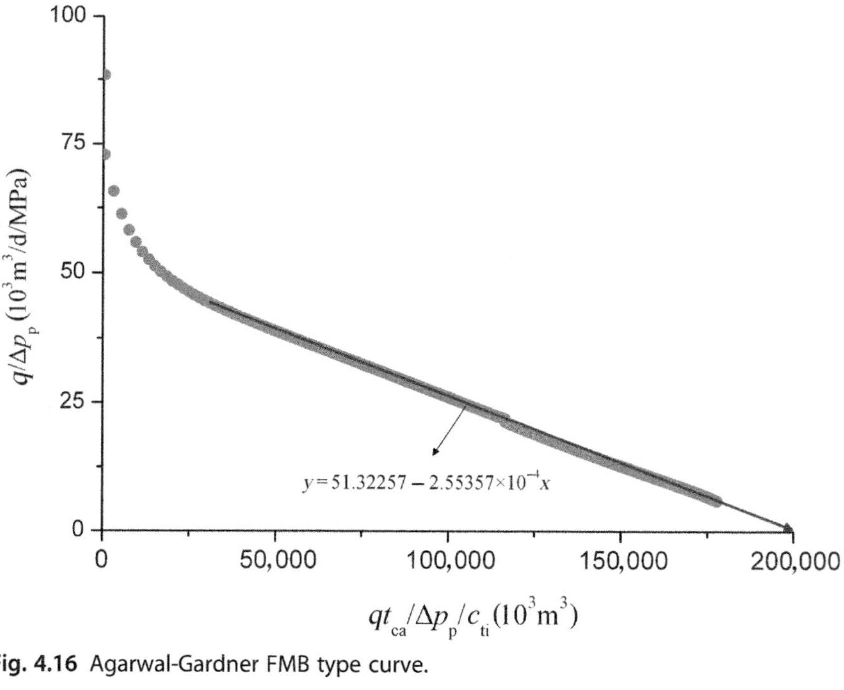

Fig. 4.16 Agarwal-Gardner FMB type curve.

(3) If the $\frac{q}{\Delta p_p} \sim \frac{q}{\Delta p_p c_{ti}} t_{ca}$ curve is plotted, the intercept of the straight line on the x-axis is the original gas in place G.
(4) Obtain the optimum match, and extrapolate G.
(5) Repeat Steps (1)–(4) based on the new G value until convergence.

4.1.3 Performance-based reserves (OGIP) estimation of fractured-vuggy carbonate gas reservoir

4.1.3.1 Selection of estimation methods

Fractured-vuggy carbonate reservoirs are diverse in types, including cave reservoirs, fractured-vuggy reservoirs, and vuggy reservoirs, and correspond to complex flow mechanisms, such as pore flow, fracture flow, and cave flow. By fluid properties, they can be divided into condensate gas reservoirs with low and medium-high condensate content, and volatile oil reservoirs. With poor operability, the reservoirs are often produced at variable pressure or variable rates. The difference between reservoir pressure and dew point pressure is minor, retrograde condensate exists in the reservoir, and multiphase flow exists in the wellbore, so the converted BHFP may suffer from significant errors. The static pressure test points are limited, and do not descend monotonically, so the traditional material balance method is not applicable. Accordingly, it is challenging to accurately evaluate the performance-based reserves or original gas in place (OGIP) of complex carbonate condensate gas wells/reservoirs in complex production conditions.

4.1.3.1.1 Adaptability of estimation methods

For the fractured-vuggy carbonate reservoirs in the Tazhong I gas field, which are complex in types and flow mechanisms, it is important to first judge whether the reserves estimation methods based on the flow theory are applicable. According to the seismic reflections of the wells, drilling breaks, and well tests and production test data, the gas field is found to have the characteristics of low production index, low formation capacity, and interstitial flow (but no tubular flow or cavity flow); and the reservoirs are mainly of medium-low permeability. Therefore, the reserves estimation methods based on the flow theory are applicable to the Tazhong I gas field (Table 4.12).

According to its definition, "performance-based reserves or original gas in place (OGIP)" is a quantity between the hydrocarbon initially in place and the estimated ultimate recovery. Therefore, it can be estimated in accordance with the Estimation Methods of Recoverable Natural Gas Reserves (SY/T 6098-2010), provided that the abandonment condition is adjusted to

Table 4.12 Data of the Tazhong I gas field

Well	Seismic reflection	Drilling break (m)	Well test model	Formation capacity (mD m)	Well testing: Nozzle (mm)	Tubing pressure (MPa)	Daily oil production (m^3)	Daily gas production (m^3)	Daily water production (m^3)	Production index: Gas well (m^3/MPa^2/d)	Oil well (m^3/MPa/d)
G14	Bead-like				5	45.7	74.4	130,921	0	154.4	
G14-2H	Flake-like				3	24.5	22.9	29,556	0	9.0	
G12	Bead-like	2.3	Homogeneous	100	6	43.4	15.3	169,537	22.6	111.2	
G22					5	21.4	40.8	70,260	1.4	20.2	
G21					5	40.4	37.7	146,082	0	80.9	
G111					5	42.7	62.9	114,366	1.9	91.5	
G11		4.7	Dual-porosity	44	5	30.0	52.5	95,440	0	32.4	
G48					4	35.8	25.1	69,485	/	40.6	
G441					5	41.4	38.1	134,935	/	435.6	
G431					4	31.2	85.4	35,890	/		30.5
G432					4	14.9	76.2	3276	/		7.8
G23	Flake-like				5	9.0	42.0	15,576	11.1		1.3
G44C		0.4			4	27.1	32.4	45,102	/	16.7	
G462			Dual-porosity	194	4	39.9	52.0	74,006	/	234.4	
G45			Composite	338	5	23.8	79.9	45,301	/	27.2	
G461	Chaotic				5	5.7	9.2	36,201	/	9.8	

"the flow rate is zero, and the pressure is 1 atm". However, the extrapolated estimation methods of recoverable gas reserves are not applicable to the Tazhong I gas field, where the reservoirs are complex, with only limited static pressure test data, and the variable pressure, variable rate, and production performance are difficult to forecast (Table 4.13).

4.1.3.1.2 Selection of optimal estimation methods

Method based on lifecycle well test As mentioned in Section 3.2 in Chapter 3, the well test analysis with the constraint of "lifecycle history matching" can greatly improve the data utilization, minimize the ambiguity of well test interpretation results, and estimate the performance-based reserves of single-wells. This technique is essentially based on material balance; that is, the short-term well test is combined with the long-term production performance to estimate the drainage area of a gas well and calculate the single-well performance-based reserves with the volumetric method.

Method based on advanced production decline analysis The advanced production decline analysis based on the type-curve match and using the daily production data of gas wells provides a new method to estimate the performance-based reserves. The advanced production decline analysis methods, represented by the Blasingame method, can deal with the complex data of fractured-vuggy carbonate gas wells at variable pressure and variable rates (Table 4.14).

For heterogeneous gas reservoirs that are characterized by strong heterogeneity, complex fluid properties, multiphase flow in wellbore, production at variable pressure or variable rate, and poor quality of performance data, a performance-based reserves estimation method integrating well test analysis and advanced production decline analysis has been proposed, depending on the geological characteristics of the reservoirs. Its process is shown in Fig. 4.17.

4.1.3.2 Advanced production decline analysis method

4.1.3.2.1 Basic data

Data preparation Reservoir, well completion, and daily production data are mainly collected, as shown in Table 4.15.

Data diagnosis Before production data analysis and interpretation, it is important to understand the common issues therein (Table 4.16).

Table 4.13 Adaptability of the extrapolated estimation methods of recoverable natural gas reserves subject to SY/T 6098-2010

No.	Estimation method	Application conditions	Well	Unit	Actual conditions	Conclusion
1	Material balance method	At least 2 static-pressure test points	✓	✓	Insufficient static-pressure test points, not decline monotonically	×
2	Elastic two-phase method	Constant-rate plus closed boundary	✓		Variable rate + fracture + multiple reservoir bodies	×
3	Production decline method	Constant-pressure	✓		Variable pressure	×
4	Improved production decline method	Cumulative production-time relationship	✓	✓	Multiple segments, large uncertainty	×
5	Prediction model method	Annual production-time relationship	✓	✓	Multiple segments, large uncertainty	×
6	Numerical simulation method	Geological model	✓	✓	Difficult modeling	×
7	Analogy method	New blocks/reservoirs that have not been put into production or that have been developed for a short period of time Old blocks that are short of performance data	✓	✓		×

Table 4.14 Adaptability of the advanced production decline analysis methods

No.	Development phase		Estimation method	Application conditions	Actual conditions	Conclusion
1	Empirical method		Arps	Constant-pressure + boundary-dominated flow	Variable pressure and variable rate	Not applicable
2	Type-curve matching	Classical	Fetkovich	Constant-rate + closed boundary		Not applicable
3		Modern	Blasingame	Boundary-dominated flow		Applicable
4			A-G (Agarwal-Gardner)			Applicable
5			NPI			Applicable
6	Numerical method		Numerical method	Solution based on the analytical matching		Applicable

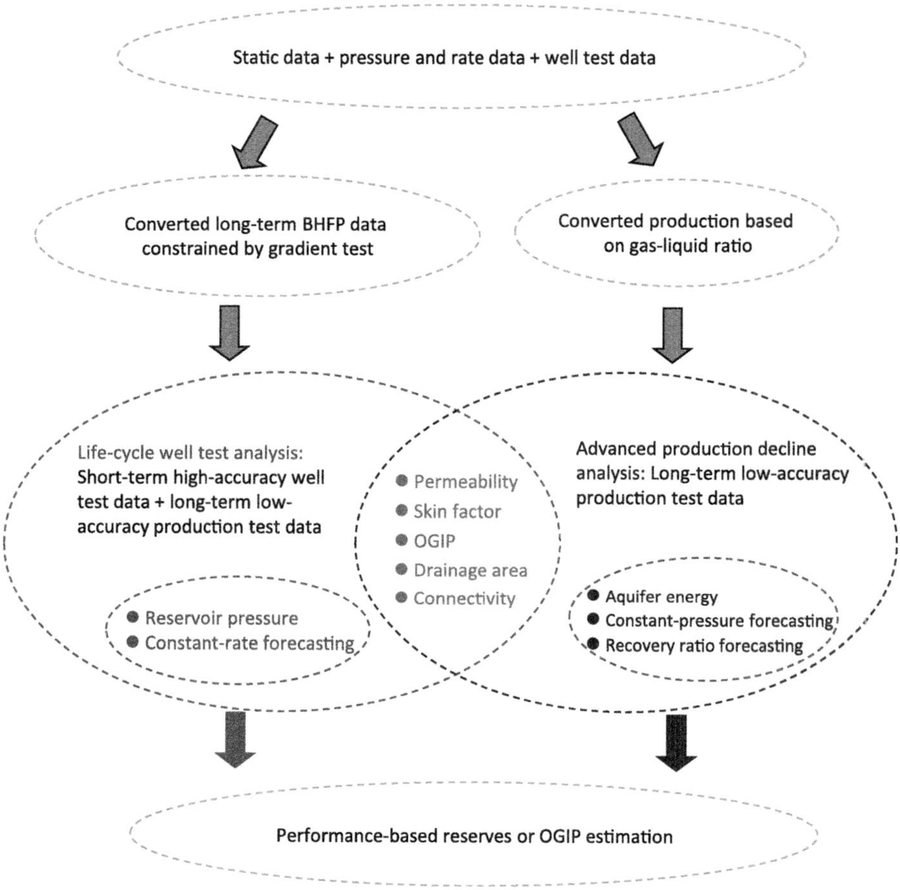

Fig. 4.17 Process of performance-based reserves, or OGIP estimation, for complex carbonate condensate gas reservoirs in the Tazhong I gas field.

Table 4.15 Basic data for advanced production decline analysis

Reservoir properties	Porosity, net pay thickness, rock compressibility, total compressibility
Physical properties of oil	Volume factor, solution gas-oil ratio (SGOR), viscosity, compressibility
Physical properties of gas	Volume factor, deviation factor, viscosity, compressibility, the relative density of gas, CO_2%, H_2S%, N_2%
Wellbore-related parameters	Mid-depth of reservoir, initial reservoir pressure, tubing/casing size, formation temperature, wellhead flowing (static) temperature
Daily production data	Wellhead tubing/casing pressure, oil/gas/water flow rate, measured flowing pressure gradient and measured static pressure gradient

Table 4.16 Common issues in production decline analysis (Anderson et al., 2010, Permission to publish by the SPE, Copyright SPE)

Issue		Severity
Pressure	No pressure measurement	High
	Incorrect initial pressure estimate	High
	Poor $p_{tf} \sim p_{wf}$ conversion	Moderate
	Liquid loading (influencing $p_{tf} \sim p_{wf}$ conversion)	Moderate
	Incorrect location of pressure measurement	Very high
Flow rate	Rate allocation (possibly causing errors)	Moderate
	Liquid loading	Moderate
Well completion	Change of horizon (new and old perforated horizons)	Very high
	Change of wellbore diameter	High
	Change of wellhead equipment	Moderate/high
	Stimulation measures	High
General	Reservoir properties	Moderate
	Physical properties of oil	Moderate
	Physical properties of gas	Moderate
	Poor time-pressure-rate synchronization	Moderate high
	Poor time-pressure-rate correlation	Very high

A simple visual inspection of the production data report (flowing pressure, rate, and time) may reveal evident or potential inconsistencies, and the time-pressure-rate data should be checked as follows:

If the rate suddenly increases or drops, the flow pressure should drop or rise accordingly. If not, the rate and pressure data may be unmatched. This may result from: a change in completion conditions, liquid supply from multiple reservoir bodies, an incorrect measuring position of flow pressure, liquid loading in the wellbore, and so forth. The discordant distribution of tubing pressure and casing pressure may result from liquid loading in a wellbore, or leaking on the tubing.

As for pressure and rate, it is usually acceptable to take the average rate, but it makes no sense to take the average pressure. The rate distribution may conform to the flow rate on site, or of the pipe manifold; but it may possibly be incapable of representing the single well performance, especially when the rate interferes in such cases as shut-in or short-term nonuniform flow. If the rate and flow pressure data are widely scattered, it indicates that the flow is unstable in the wellbore. If the measured value or split value of the rate appears as a step change for a long time, it usually indicates that the measuring frequency is low and the measurement is likely inaccurate. Such flow rate data would reduce the accuracy of calculated BHP value, even if the WHP measurement quality is quite high. If the measured value of flow pressure assumes a step change for a long time, it usually indicates that the measuring frequency is low. In such a case, the pressure data should be compared with the flow rate data to ensure that the pressure data meet the requirements.

Model diagnosis The decline type curves mainly have two functions: (a) to determine the reservoir permeability by matching the production decline curve with it, which belongs to quantitative analysis; and (b) to diagnose the production performance characteristics, that is, to qualitatively judge the flow state, change of production index, interwell interference, extraneous energy supply, and so on.

Fig. 4.18A shows how to use the production decline curves to diagnose the production performance of a gas well, and the solid line represents a normal decline curve in the constant volume reservoir. In the early transient flow period, if the real production curve gradually goes upward to approach the normal curve, it indicates that the skin factor of the well is dropping gradually, that is, the cleaning up process; otherwise, it indicates that the well is being damaged more and more seriously. In the late boundary-dominated flow period, if the real production data skews from the harmonic straight line segment and gradually moves leftward, it indicates that there is interwell interference, which reduces the original gas in place of the well; otherwise, if the real production data curve gradually shifts rightward, it indicates that there is an extraneous energy supply, as shown in Fig. 4.18B.

4.1.3.2.2 Analysis procedure

Theoretically, the advanced production decline analysis is the same as the well test analysis. Their correlation comes down to the quality of real data (the production data belong to "low frequency/low resolution" data, whereas the well test analysis data belong to "high frequency/high resolution" data) and the data acquisition (the production data are passively collected as "routine monitoring" data, whereas the well test analysis data are acquired according to the design requirements, so as to ensure the reservoir characterization is accurate and representative). To improve the results of production data analysis, the production data can be diagnosed, analyzed, and interpreted in the following procedure.

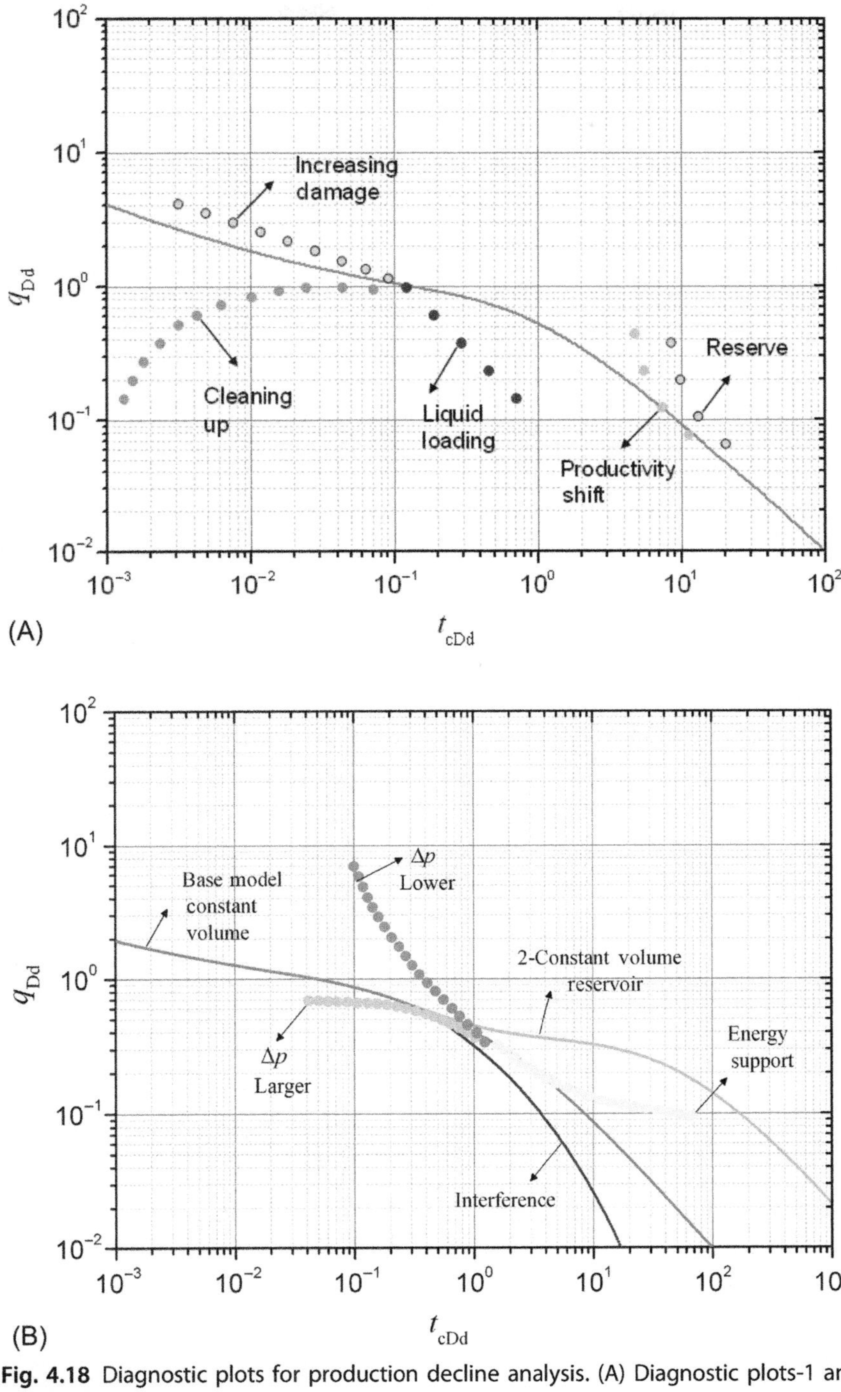

Fig. 4.18 Diagnostic plots for production decline analysis. (A) Diagnostic plots-1 and (B) diagnostic plots-2.

(1) *Flow rate conversion*: In a high gas-liquid ratio, convert the condensate oil into condensate gas, and conduct processing as a pseudo single-phase; if the flow time is less than 24h, convert the flow rate into a daily rate.
(2) *Flow pressure conversion*: If there are no BHFP data, use the related wellbore pipe flow software to convert the tubing pressure or casing pressure data into BHFP data under the constraint of the measured pressure gradient.
(3) *Data review*: review the data quality/correlativity of the production curve.
(4) *Data removal/reduction*: remove the anomalous data points on the log-log plot used by diagnosis.
(5) *Characteristic flow behavior identification*: similar to the pressure derivative curve in the well test analysis, the pressure and rate integral derivative curves can also be used to identify the flow behavior in the transient flow period, for example, the NPI method and Blasingame method.
(6) *Type-curve matching*: match the measured data with the corresponding theoretical decline curves in the reservoir model to obtain relevant parameters.
(7) *History matching*: conduct ultimate "history matching" of the model and original data of single well performance (p_{wf} and q).
(8) If the analytical solution cannot be matched very well, then make a numerical analysis on this basis to derive relevant parameters.
(9) *Production performance forecasting*: based on the performance model established previously, conduct performance forecasting at a constant pressure or rate according to the matching results, to guide the production.

As mentioned by Mattar and Anderson (2003), the production decline curve analysis technique is developing rapidly, and a variety of methods can be applied, but there is no definite method that is always capable of providing the most reliable interpretation. In practice, various methods should be combined and balanced rationally, to reduce the uncertainty of the reserves estimate.

4.1.3.2.3 Field examples

Gas well For Well A (gas well), the initial reservoir pressure is 9.20 MPa, the reservoir temperature is 48.9°C, the net pay thickness is 6.04 m, the porosity is 0.20, and the gas saturation is 0.80. The gas flow rate is much higher than the critical unloading rate. The data in the example are taken from continuous measurements of the gas flow rate and BHFP, as shown in Fig. 4.19.

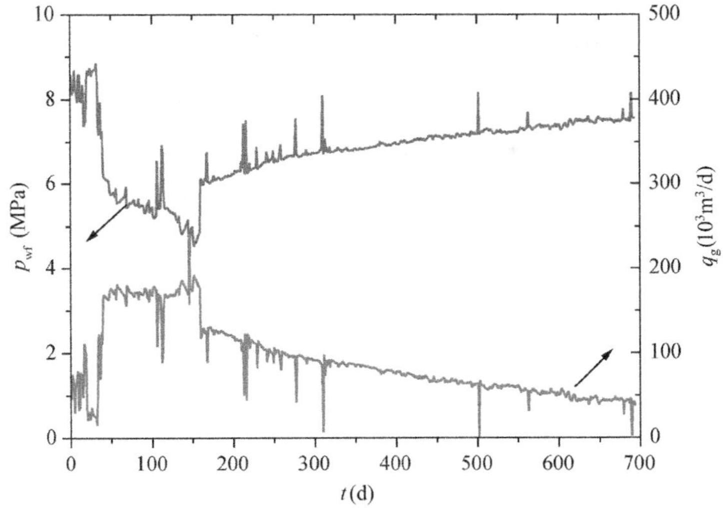

Fig. 4.19 Production history type curve of Well A.

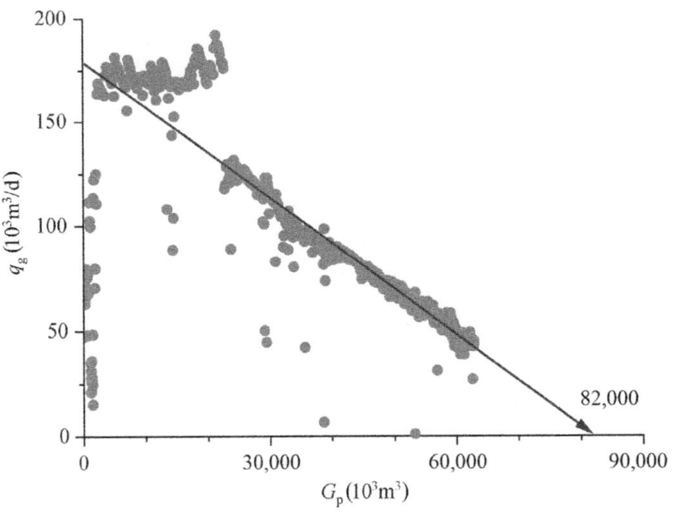

Fig. 4.20 Arps method of Well A.

Step 1: First, use the "production history curve" to review the data. In this example, the correlation between the flow rate and pressure is very good, which indicates that the data quality is higher and favorable for reliable analysis (Fig. 4.20).

Step 2: Conduct an Arps decline analysis, which shows that, if calculated according to the exponential decline, when the flow rate is zero, the cumulative production is forecasted as only 0.82×10^8 m^3 (Fig. 4.21).

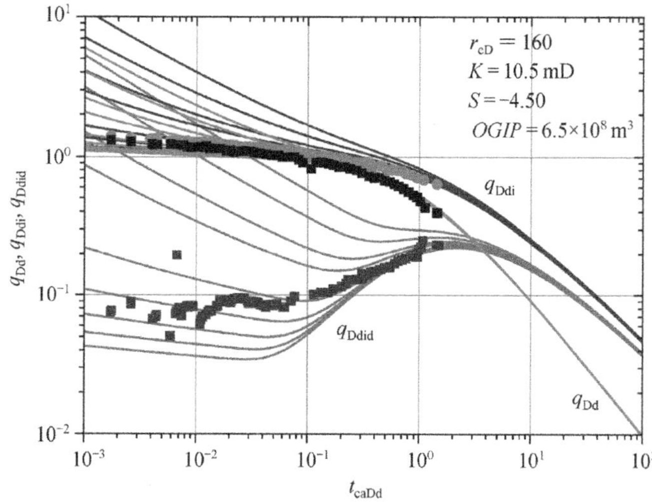

Fig. 4.21 Blasingame type curve of Well A.

Step 3: Conduct type-curve matching, which shows that the type-curve matching results of Blasingame, Agarwal-Gardner, and NPI methods are consistent, the production is in a boundary-dominated flow period, and the OGIP is close to 6.5×10^8 m^3.

Step 4: Plot the flowing material balance curve, as shown in Fig. 4.22, where the OGIP is calculated as 6.3×10^8 m^3, close to the type-curve matching result. If the abandonment pressure is assumed to be 2.1 MPa, the *EUR* will be around 5.1×10^8 m^3.

Seemingly, the results of the Arps method differ greatly from other methods. Mattar and Anderson (2003) pointed out that such a difference could be easily explained if the influence of flow pressure was taken into account. Essentially, the Arps decline analysis is associated with production conditions, and the *EUR* depends on the magnitude of pressure drawdown (which is minor). Once the influence of flow pressure (magnitude and trend) is taken into account (according to FMB), it can be noted that the actual OGIP is much higher than that forecasted by the Arps method. Therefore, under current production conditions (i.e., "no any stimulation is adopted"), the recovery ratio would be very low (0%–13%). On the contrary, the *EUR* proposed by the FMB method is not restricted by the lower pressure drawdown, indicating much larger recoverable reserves.

Oil well For Well A (oil well), the initial reservoir pressure is 70.8 MPa, the reservoir temperature is 136.4°C, the net pay thickness is 26.0 m, the

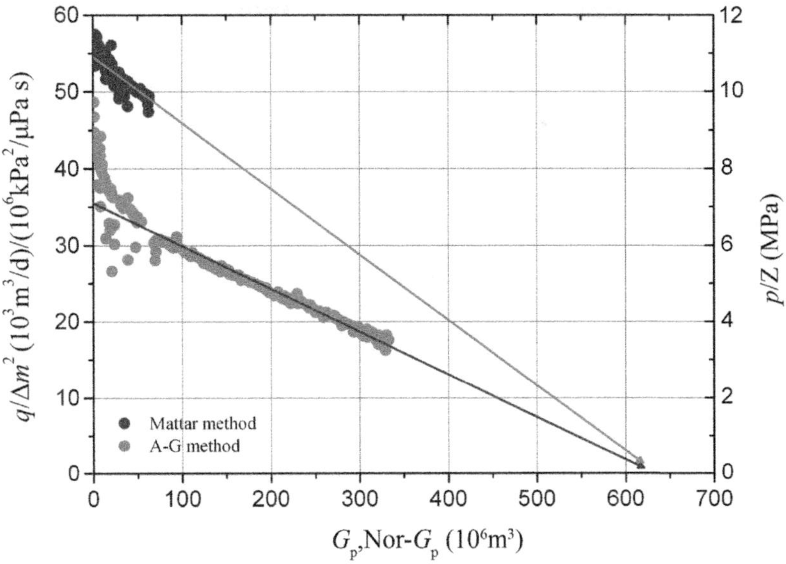

Fig. 4.22 FMB analysis of Well A.

porosity is 0.03, the gas saturation is 0.80, and there are data from 6 gradient tests. The production data are shown in Fig. 4.23.

Step 1: There is a very good correlation between flow rate and pressure. Use the wellbore pipe flow software to convert the wellhead tubing pressure into the flowing pressure at the mid-point of the reservoir, as shown in Fig. 4.23. In conversion, the gradient test data points are used for correction.

Step 2: Combine various methods for analysis to determine the reservoir parameters and the original oil in place (OOIP). The FMB type curve obtained by the Agarwal-Gardner method is shown in Fig. 4.24, and the OOIP is calculated as 9.2×10^4 m^3.

The type-curve matching results of Blasingame, Agarwal-Gardner, NPI, and Transient methods suggest a boundary-dominated flow, as shown in Fig. 4.25. Such results are basically consistent, and the OOIP is close to 9.2×10^4 m^3, as shown in Table 4.17.

A large number of analysis examples show that no single advanced production decline analysis method is applicable for treating all types of data, and flow pressure affects analysis quality almost as much as flow rate. If the flow pressure is ignored, inaccurate interpretation is highly possible. Therefore, regular monitoring of BHFP is essential for medium- and high-yield wells with complex fluid flow, to lay a foundation for the performance forecasting

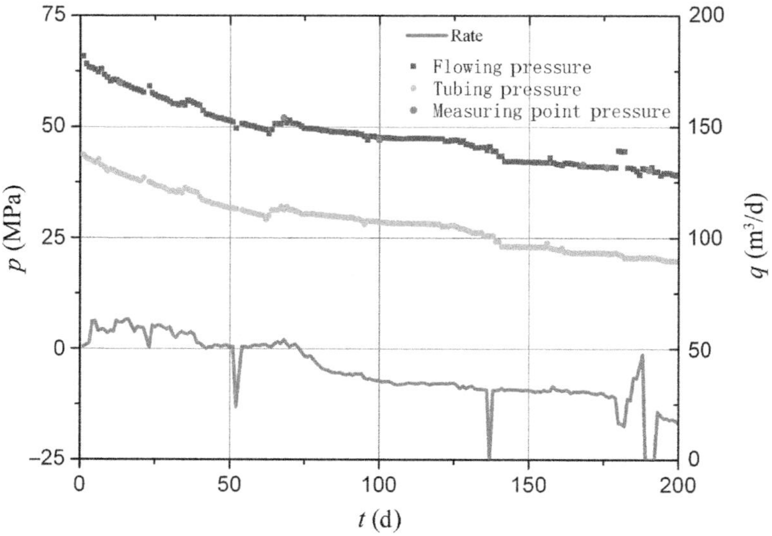

Fig. 4.23 Production data and pressure conversion of Well A.

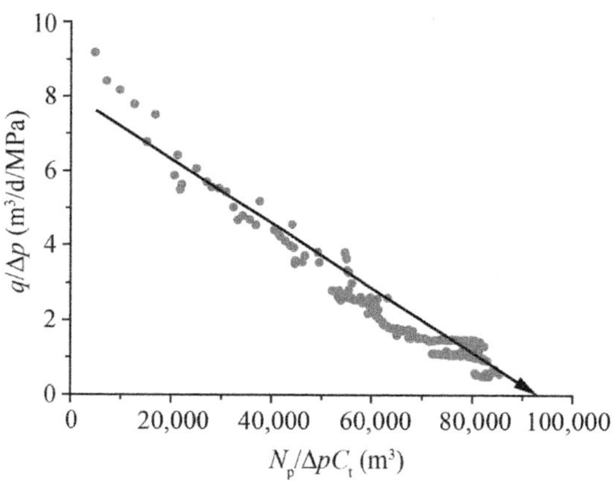

Fig. 4.24 OOIP estimation of Well A by the Agarwal-Gardner method.

and stimulation adjustment of the well. The well performance model based on the preceding process can be used to forecast the performance of a single well, as shown in Fig. 4.26.

Deconvolution analysis Applying the deconvolution algorithm in the production decline analysis can effectively reduce the influence of data noise, but also obtain more data so that the data matching effect can be significantly improved, and the uncertainty of the interpretation results can be greatly

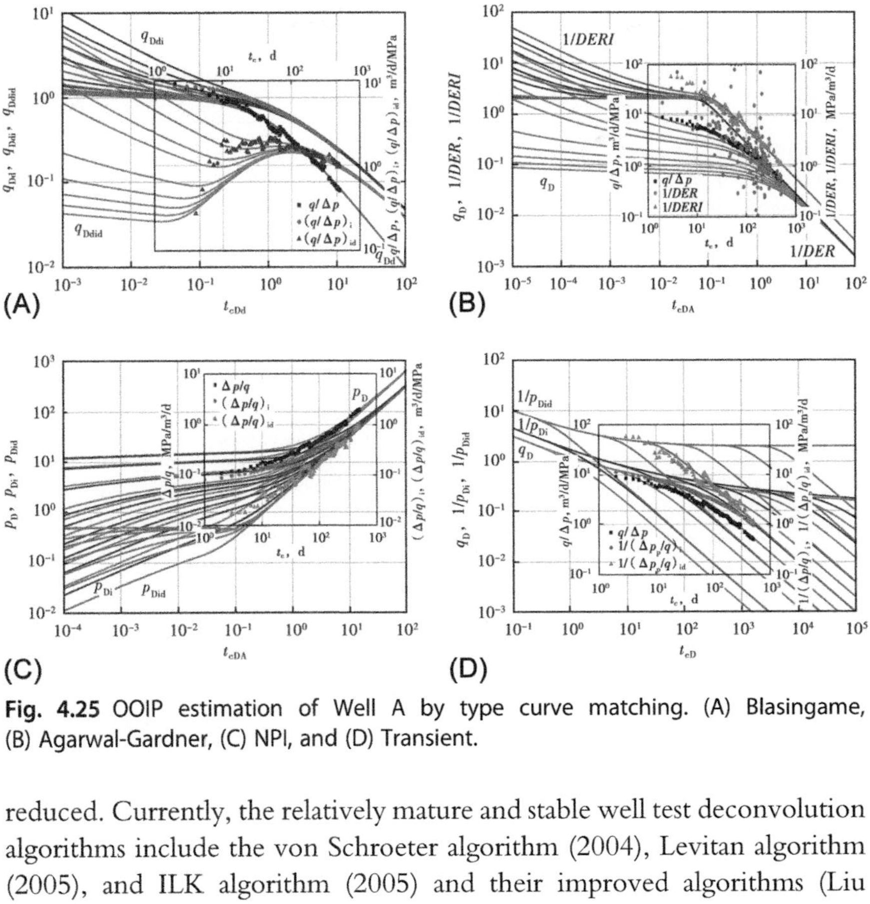

Fig. 4.25 OOIP estimation of Well A by type curve matching. (A) Blasingame, (B) Agarwal-Gardner, (C) NPI, and (D) Transient.

reduced. Currently, the relatively mature and stable well test deconvolution algorithms include the von Schroeter algorithm (2004), Levitan algorithm (2005), and ILK algorithm (2005) and their improved algorithms (Liu et al., 2018).

It can be seen from Fig. 4.27 that the result of the improved deconvolution algorithm accords well with the accurate flow data, which verifies the correctness of the improved algorithm. However, the result of the original deconvolution algorithm based on the cumulative flow data shows a large deviation in the early and late stages of the flow change. Because the flow rate changes rapidly under the unit pressure difference in the early stage, and approaches zero in the late stage, the calculations in both stages are greatly affected by the accuracy of the algorithm.

Through deconvolution, the flow data under the bottom hole pressure are converted into the flow data under the unit bottom hole pressure difference, and then the production decline analysis is performed. The characteristic parameter data points will form three clearer and smoother characteristic straight line segments with longer duration (boundary-dominated),

Table 4.17 Type-curve matching results of Well A

Method	Matched radius r_{eD}	OOIP N (10^4 m^3)	Drainage area A (10^4 m^2)	Permeability K (mD)	Skin factor S
Blasingame	7	9.27	27.8	0.21	−5.26
Agarwal-Gardner	7	8.84	26.5	0.23	−5.96
NPI	7	9.70	26.8	0.25	−5.97
Transient	7	8.94	29.1	0.19	−6.01
Average		9.19	27.6	0.22	−5.80

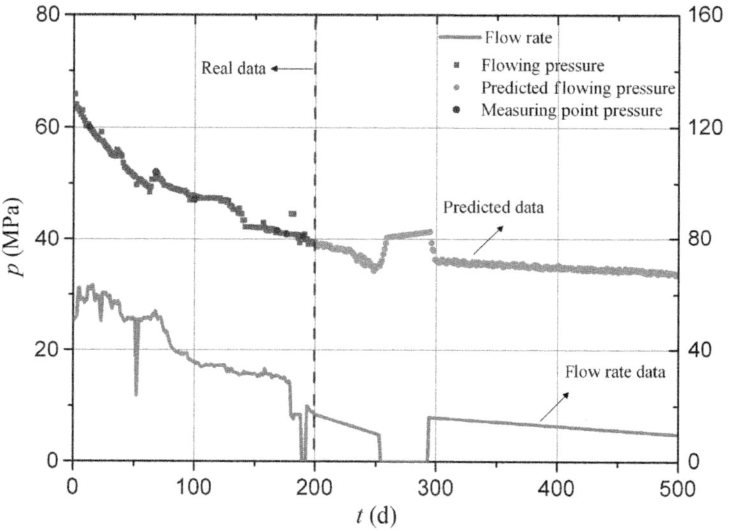

Fig. 4.26 Performance forecasting for Well A.

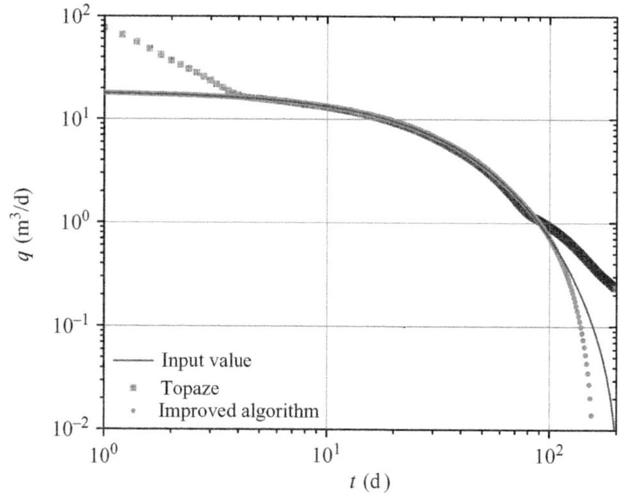

Fig. 4.27 Comparative verification for the deconvolution algorithms (Liu et al., 2018).

which can improve the matching effect of the characteristic parameter data to reduce the uncertainty of the interpretation results. The deconvolution flow rate data calculated by the improved algorithm and the original algorithm are respectively analyzed by FAST RTA software, as shown in Fig. 4.28. From the comparison between Fig. 4.28A and B, it can be seen that the data points of the characteristic parameters are more concentrated, and the data curve is smoother, when using the Blasingame method to perform production decline analysis to the flow rate data output by the improved ILK deconvolution algorithm, and it has a very high degree of matching with the type curve, which significantly improves the matching effect of the characteristic parameters. However, when the ILK original algorithm is used to calculate the output flow rate data, the calculated data points are scattered, the normalized rate and rate integral data are missing, and the length of the characteristic straight line segment of the pseudo-steady state flow is shorter, which makes interpretation more difficult, and is not conducive to data matching in the production decline analysis. This weakens the role of deconvolution in production decline analysis.

The performance-based reserves calculated by the original ILK algorithm are 8.36×10^4 m^3, and that calculated by the improved ILK algorithm are 8.78×10^4 m^3. It can be seen that the performance-based reserves obtained by the Blasingame decline analysis based on the improved ILK deconvolution algorithm are closer to the average of performance-based reserves calculated by the preceding five advanced production decline analysis methods.

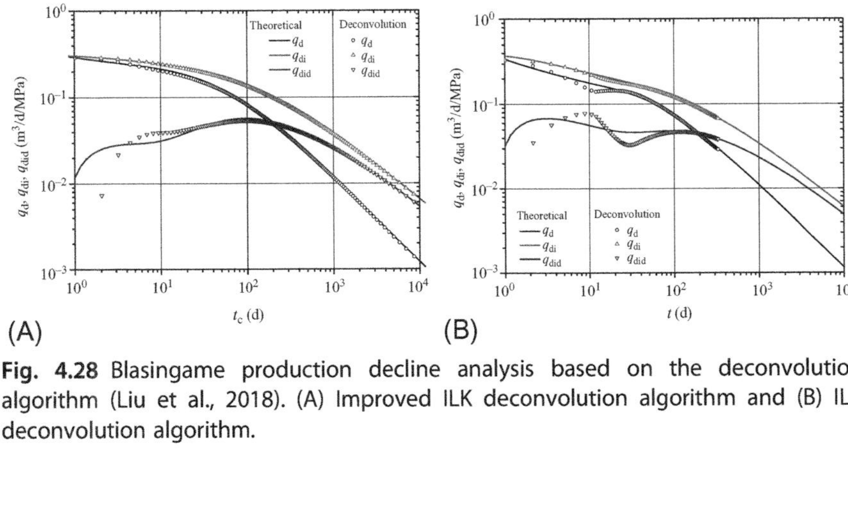

(A) (B)

Fig. 4.28 Blasingame production decline analysis based on the deconvolution algorithm (Liu et al., 2018). (A) Improved ILK deconvolution algorithm and (B) ILK deconvolution algorithm.

4.1.3.2.4 Sensitivity analysis

Basic parameters According to the analysis procedure, the accuracy of the performance-based reserves estimation is mainly sensitive to two factors: (1) the error caused by the inaccurate conversion of BHFP; and (2) the errors resulting from the initial reservoir pressure, PVT of fluids, and reservoir parameters. The latter factor will be explained herein. In the advanced production decline analysis, the OGIP is calculated by the following equation

$$G=\alpha\frac{S_g p_i}{\left(\mu_g Z c_{ti}\right)}f_{match} \tag{4.40}$$

Therefore, such parameters as initial reservoir pressure, gas saturation, porosity, and relative density of gas impact the estimation results.

A dry gas scenario is considered for illustration. A set of production data are simulated using well test software, and then the influence of each parameter on the estimation result is analyzed. It is assumed that a well is located in the midpoint of the gas reservoir, the reservoir thickness is 10 m, the drainage radius is 1000 m, the reservoir temperature is 140°C, the initial reservoir pressure is 65 MPa, the formation flow capacity is 300 mD·m, the reserves within the drainage area are 10.1×10^8 m^3, and the well is produced at the rate of 15×10^4 m^3/d for 3.5 years (Fig. 4.29).

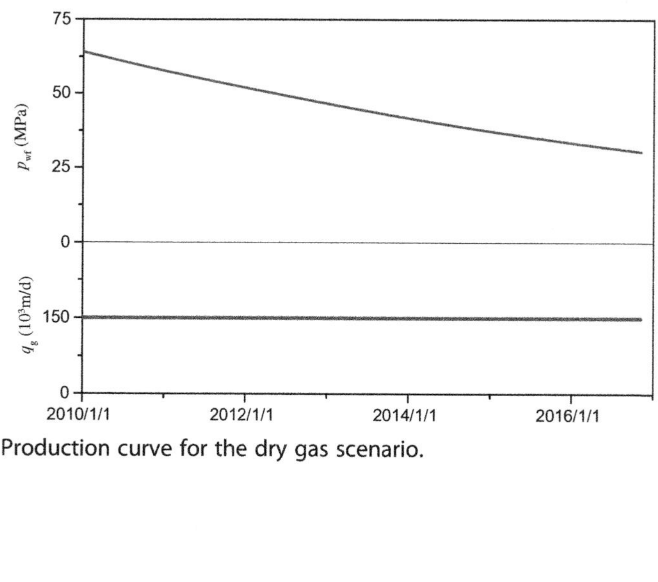

Fig. 4.29 Production curve for the dry gas scenario.

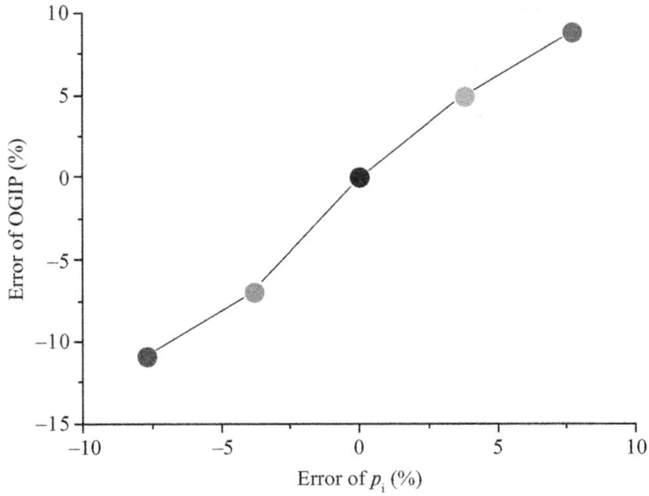

Fig. 4.30 Influence of initial reservoir pressure on OGIP estimate.

Table 4.18 Influence of porosity on the OGIP estimation

Porosity		Total compressibility (c_t)		OGIP	
Value (%)	Error (%)	Value (kPa^{-1})	Error (%)	Value (10^8 m^3)	Error (%)
5	−50	6.93E−06	3.59	9.76	−3.4
8	−20	6.76E−06	1.05	10.00	−0.97
9	−10	6.72E−06	0.45	10.06	−0.39
10	0	6.69E−06	0.00	10.10	0.00
11	10	6.66E−06	−0.45	10.14	0.39
12	20	6.64E−06	−0.75	10.18	0.78
15	50	6.58E−06	−1.64	10.27	1.65

First, the influence of the initial reservoir pressure is discussed. In Fig. 4.30, if the input value of the initial reservoir pressure is larger than the real value, the OGIP estimate is larger than the real value; otherwise, the OGIP estimate is smaller than the real value.

Porosity, a static parameter, may also influence the estimation, because it is relevant to rock compressibility (Table 4.18). Given the other parameters remain unchanged, the porosity will lead to a deviation, mainly in the total formation compressibility, with an error of less than 5%. If the compressibility is 5% larger, the OGIP will be 5% lower; otherwise, the OGIP will be 5% higher.

Gas saturation influences the estimation in total compressibility, similar to the porosity (Table 4.19). Given other parameters remain unchanged, the error in the OGIP is positively proportional to the error in the ratio of gas saturation to total compressibility.

The relative density of gas exerts influence as shown in Fig. 4.31. If the input value of the relative density of gas is greater than the real value, the OGIP estimate is greater than the real value; otherwise, the OGIP estimate is smaller than the real value.

PVT A condensate gas scenario is considered to illustrate the influence of PVT data on the performance-based reserves estimation. It is assumed that a well is located at the midpoint of the gas reservoir, and the simulation parameters are shown in Table 4.20.

In this model, the original gas in place is 1.29×10^8 m^3, and the original condensate oil in place is 10.8×10^4 t. Upon abandonment, the recovery percent of natural gas is 62.01%, and the recovery percent of condensate oil is 32.2%. The production curve is shown in Fig. 4.32.

First, the flow rate data are processed according to the pseudo-single-phase method, and the oil and gas two-phase production is converted into the condensate gas production. The empirical formula of the condensate oil molecular weight is,

$$M_o = \frac{44.29\gamma_o}{1.03 - \gamma_o} \tag{4.41}$$

The equation for the gas equivalent of condensate oil is,

$$G_e = \frac{8.3093\gamma_o T_{sc}}{M_o p_{sc}} \tag{4.42}$$

The equation for the conversion of condensate gas is

$$q_{og} = q_g \left(1 + \frac{G_e}{R_s}\right) \tag{4.43}$$

By loading the converted data to relevant software, the OGIP is calculated to be 0.93×10^8 m^3, which is 28% smaller than the real value. If the influence of PVT is considered, the OGIP is 1.34×10^8 m^3, which is 4% larger than the real value. Therefore, for condensate gas wells, the data cannot be simply treated by the pseudo single-phase method, and the influence of PVT must be considered.

Table 4.19 Influence of gas saturation on the OGIP estimation

Saturation (S_g)		Total compressibility (c_t)		S_g/c_t ratio		OGIP	
Value (%)	Error (%)	Value (kPa^{-1})	Error (%)	Value (% kPa)	Error (%)	Value (10^8 m^3)	Error (%)
60.0	−25.0	5.31E−06	−20.63	11,299,435	−5.51	9.53	−5.64
65.0	−18.8	5.66E−06	−15.40	11,484,099	−3.96	9.73	−3.69
70.0	−12.5	6.00E−06	−10.31	11,666,667	−2.44	9.83	−2.70
75.0	−6.3	6.34E−06	−5.23	11,829,653	−1.07	9.98	−1.17
76.5	−4.4	6.45E−06	−3.59	11,860,465	−0.82	10.01	−0.89
80.0	0.0	6.69E−06	0.00	11,958,146	0.00	10.1	0.00
83.5	4.4	6.93E−06	3.59	12,049,062	0.76	10.17	0.69
85.0	6.3	7.03E−06	5.08	12,091,038	1.11	10.22	1.17
90.0	12.5	7.38E−06	10.31	12,195,122	1.98	10.32	2.14

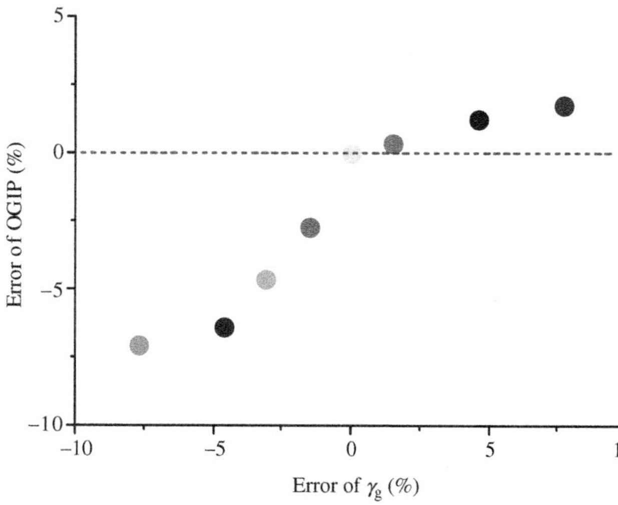

Fig. 4.31 Influence of relative density of gas on OGIP estimate.

Table 4.20 Simulation parameters for the condensate gas scenario

Parameter	Unit	Value	Parameter	Unit	Value
Top/bottom depth	m	6293–6308	Abandonment production	m^3	3000
Net pay thickness	m	15.00	Wellbore diameter	mm	89
Porosity	%	1.85	Relative density of gas		0.6210
Permeability	mD	11.00	GOR	m^3/m^3	1945.9
Saturation (S_g)	%	75.00	Separator pressure	MPa	1.5
Compressibility	kPa^{-1}	1.50E − 06	Separator temperature	°C	26.8
Initial reservoir pressure	MPa	68.11	Dewpoint pressure	MPa	51.8
Reservoir temperature	°C	151.86	Dewpoint temperature	°C	151.9
Drainage radius	m	1000	Relative density of oil		0.7681
Production proration	$10^4\ m^3$	10	Molar content of N_2	%	3.090
Abandonment pressure	MPa	18	Molar content of CO_2	%	2.664

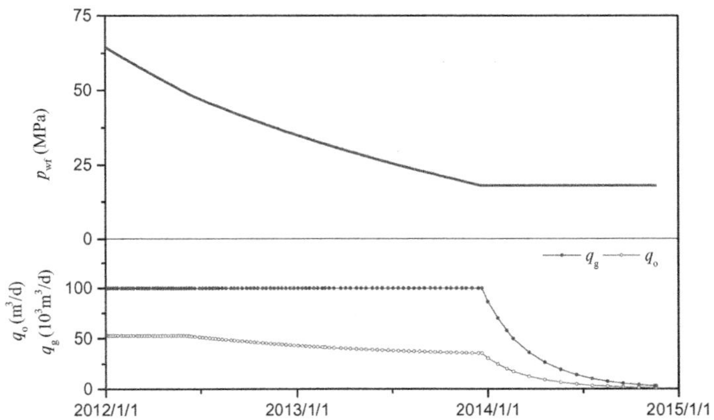

Fig. 4.32 Production curve for the condensate gas scenario.

4.1.3.2.5 Influencing factors

The influence of converted BHFP, production time, working system, water production, and other factors on the OGIP estimation are discussed herein (Table 4.21).

4.1.3.2.5.1 Converted BHFP The accuracy of BHFP conversion may significantly affect the estimation of OGIP. If the converted BHFP is smaller than the real BHFP, that is, the production pressure drawdown increases, then the estimated OGIP will be smaller. If the converted BHFP is larger than the real BHFP, that is, the production pressure drawdown decreases, then the estimated OGIP will be larger. Here, some typical wells with reliable flow rate and abundant static and flow pressure data are selected from the Tazhong I, II, and III gas fields for sensitivity analysis. The data include the entire history because these wells were put into production so that the reserves estimation for the wells, under the base case, is likely to be more reliable.

An error of ±5% in BHFP may bring an error of ±(5%–10%) in the reserves (Table 4.22). Therefore, the accuracy of BHFP conversion is crucial to the reserves estimation.

Figs. 4.33 and 4.34 show the production curves and OGIP at variable BHFP of Wells G43 and G15-5H.

Production time For fractured-vuggy carbonate reservoirs, short production time will result in a smaller OGIP of the single well when the boundary-dominated flow state is not reached.

Table 4.21 Factors influencing the OGIP estimation

Factor		Influence: High	Influence: Small	Remark
BHFP	Higher BHFP	✓		
	Smaller BHFP		✓	
Production time	Multiple fractured-vuggy units		✓	In case of short production time, the reservoir bodies far away from the wellbore are not connected
	Low permeability, fractured		✓	Gas supply near the wellbore is insufficient; the BHFP drops rapidly and becomes gentle in the late stage
Water production	Connected with large bottom aquifer	✓		Since connecting with the large bottom aquifer, the illusion of steady pressure and high yield appears in the early stage of production
	Connected with fractures		✓	Water energy is insufficient, and the flowing pressure drops immediately after water breakthrough
Working system	Large pressure difference		✓	Similar to low-permeability conditions
	Retrograde condensate		✓	Gas-liquid two-phase flow, fast flowing pressure drop

For multireservoir wells, when the production time is short, the OGIP estimated is obviously smaller. As shown in Fig. 4.35, for Well T622, after day 68 and day 123 of production, the daily oil production increased by two magnitudes consecutively in the condition of the smaller nozzle, and the calculated OGIP rose from 2.66×10^4 to 10×10^4 and 21.2×10^4 t, respectively.

Table 4.22 Influence of BHFP conversion error on OGIP estimation

Field	Typical well (gas well)	Factor	BHFP	OGIP of condensate gas (10^4 m^3)	Qty. of increase/ decrease (10^4 m^3)	Percentage of increase/ decrease (%)
TZ I	T62-6H	Base case		35,200	/	/
		BHFP decrease	5%	32,800	−2400	−7
		BHFP increase	5%	37,000	1800	5
	T62-7H	Base case		29,800	/	/
		BHFP decrease	5%	26,900	−2900	−10
		BHFP increase	5%	31,600	1800	6
TZ II	T201C	Base case		33,400	/	/
		BHFP decrease	5%	30,800	−2600	−8
		BHFP increase	5%	36,300	2900	9
	G43	Base case		31,100	/	/
		BHFP decrease	5%	28,300	−2800	−9
		BHFP increase	5%	33,800	2700	9
	G11	Base case		9966	/	/
		BHFP decrease	5%	8978	−988	−10
		BHFP increase	5%	10,810	844	8

Continued

Table 4.22 Influence of BHFP conversion error on OGIP estimation—cont'd

Field	Typical well (oil well)	Factor	BHFP	OOIP of oil (10^4 t)	Qty. of increase/decrease (10^4 t)	Percentage of increase/decrease (%)
TZ III	G15	Base case		201.0	/	/
		BHFP decrease	5%	188.0	−13	−6
		BHFP increase	5%	211.2	10.2	5
	G15-1H	Base case		90.4	/	/
		BHFP decrease	5%	82.9	−7.52	−8
		BHFP increase	5%	96.0	5.6	6
	G15-4H	Base case		53.4	/	/
		BHFP decrease	5%	50.9	−2.56	−5
		BHFP increase	5%	56.3	2.88	5
	G15-5H	Base case		65.2	/	/
		BHFP decrease	5%	61.4	−3.84	−6
		BHFP increase	5%	72.0	6.8	10

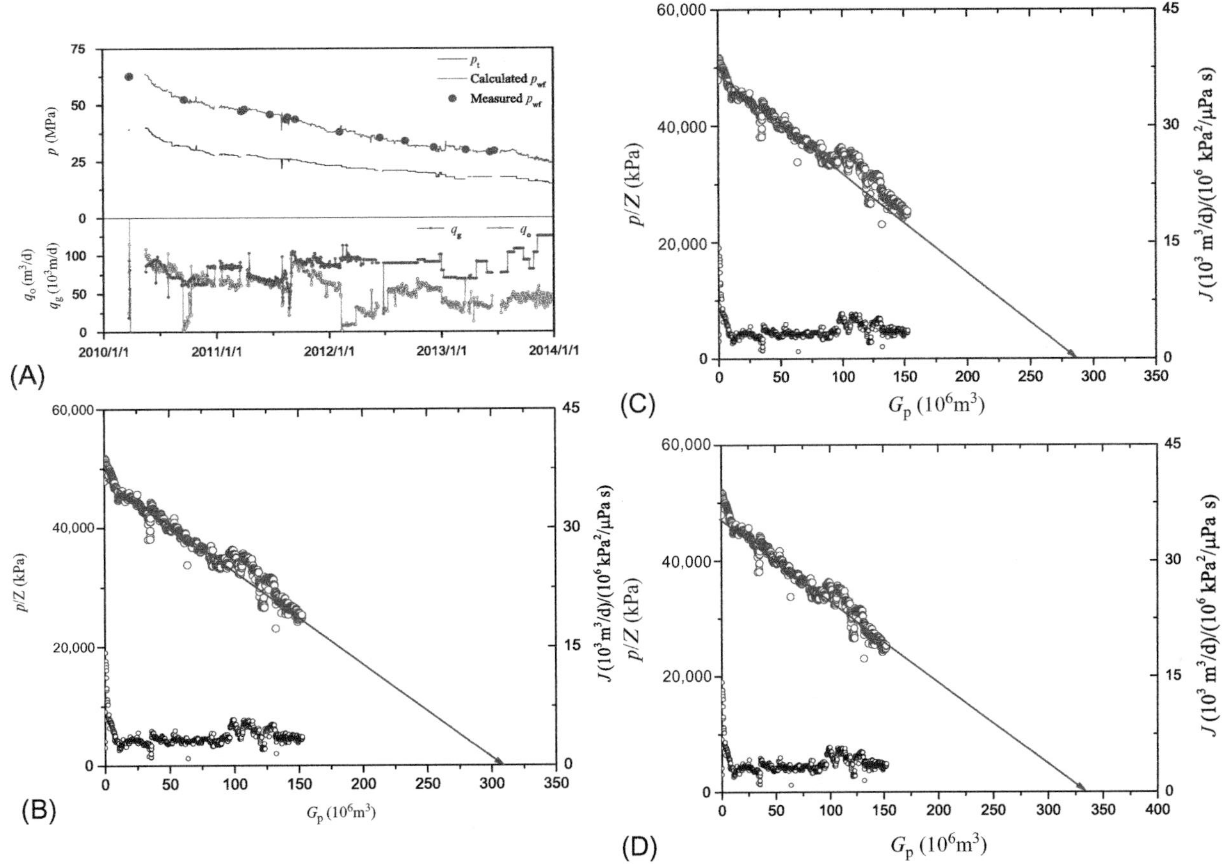

Fig. 4.33 Production curve and OGIP of Well G43. (A) Production curve, (B) OGIP in the base case, (C) OGIP when BHFP decreases by 5%, and (D) OGIP when BHFP increases by 5%.

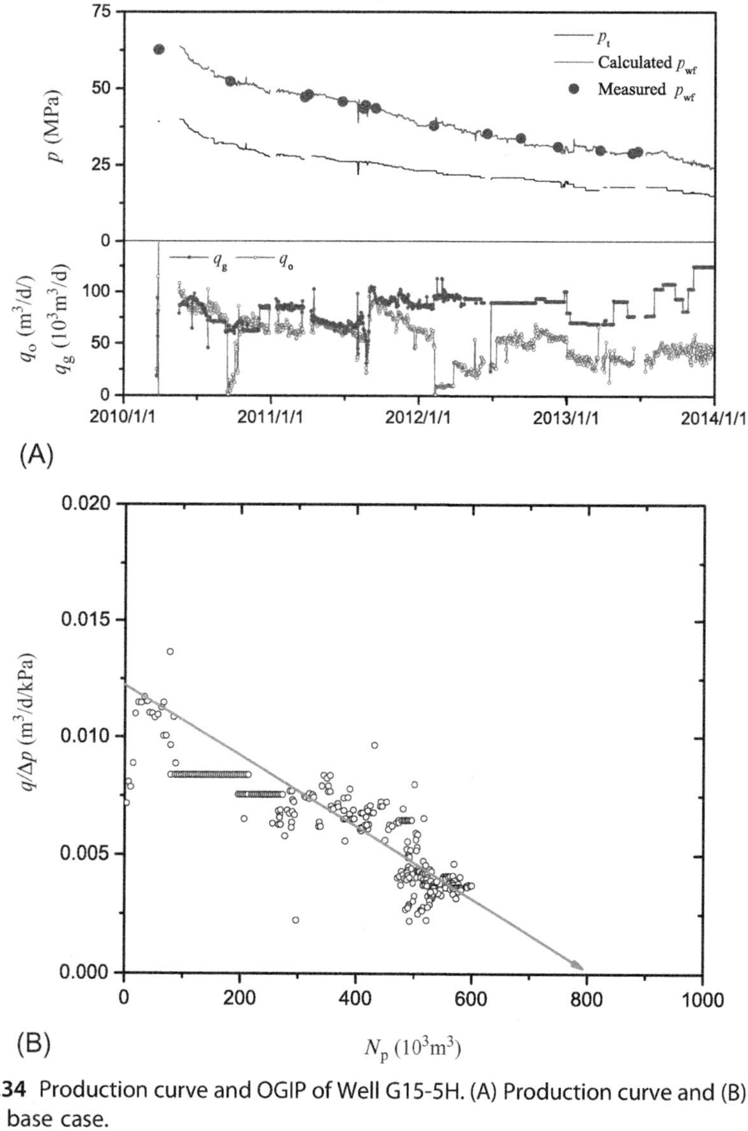

Fig. 4.34 Production curve and OGIP of Well G15-5H. (A) Production curve and (B) OGIP in the base case.

For low-permeability reservoirs, the initial pressure drops quickly; if the production time is short, the calculated OGIP will be smaller. Fig. 4.36 shows the results of a typical well (Well T242). For details, see Section 5.2 in Chapter 5.

Most wells in the Tazhong II gas field reveal the production characteristics of peripheral energy supply (Table 4.23).

For medium- and high-yield wells that are producing for less than half a year, the OGIP tends to increase with production time (Table 4.24). After

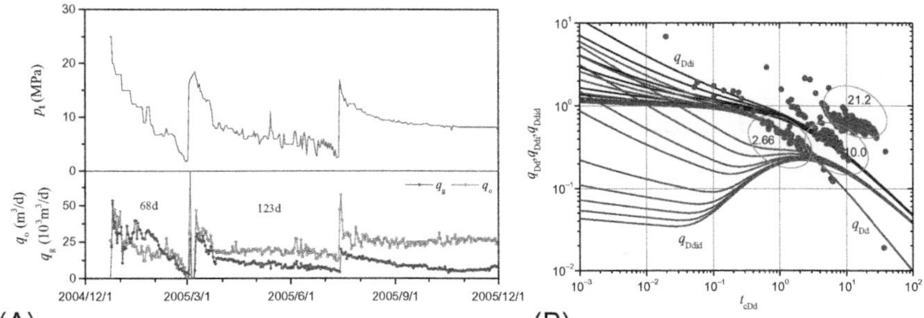

(A) (B)

Fig. 4.35 Change of OGIP of Well T622 with the production time. (A) Production history and (B) Blasingame type curve.

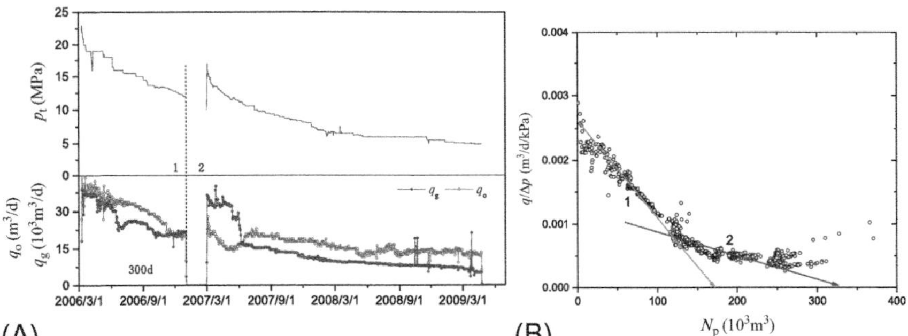

(A) (B)

Fig. 4.36 Change of OGIP of Well T242 with the production time. (A) Production history and (B) Blasingame type curve.

Table 4.23 Peripheral energy for single wells in the Tazhong II gas field

Well	Net pay thickness (m)	Porosity (%)	Energy supply	
			Well test	**RTA**
G48	9.5	2.18		Constant volume
G44-H2C	12	3.19		Constant volume
G441	9.5	2.18		✓
G46	9.5	2.18		Constant volume
G462	10.5	5.34		✓
T201C	16	2.97		Constant volume
G43	10	2.80	✓	✓
G14-1	14	3.39	✓	✓
G111	79.5	6.06		✓
G11-H7	42.16	4.96		✓
G8	14.5	4.96		✓
G22	4.5	2.50		✓

Table 4.24 Influence of production time on the OGIP estimation

Well	OGIP			Growth multiple (compared with 180 d)	
	180 d 10^8 m^3	300 d 10^8 m^3	Current 10^8 m^3	300 d Decimal	400 d+
G43	1.2	2.8	2.9	2.3	2.4
T201C	1.4	3.1	3.2	2.2	2.3
G462	1.3	2.6	2.7	2.0	2.1
G8	0.6	0.7	0.8	1.2	1.2
G111	1.8	2.2	2.2	1.2	1.6
G10	2.7	3.5	4.0	1.3	1.5
Average	1.5	2.5	2.6	1.7	1.8

the production enters the boundary-dominated flow period, the production time essentially has no impact on the OGIP estimation.

To sum up, for low-permeability or fractured reservoirs, if the production time is short and the gas supply near the wellbore is insufficient, resulting in a rapid decline in the BHFP, the OGIP will be underestimated when the early data are used; for multireservoirs, if the production time is short and the reservoirs far away from the wellbore are not communicated, the OGIP may also be underestimated.

(1) Working system

A large production pressure drawdown may cause retrograde condensate, which will result in underestimation of OGIP. The simulation results show that the OGIP is 1.65×10^8 m^3 at the flow rate of 4×10^4 m^3/d, and 1.30×10^8 m^3 at the flow rate of 8×10^4 m^3/d.

(2) Water breakthrough

Water invasion in a single well presents obvious behaviors on the production decline type curve (Fig. 4.18). Similar behaviors can also be observed on the curve of the relationship between static pressure and cumulative production (Fig. 4.37). Generally, the single-well production performance reveals three stages: (1) normal production without water invasion; (2) production with water invasion and water energy supply; and (3) deteriorating production due to water coning. The reserves are calculated for the three stages. In Stage 1, the calculation reflects the real reserves. In Stage 2, the energy supply, due to the water invasion, results in the illusion of stable pressure and high flow rate, which gives rise to the overestimation of reserves. In Stage 3, multiphase flow and a rapid drop of BHFP cause the illusion of insufficient formation energy, which leads to underestimation of reserves. For details, see Section 4.1.4.3.

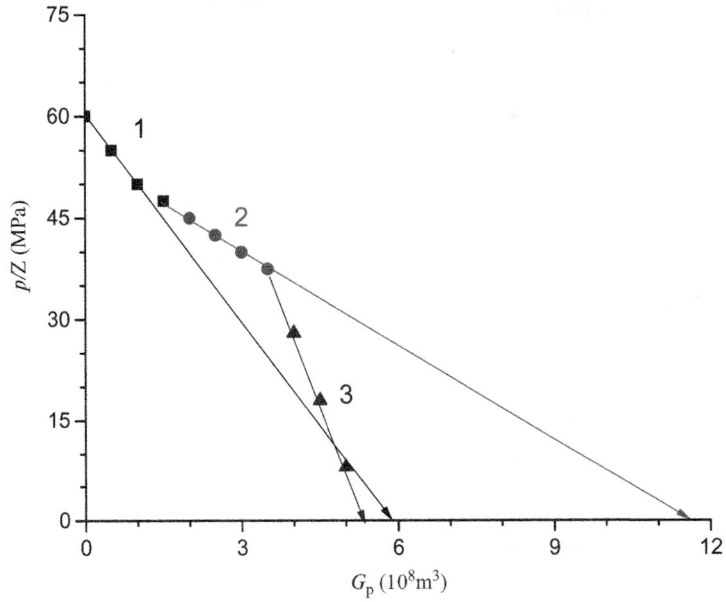

Fig. 4.37 Behaviors of water invasion on pressure drawdown curve.

4.1.3.2.6 Well interference

For a well-connected gas reservoir, the material balance curve of a single well often bends inward, revealing characteristics similar to the material balance equation of ultra-high-pressure gas reservoirs. The Blasingame type curve of a single well also bends inward during the boundary-dominated flow period (Fig. 4.18). The material balance curve of the whole gas reservoir actually manifests the characteristics of the constant-volume gas reservoir. In this case, it is not acceptable to get the OGIP of the gas reservoir simply by adding the single-well performance-based reserves (which are often overestimated). The influence of well interference should be considered, and it is only necessary to redefine the system-based material balance time for the advanced production decline analysis of the well cluster. For example, the dynamic and static analysis results show that the three wells in Cluster G15 are connected, and the Blasingame type curve of a single well suggests obvious well interference, as shown in Fig. 4.38A; the Blasingame type curve of the cluster, as shown in Fig. 4.38B, presents the characteristics of the constant-volume gas reservoir.

For a single well, the sum of OGIP of the three wells is 3.222 million tons, while for the cluster, the total OGIP is 2.576 million tons. Obviously, the former is 25% larger than the latter (Table 4.25).

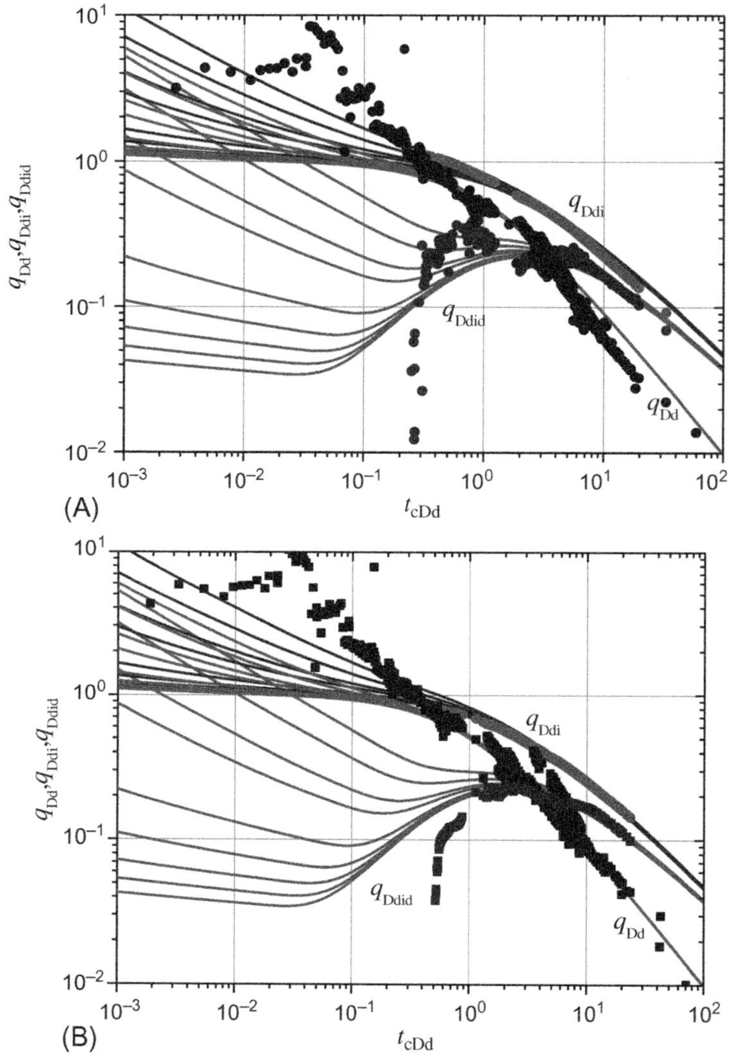

Fig. 4.38 Blasingame type curves of Cluster G15. (A) Single well and (B) cluster.

Table 4.25 OGIP with/without considering well interference

No.	Well	OGIP 10^4 t	Drainage area km^2	Cumulative production 10^4 t
1	G15	201.6	4.13	19.8
2	G15-4H	53.52	1.1	8.9
3	G15-5H	67.12	1.41	7.9
Total		322.2	6.64	36.6
4	Cluster analysis	257.6	5.06	/

4.1.3.2.7 Discussion

Advanced production decline analysis is a new technique to forecast the production performance of wells and clusters. Essentially, based on the transient flow theory, the relationship between pressure and the flow rate is analyzed using the routine production data of wells, and finally, the reservoir parameters and reserves within the drainage area are determined through matching type curves (e.g., Blasingame type curve). Hydraulic fracturing is a common measure for stimulating the single-well production of low-permeability gas reservoirs. The advanced production decline analysis technique can also be used as a fracturing evaluation method to obtain the fracture parameters. In all flow models, for the advanced production decline analysis of fractured wells, the fracture is assumed with constant flow conductivity. When they are used for fractured wells with variable conductivity, a substantial error will occur.

Variable fracture conductivity can be divided into spatially variable conductivity and temporally variable conductivity. The spatially variable conductivity means that the conductivity of the wedge-shaped fractures generated in the fracturing operation changes with the propagation of fractures due to an uneven filling of proppant. The temporally variable conductivity means that the conductivity of fractures changes with time due to factors such as proppant crushing, nesting of proppant in the formation, and blockage of cuttings. The variation of fracture conductivity is mainly related to proppant concentration, physical properties of proppant, fracture closure stress and rock hardness; but is also related to reservoir temperature, fluid properties, and salinity environment. Due to so many influencing factors, theoretically, the fracture conductivity can hardly be characterized by the general mathematical formula. Currently, the effect of variable fracture conductivity is measured through a laboratory proppant conductivity test, and then regression analysis is performed on the test data by various functions. According to the pilot tests, the spatially variable conductivity usually presents three forms of variations, that is, linear, exponential, and logarithmic; while the temporally variable conductivity demonstrates two forms of variations, that is, logarithmic and exponential. Many scholars have established transient flow models of oil and gas wells based on the variable fracture conductivity formula and applied them in well test interpretation and production performance analysis.

Here, a mathematical flow model of a fracturing vertical well considering spatially temporally variable conductivity and stress sensitivity of fractures is established, and a new exponential temporally variable conductivity formula is proposed, with the clear physical meaning of the variable conductivity, which can facilitate the type-curve matching. The Blasingame type curve under the conditions of variable conductivity and stress sensitivity is obtained by the mixed finite element method. The influences of spatially variable conductivity, temporally variable conductivity, spatially temporally variable conductivity, and stress sensitivity on the Blasingame type curve are discussed.

Mathematical model

(1) Assumptions

- **(a)** The original reservoir is homogeneous, bounded, and stress-sensitive. The permeability stress sensitivity is considered, but the porosity stress sensitivity is ignored.
- **(b)** The gas reservoir is a dry gas reservoir, or a low-water-saturation reservoir. The water is in a bound water state, and there is no threshold pressure gradient effect. The flow of gas in the reservoir is single-phase flow and meets Darcy's law.
- **(c)** Through hydraulic fracturing, a finite conductivity fracture is induced. The fracture features spatially variable conductivity and temporally variable conductivity. The flow of gas in the fracture is a one-dimensional flow. The damage of the fracturing fluid to the reservoir is considered, and the degree of damage is measured by the skin factor of the fracture.
- **(d)** The fracture volume is very small relative to the reservoir pore volume. The permeability of the fracture is much larger than the permeability of the reservoir. The fracture pressure drawdown causes the gas volume expansion in the fracture to have little effect on the whole flow, so the derivative of the pseudo-pressure versus time in the fracture control equation can be ignored.
- **(e)** The gas compressibility and viscosity changed with pressure are considered, the gas compressibility and the deviation factor are calculated by the DPR method (Dranchuk et al., 1973), and the gas viscosity is determined by the Lee method (Lee et al., 1966).
- **(f)** The influence of the wellbore storage effect, temperature change, and other factors on the flow are ignored.

(2) Building of the mathematical model

Reservoir control equation:

$$\frac{\partial}{\partial x}\left[K_{\mathrm{r}}(p)\frac{\partial m}{\partial x}\right]+\frac{\partial}{\partial y}\left[K_{\mathrm{r}}(p)\frac{\partial m}{\partial y}\right]-\alpha_3 Tq_{\mathrm{f}}=\alpha_1\phi c_{\mathrm{t}}(p)\mu(p)\frac{\partial m}{\partial t} \tag{4.44}$$

Fracture control equation:

$$\frac{\partial}{\partial l}\left[K_{\mathrm{f}}(l,t)\frac{\partial m}{\partial l}\right]+\alpha_3 T q_{\mathrm{f}}=0 \tag{4.45}$$

Initial condition:

$$m(x,y,0)=m_{\mathrm{i}} \tag{4.46}$$

Inner boundary condition:

$$\text{Known wellhead production}: \left.\frac{\partial m}{\partial l}\right|_{\Gamma_{\mathrm{in}}}=\frac{\alpha_2 q_{\mathrm{sc}}(t)T}{w_{\mathrm{f}}K_{\mathrm{f}}(0,t)h} \tag{4.47}$$

$$\text{Known bottom hole pressure}: m|_{\Gamma_{\mathrm{in}}}=m_{\mathrm{w}}(t) \tag{4.48}$$

Closed outer boundary:

$$\left.\frac{\partial m}{\partial n}\right|_{\Gamma_{\mathrm{out}}}=0 \tag{4.49}$$

Material balance equation:

$$\frac{\overline{p}}{Z}=\left(\frac{p}{Z}\right)_{\mathrm{i}}\left(1-\frac{G_{\mathrm{p}}}{G}\right) \tag{4.50}$$

Pseudo-pressure definition:

$$m(p)=2\int_0^p\frac{p}{\mu Z}dp \tag{4.51}$$

(3) Variable fracture conductivity
Spatially variable conductivity: The fracture conductivity ($K_{\mathrm{f}}w_{\mathrm{f}}$) changes with the fracture length in a linear, exponential, or logarithmic manner (Mou and Fan, 2006). The calculation equations are:

$$F_{\mathrm{cl}}=F_{\mathrm{c0}}(1-a_{\mathrm{s}}l_{\mathrm{D}}) \tag{4.52}$$

$$F_{\mathrm{cl}}=F_{\mathrm{c0}}\mathrm{e}^{-b_{\mathrm{s}}l_{\mathrm{D}}} \tag{4.53}$$

$$F_{\mathrm{cl}}=F_{\mathrm{c0}}[1-c_{\mathrm{s}}\ln(1+l_{\mathrm{D}})] \tag{4.54}$$

Temporally variable conductivity: The fracture conductivity changes with the production time in a logarithmic or exponential manner (Wen et al., 2005). The calculation equations are:

$$F_{\mathrm{ct}}=F_{\mathrm{ci}}[1-\chi\ln(1+t)] \tag{4.55}$$

$$F_{\mathrm{ct}}=F_{\mathrm{ci}}\mathrm{e}^{-\gamma t}+F_{\mathrm{cr}} \tag{4.56}$$

The commonly used exponential variation formula (Eq. 4.56) is modified to Eq. (4.57), which can derive a conductivity with clearer physical meaning, thus being conducive to the production data interpretation and analysis.

$$F_{ct} = F_{ci}\left(\eta e^{-\frac{t}{8760C}} + 1 - \eta\right) \tag{4.57}$$

Spatially temporally variable conductivity: The spatially variable conductivity and the temporally variable conductivity are simultaneously considered; that is, the fracture conductivity not only changes with the position of the fracture, but also changes with time. Here, taking the exponential variation scenario as an example, the calculation equation of spatially temporally variable conductivity is derived (Eq. 4.58). Similarly, the calculation equations of spatially temporally variable conductivity under other scenarios can be obtained.

$$F_c(l, t) = F_{ci0}e^{-bl}\left(\eta e^{-\frac{t}{8760C}} + 1 - \eta\right) \tag{4.58}$$

(4) Stress sensitivity
Tight gas reservoirs may have nonlinear flow characteristics, such as stress sensitivity, threshold pressure gradient, and slippage effect. As to the stress sensitivity, the reservoir permeability is no longer constant, but decreases with the increase in effective stress. Numerous experiments have shown that tight gas reservoirs have strong stress sensitivity, which must be considered in the production decline analysis for wells with long production history. The threshold pressure gradient in the gas reservoir is mainly generated due to the gas-water reaction, which is essentially different from the threshold pressure gradient in the oil reservoir. Therefore, water existing in the gas reservoir is a necessary condition for the generation of the threshold pressure gradient. Moreover, the slippage effect may occur in tight gas reservoirs. The slippage effect will cause the increase of the apparent permeability. The tighter the reservoir is, the smaller the reservoir pressure, and the stronger the slippage effect will be. The flow model considering the slippage effect is mainly applied to shallow coalbed methane reservoirs. The influence of the slippage effect on tight sandstone gas reservoirs with generally high reservoir pressure needs to be investigated.

According to the Klinkenberg (1941) formula (Eq. 4.59), the slippage effect is mainly determined by the slippage factor and average reservoir pressure. In order to identify the influence of the slippage effect on flow in a tight sandstone gas reservoir, 96 slippage effect experiments were conducted on

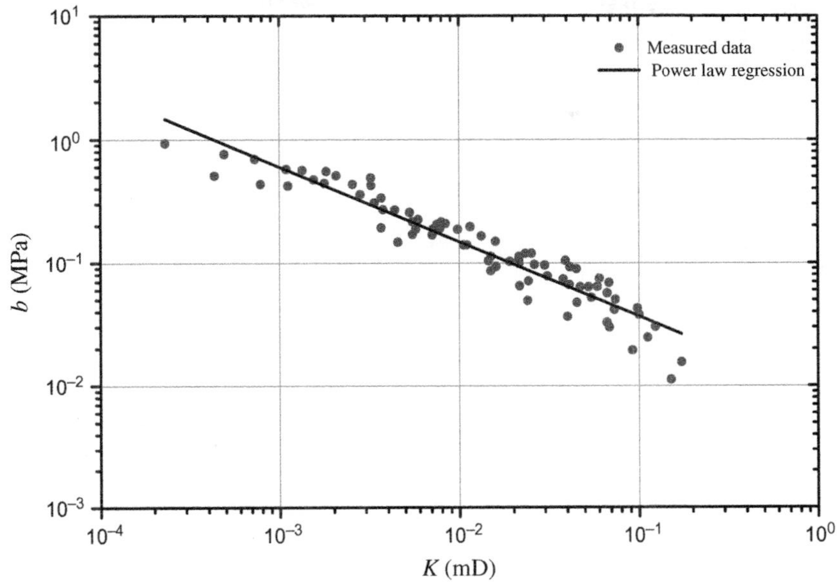

Fig. 4.39 Results of slippage effect experiments.

the core samples from the Sulige gas field, and the relationship between the slippage factor and the absolute permeability was obtained (Eq. 4.60), as shown in Fig. 4.39.

$$K_g = K_\infty\left(1 + \frac{b}{p_m}\right) \tag{4.59}$$

$$b = 0.00911K_\infty^{-0.6061} \tag{4.60}$$

It is assumed that the absolute permeability of the reservoir is 0.1 mD, the initial reservoir pressure is 30 MPa, and the average reservoir pressure in the late stage of production is 10 MPa. According to Eqs. (4.59), (4.60), it is determined that the slippage effect influences the early and late apparent permeability of the reservoir by 0.12% and 0.37%, respectively. Even if the absolute permeability is 0.01 mD, the influence is only 0.50% and 1.50%, respectively. Fig. 4.40 shows the influence of the slippage effect on the Blasingame type curve. It can be observed that the slippage effect has little influence on the production decline curve. Thus, the slippage effect can be neglected in the production decline analysis.

There are many methods for evaluating reservoir stress sensitivity (Luo et al., 2008). The exponential relation and power-law relation are the most popular. Considering the higher relevance of the power-law regression test

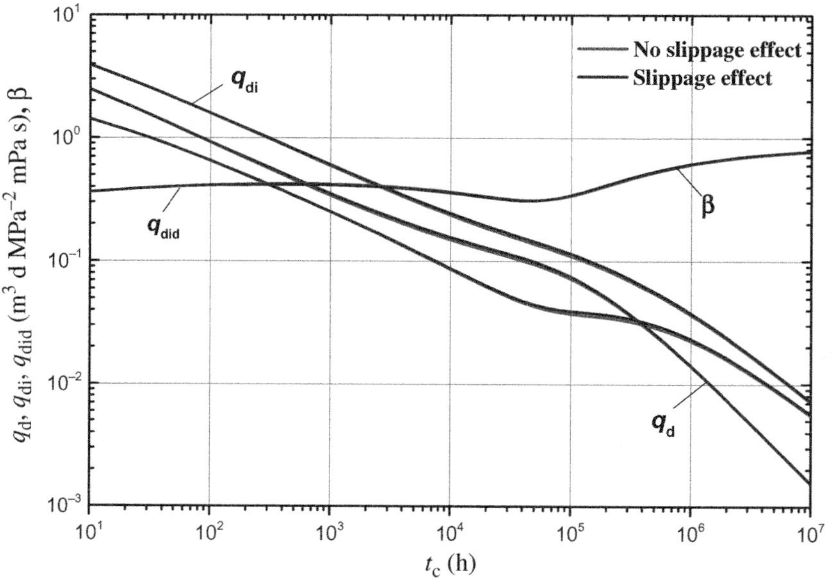

Fig. 4.40 Influence of the slippage effect on a Blasingame type curve.

data, the formula (Eq. 4.61) reflecting the permeability changes with the effective stress in a power-law relationship is adopted herein.

$$K_r(p) = K_{ri}\left(\frac{\sigma_e}{\sigma_{ei}}\right)^{-S_p} = K_{ri}\left(\frac{\sigma - p}{\sigma - p_i}\right)^{-S_p} \tag{4.61}$$

(5) Model solution

The model is solved by the mixed finite element method (Wan et al., 2016). As shown in Eq. (4.62), the whole computational domain is divided into two parts. One is a two-dimensional flow reservoir region, and the other is a one-dimensional flow fracture region.

$$\iint_{\Omega} FEQd\Omega = \iint_{\Omega_m} FEQd\Omega_m + w_f \cdot \int_{\bar{\Omega}_f} FEQd\bar{\Omega}_f \tag{4.62}$$

Where, FEQ represents the fluid flow equation, Ω represents the whole flow region, Ω_m represents the reservoir flow region, and $\bar{\Omega}_f$ represents the fracture flow region.

Using the Galerkin weighted residual method (Zhang et al., 1991) to discretize the control equations of reservoir and fracture, we obtain the two-dimensional finite element equation of the reservoir region as follows:

$$
\begin{aligned}
&AK_r(p)\left[b_ib_i+c_ic_i+\frac{\alpha_1\phi\mu(p)c_t(p)}{6\Delta tK_r}\right]m_i^{n+1}\\
&+AK_r(p)\left[b_ib_j+c_ic_j+\frac{\alpha_1\phi\mu(p)c_t(p)}{12\Delta tK_r}\right]m_j^{n+1}\\
&+AK_r(p)\left[b_ib_k+c_ic_k+\frac{\alpha_1\phi\mu(p)c_t(p)}{12\Delta tK_r}\right]m_k^{n+1}\\
&-\frac{K_r(p)L}{3}\frac{\partial m_i^{n+1}}{\partial n}-\frac{K_r(p)L}{6}\frac{\partial m_{j(k)}^{n+1}}{\partial n}=\frac{\alpha_1\phi\mu(p)c_t(p)}{6\Delta t}m_i^n\\
&+\frac{\alpha_1\phi\mu(p)c_t(p)}{12\Delta t}m_j^n+\frac{\alpha_1\phi\mu(p)c_t(p)}{12\Delta t}m_k^n
\end{aligned}
\tag{4.63}
$$

The one-dimensional finite element equation of the fracture region is:

$$
\frac{F_c(l,t)}{L}m_i^{n+1}-\frac{F_c(l,t)}{L}m_j^{n+1}+F_c(l,t)\frac{\partial m_i}{\partial l}=0 \tag{4.64}
$$

where, A represents the area of the triangular grid, m^2; b_i, b_j, b_k, c_i, c_j and c_k represent the coefficients of finite element units; i, j, and k represent the node numbers of the triangular grid; Δt represents the time step, h; L represents the one-dimensional grid length of fracture, m.

The finite element stiffness matrix of the reservoir region and the stiffness matrix of the fracture region are established by Eqs. (4.63), (4.64), and then they are combined into a system stiffness matrix. According to the interporosity flow expression, the boundary between the fracture and the reservoir is eliminated by superimposing the reservoir and the fracture unit to obtain the last two terms of the left side of Eq. (4.63) and the last term of Eq. (4.64) (see the literature for details). As the fracture conductivity, reservoir permeability, gas viscosity, and compressibility change with time, each time step is calculated according to the specific calculation method for each parameter in practice. In addition, the model adopts two methods to calculate the average reservoir pressure; namely, the combined material balance Eq. (4.50) when the OGIP G is iteratively obtained from the production data, and the weighted averaging of grid unit pressure in case of the whole history matching. Finally, the linear equations solver is used to solve the combined linear equations to obtain the solution of the model.

Theoretical analysis (1) Type curve under spatially variable conductivity

It is assumed that the reservoir permeability is 0.1 mD, the initial reservoir pressure is 30 MPa, the reservoir temperature is 100°C, the net pay thickness is 8 m, the porosity is 10%, the circular closure radius is 500 m, the relative density of gas is 0.7, the fracture half-length is 100 m, and the spatially variable conductivity of the fracture is considered separately, which is assumed to be 200 mD·m. For the purpose of comparative analysis, the values of the variable conductivity coefficients a_s, b_s, and c_s are set to ensure the fracture gap conductivity is 20 mD m. The Blasingame type curves under different spatially variable conductivities are obtained, as shown in Fig. 4.41. According to the comparison, it is found that the variable conductivity mainly affects the production decline type curve in the early stage. The spatially variable conductivity will reduce the value of the early-stage type curve, similar to the influence of the fracture skin factor; the interpretation using conventional models without considering variable conductivity will give rise to a larger skin factor. The late-stage type curves are almost coincidental under different variable conductivities, indicating that the variable conductivity does not affect the pressure diffusion rate, nor the occurrence time of the pseudo-steady state. In addition, given the same conductivity at fracture tip and fracture gap, a descending order of influence on the decline of the type curves is found among the exponential variation, the logarithmic change, and the linear variation.

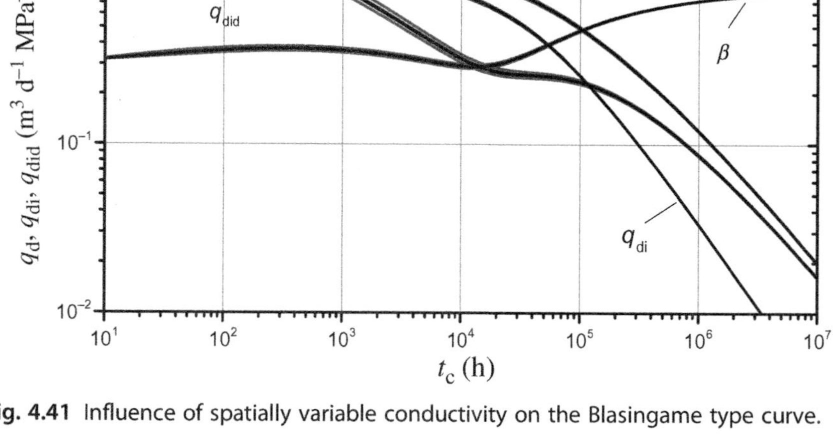

Fig. 4.41 Influence of spatially variable conductivity on the Blasingame type curve.

(2) Type curve under temporally variable conductivity

With the same reservoir parameters as those listed in the previous section, the influence of temporally variable conductivity on the Blasingame type curve is discussed. It is assumed that the fracture conductivity is 200 mD·m initially, and decays to 20 mD·m after 1 year of production, and thereafter the conductivity is basically stable. The influence is calculated under the logarithmic change and exponential variation scenarios, as shown in Fig. 4.42. It can be seen that compared with the constant conductivity, the temporally variable fracture conductivity has a more obvious influence on the production decline curve, especially in the middle stage, when the flow rate curve and rate integral curve "downwarp" to form S-shaped curves. Similar to the spatially variable conductivity, the temporally variable conductivity has little influence on the late-stage type curve, because the temporally variable conductivity also does not affect the pressure diffusion rate of the reservoir. In addition, the type curve under the logarithmic change is not as smooth as that under the exponential variation, and the influence of the logarithmic change is less than that under the exponential variation. Essentially, the logarithmic change is relatively gentle in the whole process, while the exponential variation descends.

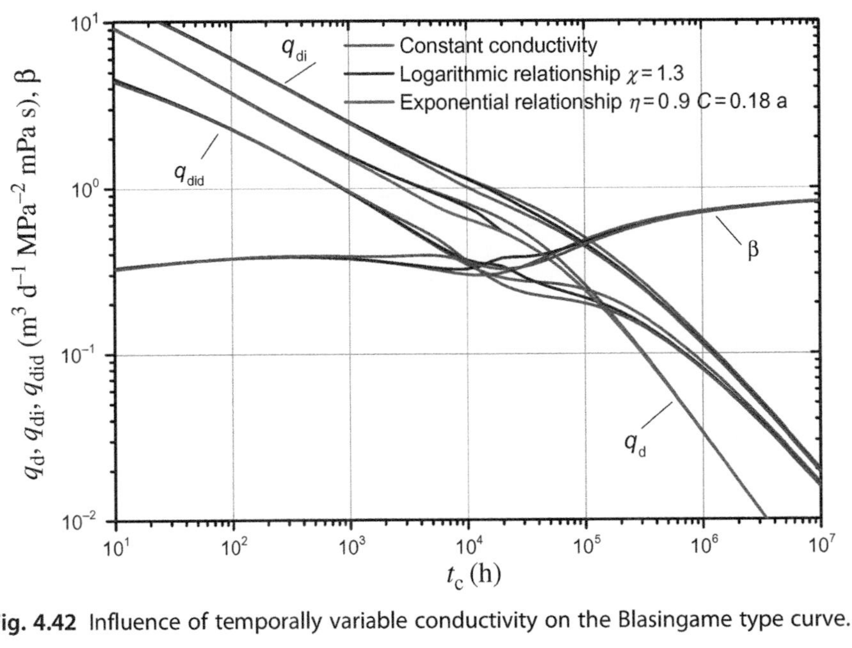

Fig. 4.42 Influence of temporally variable conductivity on the Blasingame type curve.

With the logarithmically variable conductivity as an example, the influence of the variable conductivity coefficient on the type curve is analyzed. Fig. 4.43 shows the type curves at the same variable conductivity time coefficient of 0.1 a, and different conductivity decay coefficients η. Fig. 4.44 shows the type curves at the same conductivity decay coefficient of 0.9, and different variable conductivity time coefficients. It can be seen that the larger the conductivity decay coefficient η, the larger the "downwarping" amplitude of the flow rate curve and the rate integral curve, and the more obvious the S-shaped curve. The larger the conductivity time coefficient, viz. the smaller the decline rate of the fracture conductivity, the later the downwarping of the flow rate curve and rate integral curve, and the smaller the influence on the type curve. When the conductivity time coefficient is very large (as shown in the fourth set of curves in Fig. 4.44), the influence is obscure or negligible. Clearly, the conductivity decay coefficient in the temporally variable conductivity model controls the downwarping amplitude of the type curve, while the conductivity time coefficient controls the downwarping start time of the type curve. This is significant for the type curve matching of fractured wells with temporally variable conductivity.

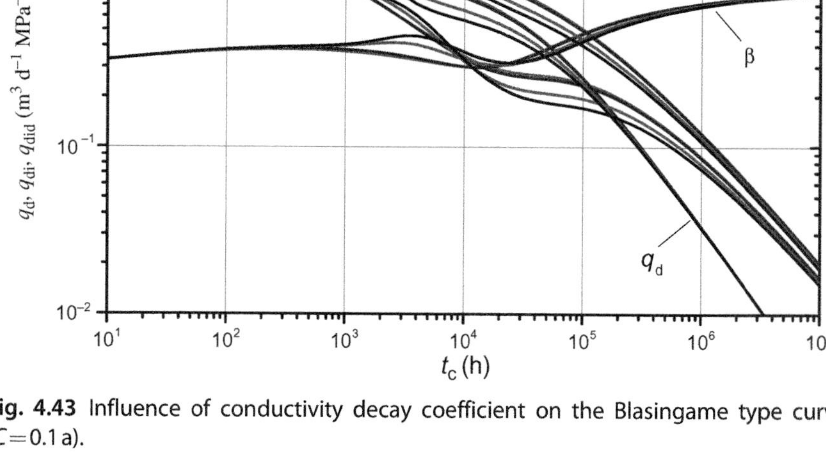

Fig. 4.43 Influence of conductivity decay coefficient on the Blasingame type curve (C=0.1 a).

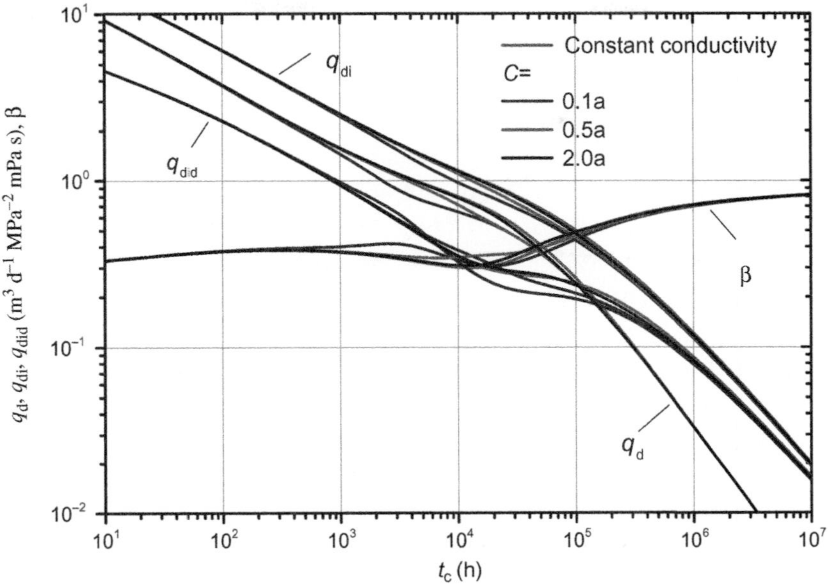

Fig. 4.44 Influence of conductivity time coefficient on the Blasingame type curve ($\eta=0.9$).

(3) Type curve under spatially temporally variable conductivity

Fig. 45 shows the Blasingame type curves under different variable conductivity conditions. The spatially variable conductivity and the temporally variable conductivity are both exponentially changing. The variable conductivity coefficients b_s, η, and C are 2, 0.9, and 0.2a, respectively. It is seen that, in the early stage, curve 1 and curve 3 coincide, while curve 2 and curve 4 coincide, which indicates that the type curve is mainly subject to the influence of the spatially variable conductivity in the early stage. In the middle stage, curve 2 and curve 4 differ greatly, while curve 3 and curve 4 do not vary so much, which indicates that the type curve is subject to the influence of both temporally variable conductivity and spatially variable conductivity in the middle stage, with the temporally variable conductivity in dominance. Comparison of curves 1–4 reveals that the type curve is generally not subject to the influence of the variable conductivity, and all the curves are basically coincident. Through mono-factor analysis of spatially variable conductivity and temporally variable conductivity, it can be found that the Blasingame type curve, considering both spatially variable conductivity and temporally variable conductivity, is a superposition of the two effects.

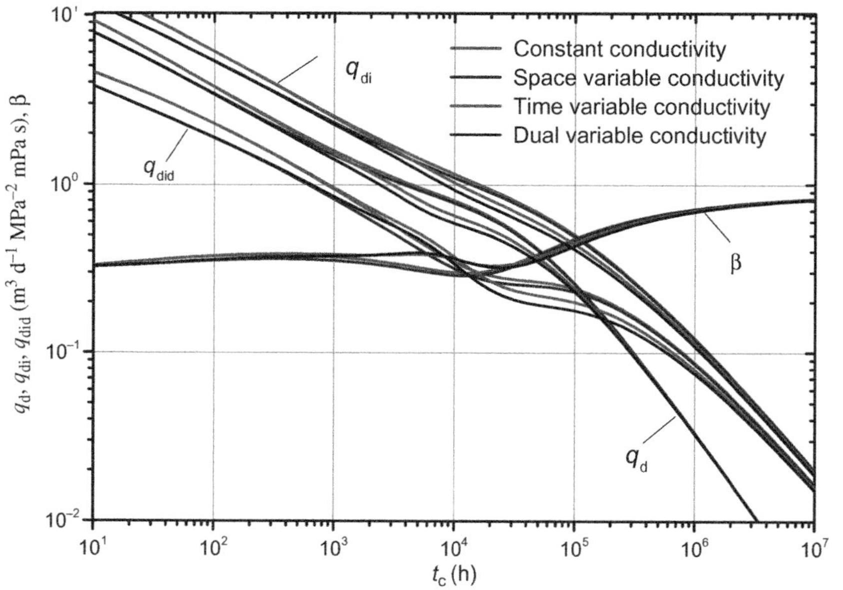

Fig. 4.45 Influence of spatially temporally variable conductivity on the Blasingame type curve.

(4) Type curve under stress sensitivity and spatially temporally variable conductivity

In practice, the wells in tight gas reservoirs may be exposed to both spatially temporally variable conductivity and stress sensitivity, which should be considered in the model. Under the assumption that the overburden pressure is 60 MPa, the Blasingame type curves at different stress sensitivity coefficients under the spatially temporally variable conductivity are calculated by using spatially temporally variable conductivity coefficient (Fig. 4.46). It can be seen that the curves at different stress sensitivity coefficients basically coincide in the early stage, but vary gradually after a period of time. Similar to the temporally variable conductivity, the stress sensitivity will reduce the value of the type curve in the middle and later stages; the larger the stress sensitivity coefficient is, the greater the reduction will be. Different from the temporally variable conductivity, the stress sensitivity will reduce the formation permeability, thereby slowing down the pressure diffusion and delaying the boundary response. The β-curve can clearly reflect this behavior. The stronger the stress sensitivity is, the later the pseudo-steady state will occur. This will be a primary measure to distinguish stress sensitivity and temporally variable conductivity.

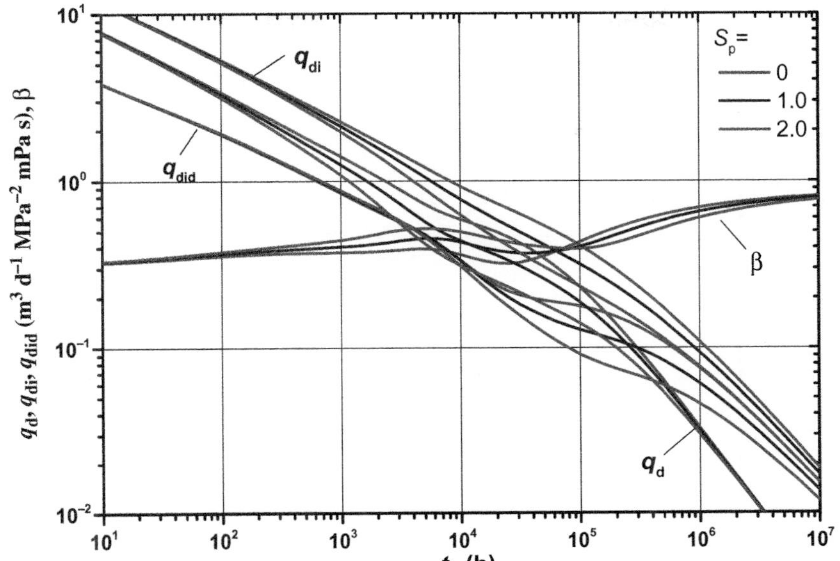

Fig. 4.46 Blasingame type curve under stress sensitivity and spatially temporally variable conductivity.

After the production enters the pseudo-steady state in the late stage, the effects of the variable conductivity and the stress sensitivity are minimal; that is, the variable conductivity and the stress sensitivity do not affect the matching of OGIP.

Field example The spatially temporally variable conductivity flow model proposed herein is used to analyze the production data a gas well, and compared with the conventional model. For this well, the initial reservoir pressure is 32 MPa, the reservoir temperature is 107°C, the porosity is 0.137, the net pay thickness is 5.3 m, the gas saturation is 74%, the relative density of gas is 0.6, the production time is about 2.5 years, and the cumulative production is 848×10^4 m^3. The results of the conventional model and the proposed model are shown in Figs. 4.47 and 4.48, and the history matching results of daily production and cumulative production are shown in Fig. 4.49. The reservoir parameters and fracture parameters obtained by type curve matching are given in Table 4.26. According to the Blasingame type curve of measured production, the production curve of this well suddenly "downwarps" after producing for not more than 1000 h, thus presenting an obvious S-shape, which is very consistent with the behavior caused by

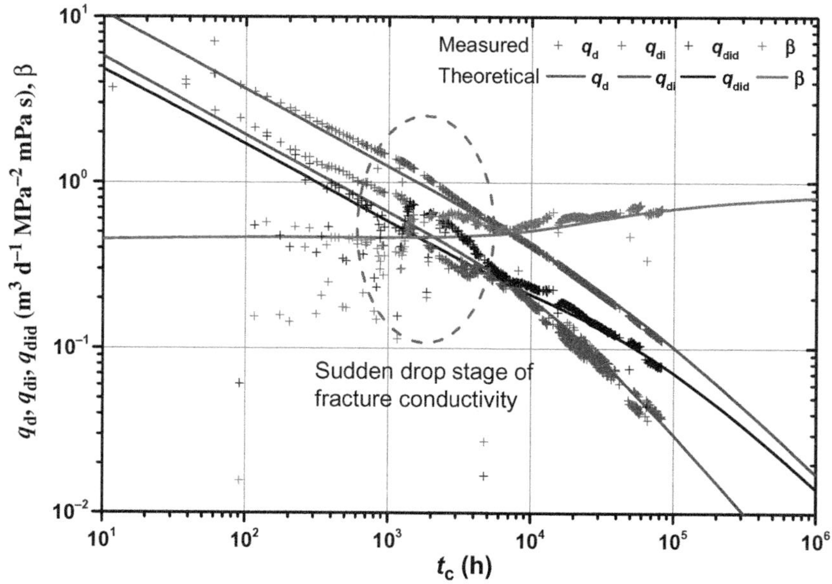

Fig. 4.47 Blasingame type curve based on conventional model.

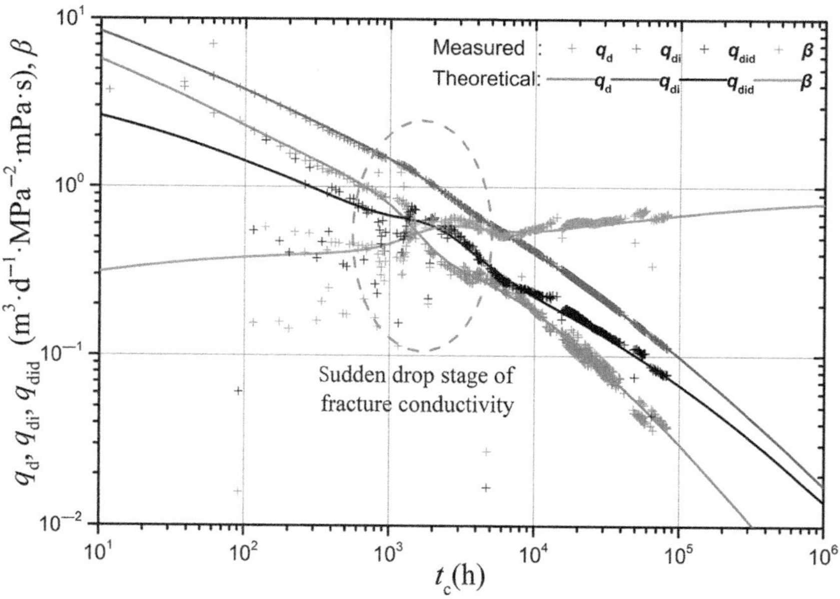

Fig. 4.48 Blasingame type curve based on the proposed model.

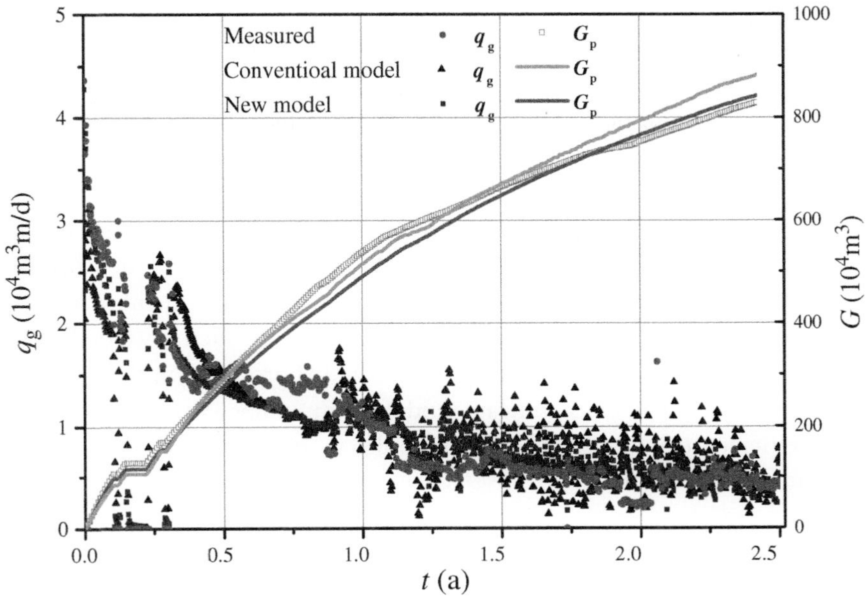

Fig. 4.49 History matching curve of the well.

Table 4.26 Production decline analysis results of the well

No.	Parameter	Conventional model	The proposed model
1	Reservoir permeability (mD)	0.0235	0.1088
2	Fracture half-length (m)	165	98
3	Initial fracture conductivity at the gap (mD·m)	860	580
4	Skin factor of fracture	0	0.03
5	Spatially variable conductivity coefficient, b	/	1.6
6	Fracture conductivity decay coefficient, η	/	0.989
7	Temporally variable conductivity coefficient, C (a)	/	0.045
8	Stress sensitivity coefficient	/	1.15
9	Drainage radius (m)	197	203
10	OGIP (10^8 m^3)	0.162	0.172
11	Current average reservoir pressure (MPa)	13.8	14.8
12	Recovery percent (%)	52.5	49.4

temporally variable conductivity. The proposed model can efficiently match the type curve and the production curve. However, the conventional model cannot match the sudden decline of fracture conductivity, but can roughly match the variation of the early-stage type curve by increasing the fracture half-length and fracture conductivity, and reducing the reservoir permeability, which apparently corresponds to larger fracture half-length and fracture conductivity. In addition, the interpretation results of both models on drainage radius and OGIP are basically consistent, which further clarifies that the variable conductivity and stress sensitivity will not substantially influence the estimation of OGIP. According to the variable conductivity coefficient interpreted, it can be seen that in the early stage of production, the conductivity decay of the well is significant, and the temporally variable conductivity coefficient reaches 98.9%, which is also a major driver for the rapid decline of production.

Conclusion The spatially variable conductivity affects the Blasingame type curve mainly by reducing the value of the curve in the early stage, which is similar to the influence of the skin factor. The exponential variation, logarithmic variation, and linear variation stay in a descending order of influence.

The temporally variable conductivity will make the flow rate and rate integral curves "downwarp" to form S-shaped curves, which is a key indicator to judge whether the temporally variable fracture conductivity exists. The influence of spatially temporally variable conductivity on the type curve is the superposition of the influence of spatially variable conductivity and temporally variable conductivity. The stress sensitivity will also reduce the value of the Blasingame type curve, but will not make the curve "downwarp" suddenly; moreover, the stress sensitivity may delay the start of the pseudo-steady state, which is obviously different from the variable conductivity effect. If variable conductivity and stress sensitivity are neglected, the estimation of performance-based reserves will not be affected greatly, but big errors may appear in the evaluation of fracture parameters and reservoir permeability.

4.1.4 Application of reserves estimation results

4.1.4.1 Performance-based reserves or OGIP estimation

The combination of advanced production decline analysis with a well test can accurately evaluate the performance-based reserves, or OGIP, of a single well (Fig. 4.50). The two methods are combined and balanced rationally to reduce the uncertainty of the reserves estimate.

As shown in Fig. 4.51, the seismic amplitude energy RMS, porosity, fluid volume, and hydrocarbon accumulation coefficient of the highly efficient well, efficient well, and inefficient well tend to decrease.

As shown in Fig. 4.52, the OGIP is basically proportional to RMS, and linearly related to the delineated volume (except for outliers). When RMS is greater than 4000, the probability of highly efficient and efficient wells is 88%.

4.1.4.2 Multiple reservoirs evaluation

For fractured-vuggy carbonate reservoirs, the reservoir heterogeneity is strong. If multiple reservoirs are communicated during production, the production decline curve shows a stepped rise (Fig. 4.35). The reserves are 2.66×10^4, 10.0×10^4, and 21.2×10^4 m^3, respectively, when the data of the first, second, and third segments are matched with the normalized rate curves.

4.1.4.3 Early warning of water breakthrough

For oil and gas wells with edge and bottom water, the production performance of a well presents a trend in four stages: normal production, water energy supply, deteriorating production due to water coning, and liquid carrying (Fig. 4.53). The behaviors of the four stages of the production decline analysis curve and the FMB curve are as follows:

(1) In the stage of normal production, the normalized rate curve is consistent with the type curve, and the FMB curve is an initial straight line.

(2) In the stage of production with water energy supply, the normalized rate curve deviates from the boundary-dominated flow straight line segment to the upper right, and the FMB curve deviates from the initial straight line segment to the upper right.

(3) In the stage of deteriorating production due to water coning, the normalized rate curve deviates from the boundary-dominated flow straight line segment to the lower left, and does not intersect with theoretical type curve; the FMB curve deviates from the initial straight line segment to the lower left.

(4) In the stage of production with liquid carrying, the normalized rate curve falls below the theoretical type curve.

Based on the comprehensive analysis of flow rate, pressure, and production decline curves, the wells can be divided into two classes by water invasion,

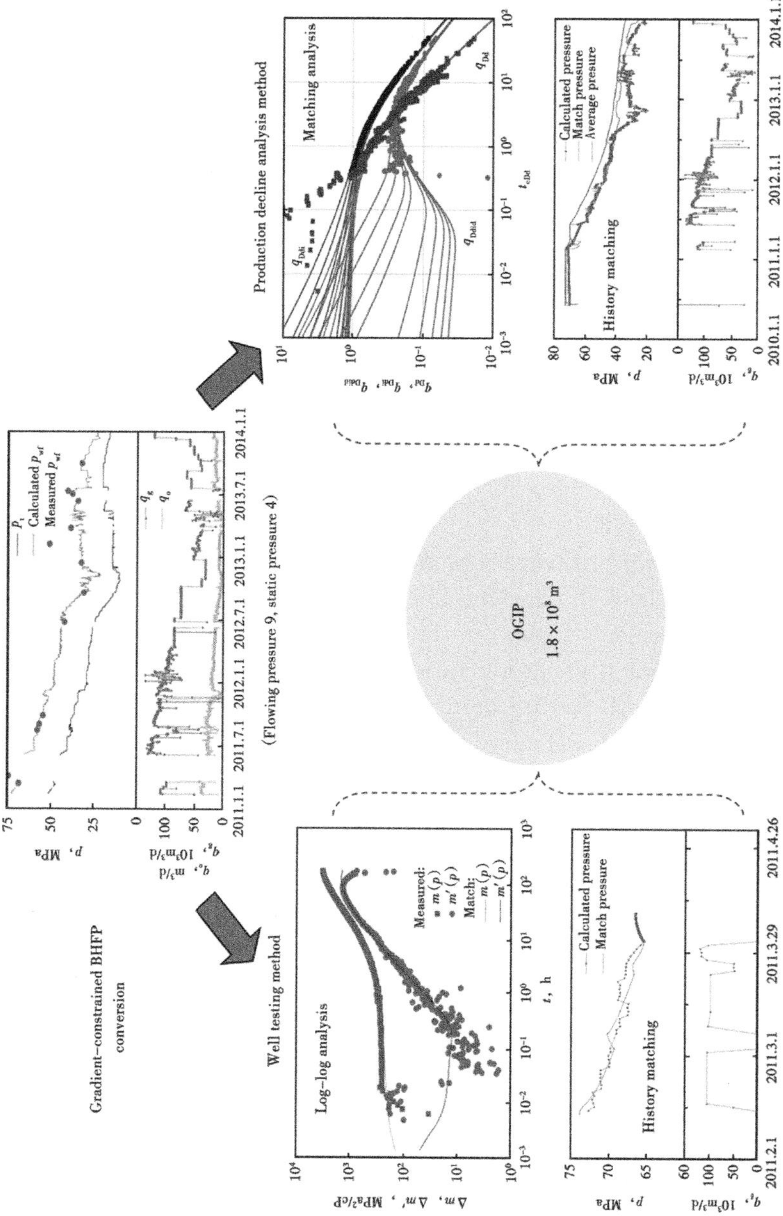

Fig. 4.50 Determination of single-well OGIP through the combination of production.

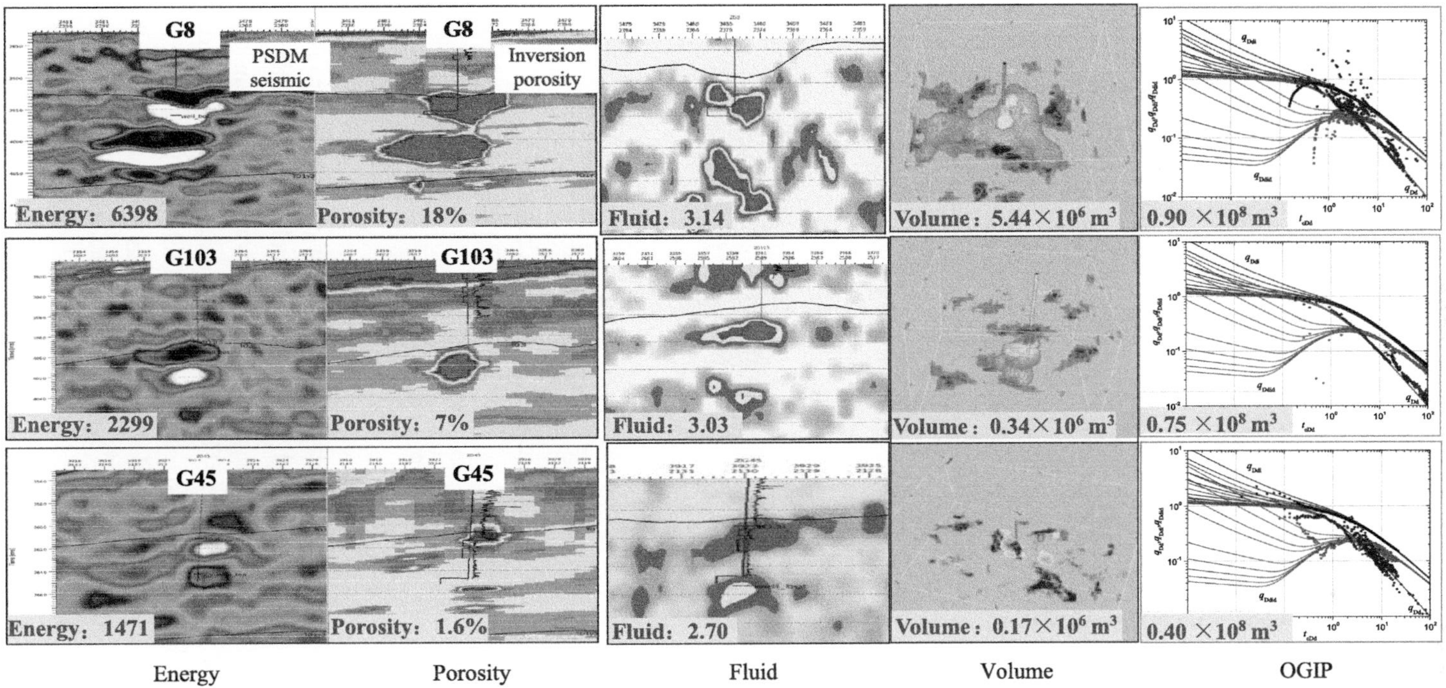

Fig. 4.51 Dynamic and static parameters of highly efficient well, efficient well, and inefficient well.

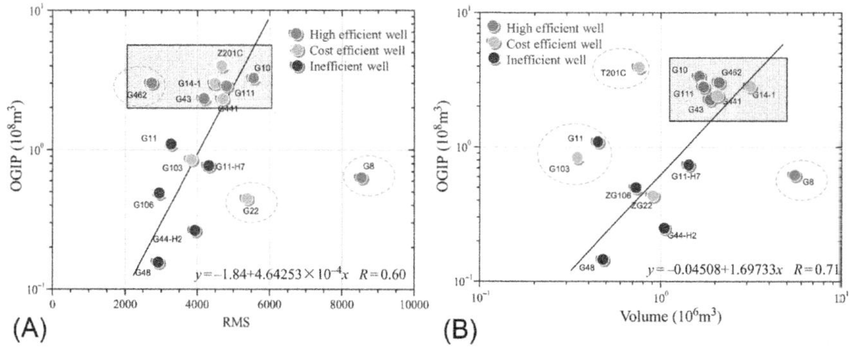

Fig. 4.52 Relationship between OGIP and RMS or volume. (A) OGIP vs. RMS and (B) OGIP vs. volume.

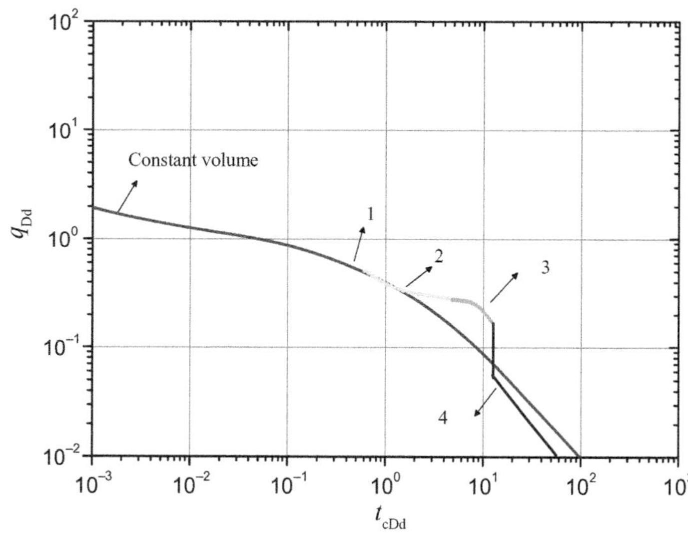

Fig. 4.53 Blasingame type curve in case of water invasion.

namely, wells with water breakthrough and wells without water breakthrough. The wells without water breakthrough fall into three subclasses: wells without water invasion, wells in the early stage of water invasion, and wells in the middle-late stage of water invasion. The flow rate, pressure, and production decline curves are different for these classes of wells (Table 4.27).

A well (T63-C) with water breakthrough in the gas field was analyzed. According to the production decline type curve, the data points shift to the upper right, and then to the lower left. The upward shift represents the

Table 4.27 Classification of wells by water invasion

No.	Class		Production characteristics	Production decline analysis curve behaviors
1	Without water breakthrough	Without water invasion	Flow rate: stable Pressure: a decline in consistent trend	Type curve: coincide with a type curve FMB curve: straight line
2		In the early stage of water invasion	Flow rate: stable or rises Pressure: drop at a lower rate than before, with significant inflection	Type curve: the data points deviate from the type curve to the upper right FMB curve: deviate from the straight line and upwarp
3		In the middle-late stage of water invasion	Flow rate: drop gradually or keep stable Pressure: drop at a higher rate than before	Type curve: deviate from the type curve to the lower left FMB curve: deviate from the straight line and downwarp
4	With water breakthrough	With water breakthrough	Flow rate: drop gradually Pressure: drop	Type curve: data points migrate downward FMB curve: deviate from the straight line and upwarp and then downwarp

energy supply, and the downward shift represents the poorer gas production capacity, which is a manifestation of water breakthrough (Fig. 4.54).

The type curve is used to judge the water breakthrough, and proper measures are taken in a timely manner to prolong the water-free production period. Fig. 4.55 shows the analysis of water breakthrough for Well G503 in the Tazhong I gas field. It is suggested that there is water invasion, and water breakthrough may occur. Accordingly, proper measures, such as managed-pressure production, are taken, so that the cumulative recoverable reserves are increased.

For the purpose of OGIP estimation, matching should be carefully conducted, and the early data should be preferred; otherwise, the OGIP may be overestimated. As shown in Fig. 4.56, the OGIP is 1.1×10^8 m^3 when the early data are used, and 1.4×10^8 m^3 when the latter data are used, suggesting a difference of 27%.

4.1.4.4 Estimation of production pressure drawdown

Because there are only limited static pressure data in the Tazhong I gas field, the reservoir pressure is predicted according to the result of single well decline curve analysis, so as to derive the production pressure drawdown of the single well (Fig. 4.57).

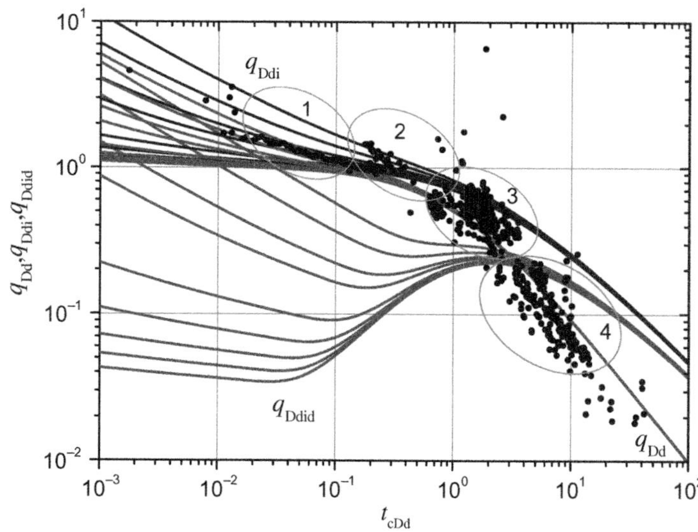

Fig. 4.54 Type curve of Well T63-C with water breakthrough.

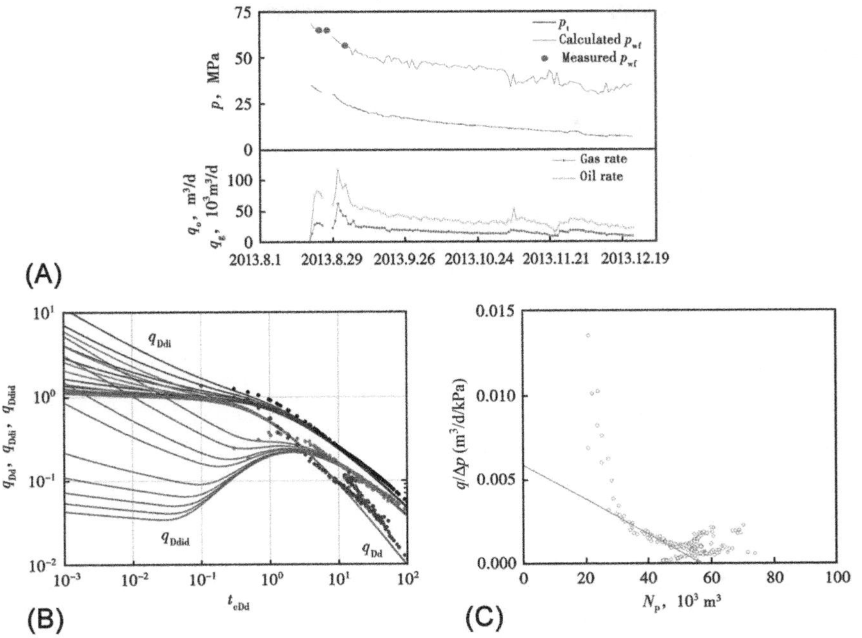

Fig. 4.55 Water breakthrough analysis for Well G503. (A) Production curve, (B) Blasingame type curve, and (C) FMB curve.

4.1.4.5 Estimation of new well reserves

According to the result of performance-based reserves estimation for wells in the Tazhong I gas field, the production index (the production per unit of pressure drop) is not highly correlated to the OGIP (Fig. 4.58); while the cumulative gas production per unit of pressure drop is well correlated to the OGIP (Fig. 4.59). For wells with shorter production time, the performance-based reserves, or OGIP, can be preliminarily evaluated according to Fig. 4.59.

4.2 Recoverable reserves estimation

4.2.1 Recoverable reserves (or ultimate recovery) estimation of single well

The performance-based estimation of single-well recoverable reserves (or ultimate recovery) will be expounded upon in Chapter 5. Fig. 4.60 shows the relationship between recoverable reserves and performance-based reserves in the Tazhong gas field. The average recoverable reserves per well

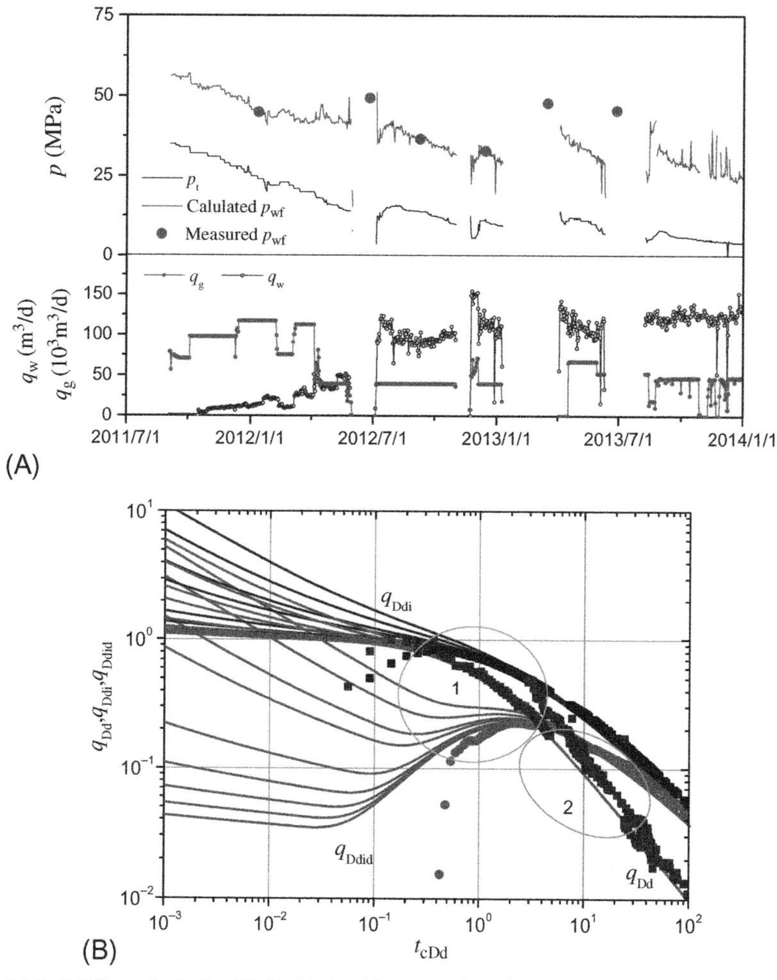

Fig. 4.56 OGIP analysis for Well G441 with water breakthrough. (A) Production curve and (B) Blasingame type curve.

are 0.62×10^8 m^3 and the average condensate oil reserves per well are 2.0×10^4 t. The recovery percent of natural gas is 57.3% and that of condensate oil is 28.8%.

4.2.2 Relationship between initial flow rate and recoverable reserves

Fig. 4.61 shows a poor relationship between initial flow rate and recoverable reserves of gas wells in the Tazhong gas field. This reflects the production

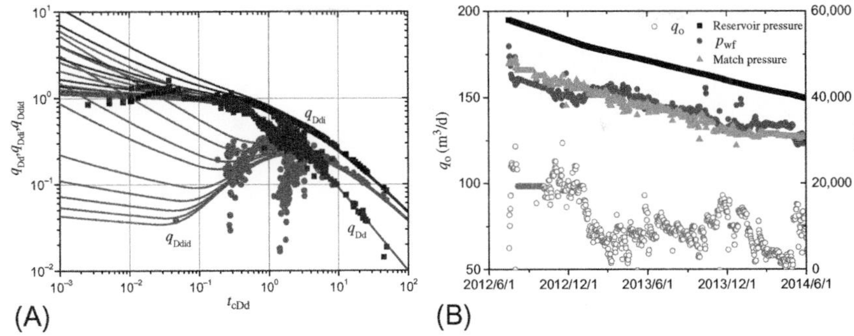

Fig. 4.57 Estimation of production pressure drawdown for Well G15-5H. (A) Blasingame and (B) pressure forecasting.

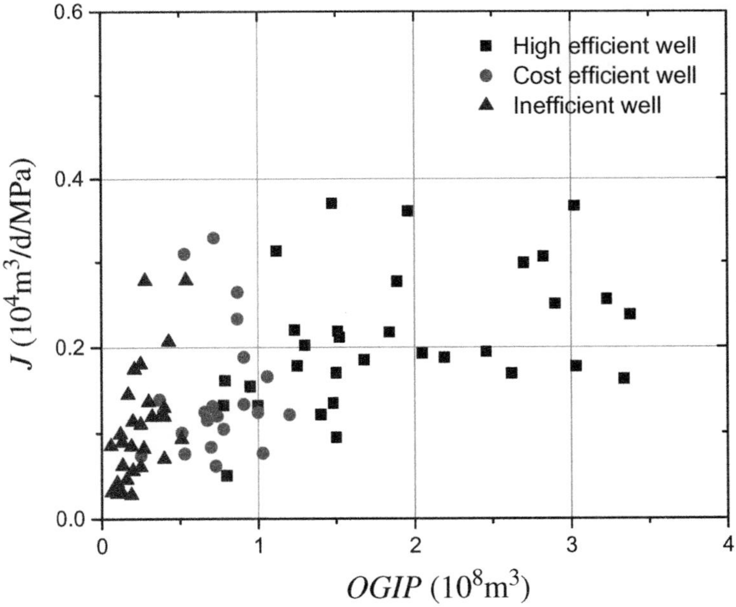

Fig. 4.58 Production index vs. OGIP.

characteristics of carbonate reservoirs, that is, high yield, but not necessarily highly efficient.

Fig. 4.62 shows the correlation among decline rate, recoverable reserves, and flow rate for different blocks. Highly efficient wells are characterized by high initial flow rate, low decline rate, and large recoverable reserves; and they are mainly distributed in the upper left of the figure. Inefficient wells are characterized by high decline rate and small recoverable reserves, and

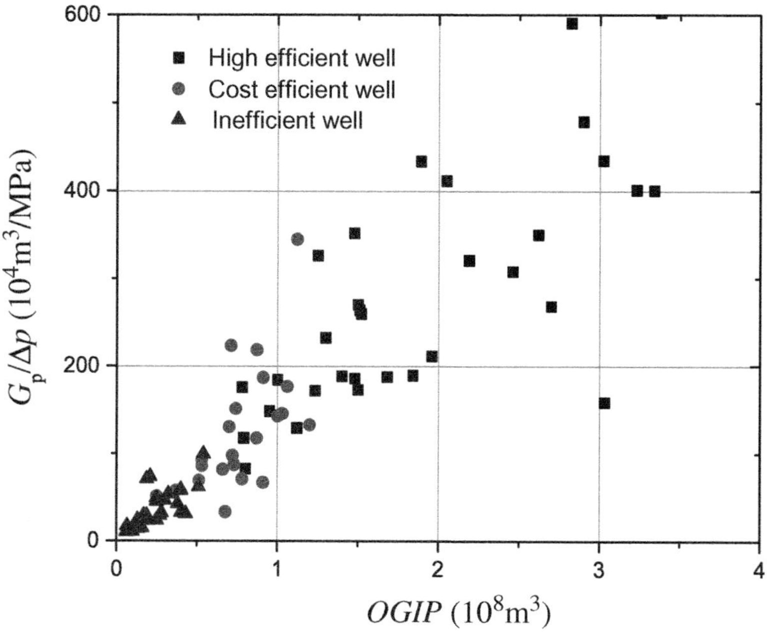

Fig. 4.59 Cumulative gas production per unit of tubing pressure drop vs. OGIP.

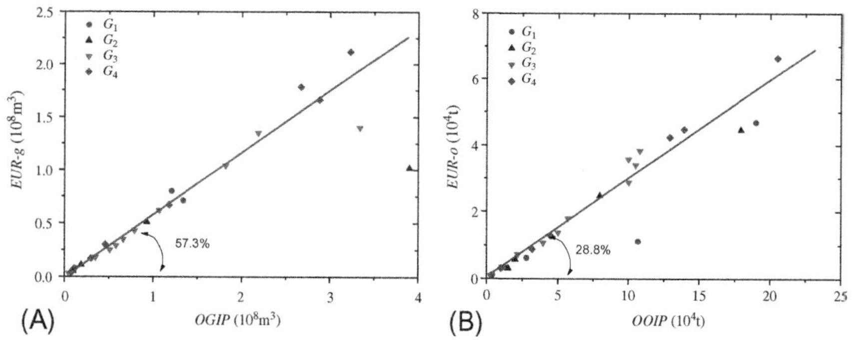

Fig. 4.60 Relationship between performance-based reserves and recoverable reserves. (A) Natural gas and (B) condensate oil.

they are mainly distributed in the lower right of the figure. Efficient wells are located in the middle. Therefore, it can be preliminarily inferred that the wells with high initial flow rate and slow production decline are basically highly efficient wells, and the wells with high initial flow rate but fast production decline or low initial flow rate and fast production decline are basically efficient or inefficient wells.

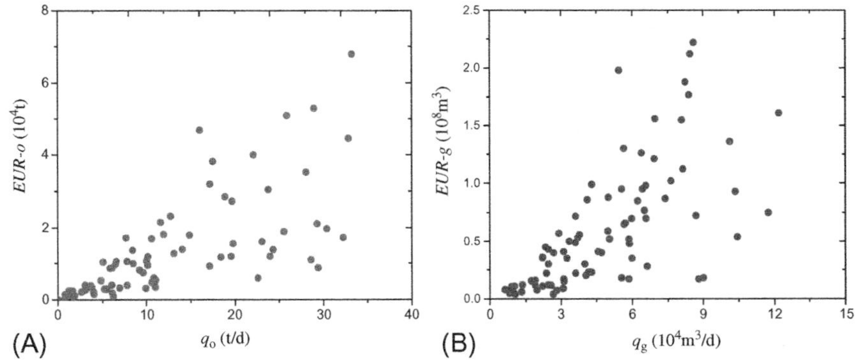

Fig. 4.61 Relationship between initial flow rate and recoverable reserves of gas wells in the Tazhong gas field. (A) Natural gas and (B) condensate oil.

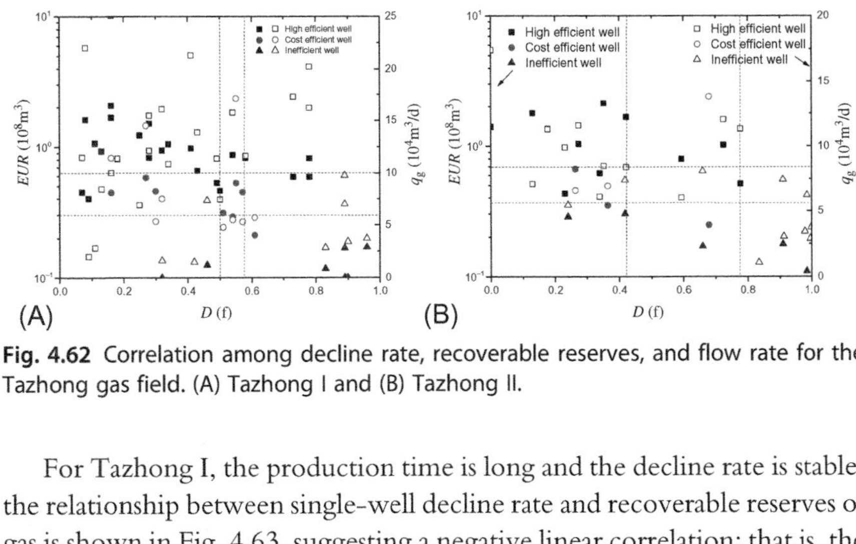

Fig. 4.62 Correlation among decline rate, recoverable reserves, and flow rate for the Tazhong gas field. (A) Tazhong I and (B) Tazhong II.

For Tazhong I, the production time is long and the decline rate is stable; the relationship between single-well decline rate and recoverable reserves of gas is shown in Fig. 4.63, suggesting a negative linear correlation; that is, the larger the single-well decline rate is, the smaller the single-well recoverable reserves are. For Tazhong III, the relationship between decline rate and recoverable reserves of oil is shown in Fig. 4.63, suggesting a negative linear correlation, which is consistent with the law in Tazhong I.

4.3 Producing reserves estimation

A large number of proven reserves have been booked. To estimate the producing reserves and the productivity thereof, static data are not valid, but only dynamic data are applicable. Here, the producing reserves estimation is discussed with the Tazhong I as an example.

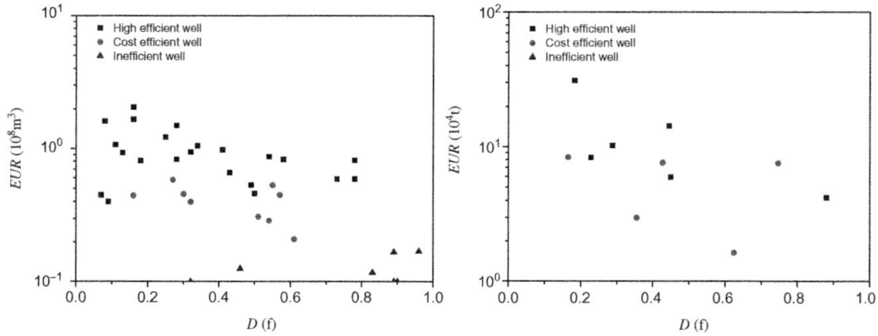

Fig. 4.63 Relationship between decline rate and recoverable reserves of gas wells in Tazhong I (*left*) and Tazhong III (*right*).

4.3.1 Analysis procedure

According to the type of fluid production, the data preprocessing flow is optimized. First, the single-well reserves are estimated. Then, the producing reserves of the block are evaluated according to the area of different reflection types (Fig. 4.64).

Based on single-well production test data, the single-well performance-based reserves and single-well reserves abundance (the abundance of reserves with bead-like and flake-like reflections) are determined by reservoir engineering methods. Based on static data, the reserves abundance and area of different reflection types are determined, the proved reserves are classified and evaluated, and the producing reserves are worked out. Using single-well

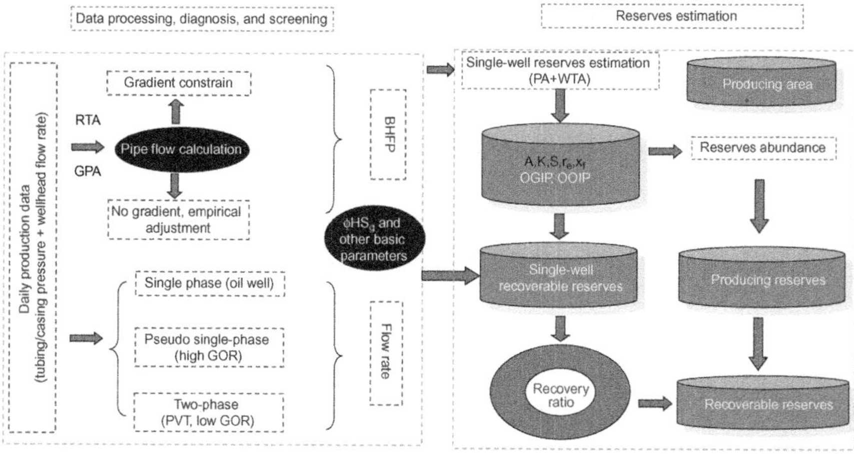

Fig. 4.64 Workflow of producing reserves estimation.

production data, the single-well recoverable reserves are determined, and then combined with the performance-based reserves to determine the average recovery factor of the block. Depending on the recovery factor and producing reserves of the block, the recoverable reserves in the producing area are determined.

4.3.2 Reserves abundance

If the single-well performance-based reserves have been estimated, the producing reserves of the block can be calculated by combining the average reserves abundance of different reflection types with the corresponding area. The average reserves abundance of a single well are determined comprehensively based on the results of well test analysis (WTA) and rate transient analysis (RTA).

4.3.2.1 Reserves abundance based on WTA

If the gas saturation is considered, the average drainage radius is 473 m, and reserves abundance is 3.6×10^8 m^3/km^2 in Block II of the Tazhong I gas field, according to the WTA interpretation. The vertical wells commonly present bead-like and flake-like reflections, so the calculated abundance is for the bead-like and flake-like sections (Table 4.28).

4.3.2.2 Reserves abundance based on RTA

The reserves abundance of these wells based on RTA is shown in Table 4.29. It is found that the reserves abundance based on RTA is 1.50 times of that based on WTA.

Why is the reserves abundance based on RTA larger? According to the WTA results, most of the wells show the characteristics of the composite model with deteriorating periphery (Fig. 4.65); that is to say, the WTA method takes into account the heterogeneity of the reservoir. However, the RTA method cannot take into account the heterogeneity. Accordingly, the reserves abundance based on RTA is higher than that based on WTA, given the same reserves (Fig. 4.66).

Because there are few well test data, the reserves abundance based on RTA are corrected by this coefficient, and then the reserves abundance in the case of bead-like and flake-like reflections is $(3.4–3.6) \times 10^8$ m^3/km^2. According to the results of geological studies, the reserves abundance in cases of "pure flake-like" reflection is 0.95 times of the reserves abundance in cases of "bead-like and flake-like" reflections.

Table 4.28 Reserves abundance of Block II based on WTA

Well	Seismic attribute	WTA				Gas saturation considered			
		Drainage radius (m)	Area (km^2)	Performance-based reserves (10^8 m^3)	Reserves abundance (10^8 m^3/km^2)	Saturation (f)	Drainage radius (m)	Area (km^2)	Reserves abundance (10^8 m^3/km^2)
G14-1	Bead-like	1500	7.07	3.67	0.52	0.75	1730	9.42	0.39
G11	Bead-like	310	0.3	1.13	3.75	0.68	370	0.44	2.55
G462	Flake-like	330	0.34	2	5.85	0.73	380	0.47	4.27
T201C	Flake-like	500	0.79	3.06	3.9	0.73	580	1.08	2.85
G43	Bead-like	490	0.75	2.93	3.88	0.76	560	0.99	2.95
Average		408	0.55	2.28	4.34		473	0.75	3.06

Table 4.29 Reserves abundances based on WTA and RTA

Well	WTA			RTA			
	Drainage radius (m)	Area (km^2)	Reserves abundance (10^8 m^3/km^2)	Performance-based reserves (10^8 m^3)	Drainage radius (m)	Area (km^2)	Reserves abundance (10^8 m^3/km^2)
G14-1	1730	9.42	0.39	3.4	480	0.72	4.7
G11	370	0.44	2.55	1.0	320	0.32	3.1
G462	380	0.47	4.27	2.0	300	0.28	7.1
T201C	580	1.08	2.85	2.5	400	0.5	5.0
G43	560	0.99	2.95	2.3	450	0.64	3.6
Average	470	0.75	3.06	1.95	370	0.44	4.5

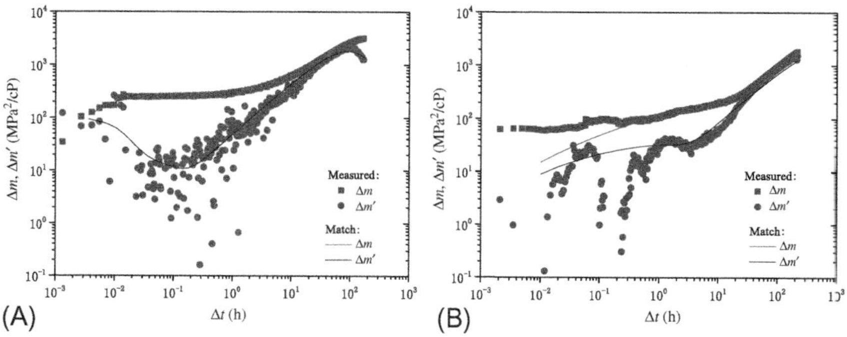

Fig. 4.65 Typical log-log plot of Tazhong II block. (A) Log-log plot for Well G14-1 and (B) log-log plot for Well G43.

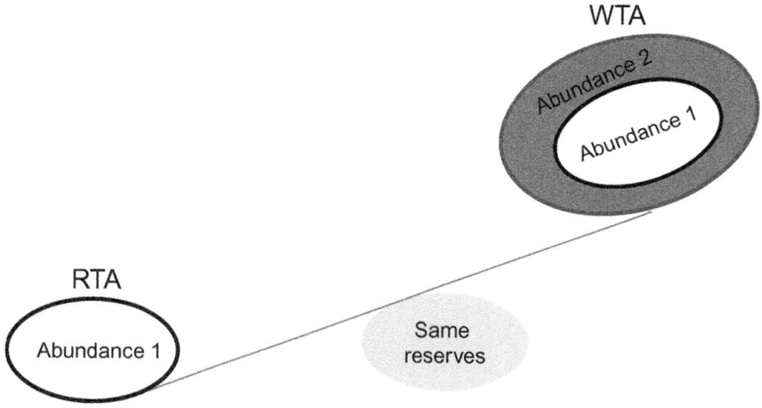

Fig. 4.66 Reserves abundance based on WTA and RTA.

4.3.3 Producing reserves estimation

If the area of different reflection types is known, the producing reserves of the whole block can be calculated according to the reserves abundance.

4.4 Summary

For fractured-vuggy carbonate reservoirs that are characterized by strong heterogeneity, complex fluid properties, multiphase flow in a wellbore, production at variable pressure or variable rate, and poor quality of dynamic data, the traditional material balance method fails to estimate the single-well performance-based reserves. Therefore, a performance-based reserves estimation method integrating well test analysis and advanced production decline analysis has been proposed. The new method can be used to effectively estimate the single-well performance-based reserves in fractured-vuggy gas reservoirs.

Main factors affecting the accuracy of single-well performance-based reserves estimation include BHFP conversion, production time, working system, water production, interwell interference, initial reservoir pressure, PVT, and reservoir parameters. If variable conductivity and stress sensitivity are neglected, the estimation of performance-based reserves will not be affected greatly, but significant errors may appear in the evaluation of fracture parameters and reservoir permeability.

The seismic amplitude energy RMS, porosity, and delineated volume of highly efficient, efficient, and inefficient wells tend to decrease generally. The performance-based reserves are proportional to RMS and linearly related to the delineated volume.

The performance-based reserves estimation technique of condensate gas reservoirs proposed considering the characteristics of carbonate reservoirs and the complexity of production conditions in the Tazhong area achieves a lower uncertainty of reserves estimation. In combination with static data, the technique can effectively evaluate reservoir parameters and producing reserves, thereby providing a scientific basis for the compilation of a development plan and recovery enhancement in the middle and later stages.

CHAPTER 5

Performance forecasting method of fractured vuggy carbonate gas reservoir

Contents

5.1 Overview of performance forecasting methods

Fractured vuggy carbonate condensate gas reservoirs contain abundant pores, vugs, and fractures, and feature extremely strong heterogeneity. They are often characterized by complicated development patterns and fast changes in production and pressure. Generally, the reservoirs are poorly connected, and usually one reservoir is deployed with one well. More than 90% of the producing wells are not communicated, except for certain clusters.

The material balance method and numerical simulation method are often used for early performance forecasting of gas reservoirs. However, for fractured vuggy carbonate gas reservoirs, it is hard to perform three-dimensional geologic modeling and numerical simulation for the entire reservoir. Only the gas reservoir engineering method can be used to forecast the production performance of a well or clusters. As mentioned previously, the production performance technique of gas reservoirs based on modern well testing analysis and advanced production decline analysis can be used to forecast single

Dynamic Description Technology of Fractured Vuggy Carbonate Gas Reservoirs
https://doi.org/10.1016/B978-0-12-818324-3.00005-X

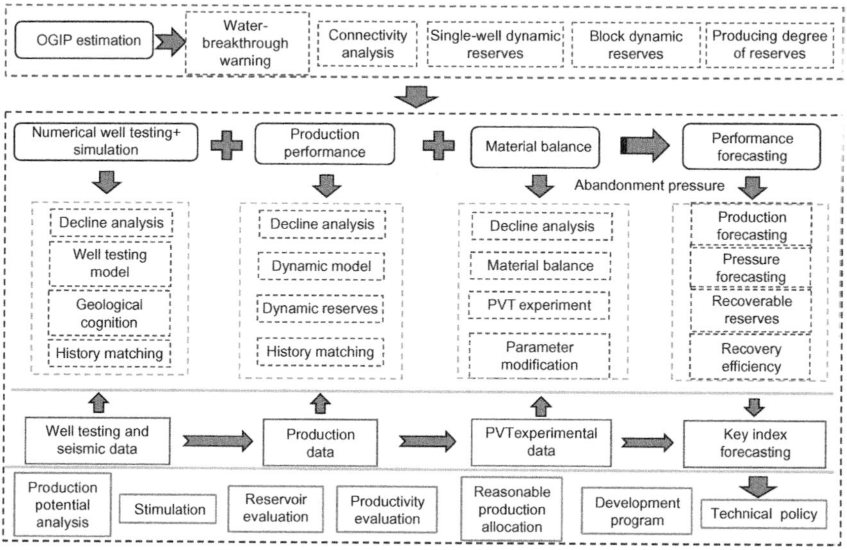

Fig. 5.1 Process of index forecasting.

well development indices, such as oil and gas production, reservoir pressure, and recoverable reserves. The forecasting process is shown in Fig. 5.1.

First, the dynamic oil and gas reserves of single wells or clusters are determined as the basis of the development indices forecasting. Then the numerical well testing, or simulation method, of single wells or clusters, the advanced production decline analysis method, or the gas reservoir engineering method are adopted, depending on the availability of data. Finally, the recoverable reserves of natural gas and condensate oil are determined under the constraint of the abandonment pressure.

5.2 Performance forecasting method of single wells

This section introduces four performance forecasting methods of single wells, including the material balance method, the numerical well testing method, the advanced production decline analysis method, and the single-well numerical simulation method.

5.2.1 Material balance method

This method is often used for four cases; that is, stress-sensitive reservoirs, low-permeability reservoirs, gas reservoirs with recharge capacity, and condensate gas reservoirs.

5.2.1.1 *Stress-sensitive gas reservoirs*

5.2.1.1.1 Deliverability evaluation method

The pseudo-pressure, considering stress sensitivity, is defined as:

$$m(p) = p_i + \left(\frac{\mu Z}{pK}\right)_i \int_{p_i}^{p} \frac{pK}{\mu Z} dp \tag{5.1}$$

Where

$$K = K_i \exp(-\gamma \Delta p)$$

γ refers to the permeability modulus, and the preceding pseudo-pressure refers to the standardized pseudo-pressure, featuring the dimension of pressure. For high-speed non-Darcy gas flow, the pressure gradient equation can be expressed as:

$$\frac{dp}{dr} = \frac{\mu}{K} v + \beta \rho_{sc} v^2 \tag{5.2}$$

The gas flow velocity in any section of the reservoir is expressed as:

$$v = -\frac{q_{sc} p_{sc} TZ}{2\pi r h T_{sc} p} \tag{5.3}$$

After differentiating Eq. (5.1), we have

$$\frac{dm}{dr} = \left(\frac{\mu Z}{pK}\right)_i \frac{T p_{sc} q_{sc}}{2\pi h T_{sc}} \frac{1}{r} + \left(\frac{\mu Z}{pK}\right)_i \frac{K \beta M p_{sc}^2 q_{sc}^2}{\mu \quad R} \frac{T}{T_{sc}^2} \frac{1}{4\pi^2 h^2} \frac{1}{r^2} \tag{5.4}$$

Converting the preceding equation to SI units, and according to the relationship between the D value and β value, we have

$$D = \frac{K \gamma_g p_{sc}}{1.8738 \times 10^{12} \mu h r_w T_{sc} R} \beta \tag{5.5}$$

After integration, we have

$$\Delta m = Aq + Bq^2 \tag{5.6}$$

Where

$$\Delta m = \left(\frac{\mu Z}{p}\right)_i \int_{p_w}^{p_i} \frac{\exp[-\gamma(p_i - p)]p}{\mu Z} dp$$

$$A = 6.345 \left(\frac{\mu Z}{pK}\right)_i \frac{T}{h} \left(\ln \frac{r_e}{r_w} + S \right)$$

$$B=6.345\left(\frac{\mu Z}{Kp}\right)_{\mathrm{i}}\frac{T}{h}D$$

After sorting out the preceding equation, we have

$$\frac{\Delta m}{q}=A+Bq \tag{5.7}$$

$\frac{\Delta m}{q}\sim q$ meets the linear relation. Their relationship is plotted, with the line slope as B and the line intercept as A. After A and B are obtained, the absolute open flow potential of gas well q_{AOF} can be obtained according to the following equation:

$$q_{\mathrm{AOF}}=\frac{-A+\sqrt{A^2+4B\Delta m}}{2B} \tag{5.8}$$

5.2.1.1.2 Performance forecasting method

The material balance equation of ultra-high pressure gas reservoirs and the turbulent flow deliverability equation considering stress sensitivity are combined to establish a performance forecasting model of stress-sensitive ultra-high pressure gas reservoirs, which is expressed as:

$$\frac{p}{Z}=\frac{p_{\mathrm{i}}}{Z_{\mathrm{i}}}\left(1-\frac{G_{\mathrm{p}}}{G}\right)\frac{1}{1-c_{\mathrm{e}}(p_{\mathrm{i}}-p)} \tag{5.9}$$

Where

$$c_{\mathrm{e}}=\frac{c_{\mathrm{w}}S_{wi}+c_{\mathrm{f}}}{1-S_{wi}} \tag{5.10}$$

According to experimental results, the empirical equation based on the matching of pore compressibility and effective stress is

$$\frac{c_{\mathrm{f}}}{c_{\mathrm{fi}}}=7.4945p_{\mathrm{eff}}^{-0.8435} \tag{5.11}$$

According to the actual deliverability test data, we can obtain the turbulent flow deliverability equation through the regression of Eq. (5.7). Eqs. (5.7), (5.9), (5.10), and (5.11) are solved simultaneously to forecast the long-term production performance of a single well (Sun, 2007).

5.2.1.1.3 Field Example

Well X is an ultra-high pressure gas well in the Tarim Oilfield. The permeability modulus is 0.01 MPa^{-1}, according to the well testing interpretation results. The initial reservoir pressure is 105.06 MPa. The minimum bottom-hole flow pressure is 10.0 MPa. The economic limit production is 10×10^4 m^3/d and the production allocation is 45×10^4 m^3/d. The formation temperature is 405 K. The geological reserves are 50×10^8 m^3. The relative density of natural gas is 0.63. The pseudo-critical temperature is 203 K, and the pseudo-critical pressure is 4.601 MPa. The water saturation is 0.33. The rock compressibility is 4.3659×10^{-4} MPa^{-1} and the water compressibility is 3.05×10^{-4} MPa^{-1}. Table 5.1 provides the deliverability well testing data under four working systems. Before the absolute open flow potential is calculated, the condensate oil flow rate needs to be converted to the condensate gas flow rate.

The absolute open flow potential is 147.97×10^4 m^3/d, according to the conventional turbulent flow pseudo-pressure method, and it is 127.3×10^4 m^3/d when $\gamma = 0.01$ MPa^{-1}. The normalized turbulent flow pseudo-pressure deliverability equation is

$$\Delta m = 0.14954q + 0.00246q^2$$

The absolute open flow potential when $\gamma = 0.01$ MPa^{-1} is about 86% of that, without considering stress sensitivity. The IPR curves under different permeability moduli are shown in Fig. 5.2. The results show that the permeability modulus γ has a significant influence on deliverability forecasting.

The performance is forecasted by taking γ as 0, 0.01, and 0.02 MPa^{-1}, respectively. The results are shown in Fig. 5.3. It is seen that stress sensitivity significantly impacts all indices of single well performance forecasting. The bigger the permeability modulus γ is, the shorter the stable production period is, and the lower the forecasted recovery degree at the end of the period is. Under the three γ cases, the stable production period is 19 years, 13.5, years and 9 years, respectively, and the forecasted recovery degree is 75.77%, 67.44%, and 57.73% respectively.

Table 5.1 Deliverability well testing data of Well X

Working systems	Flow pressure (MPa)	Condensate oil rate (m^3/d)	Gas flow rate (10^4 m^3/d)	Condensate gas flow rate (10^4 m^3/d)
1	99.2272	26.88	26.28	26.62
2	94.4251	43.68	40.34	40.89
3	89.0139	47.52	54.68	55.28
4	80.4079	60.00	65.00	65.76

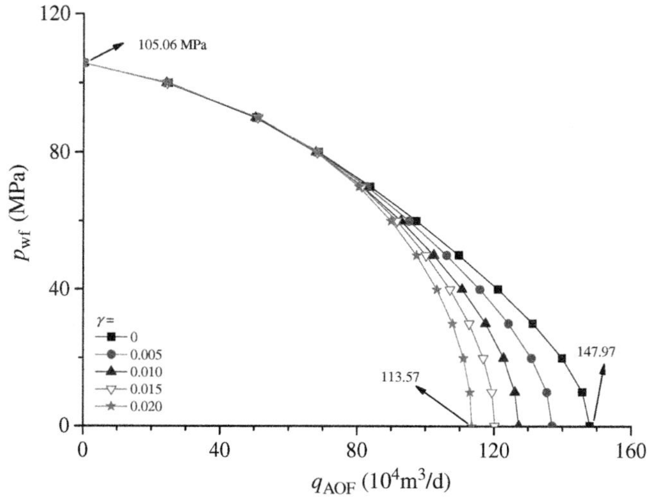

Fig. 5.2 IPR curves of Well X under different γ values.

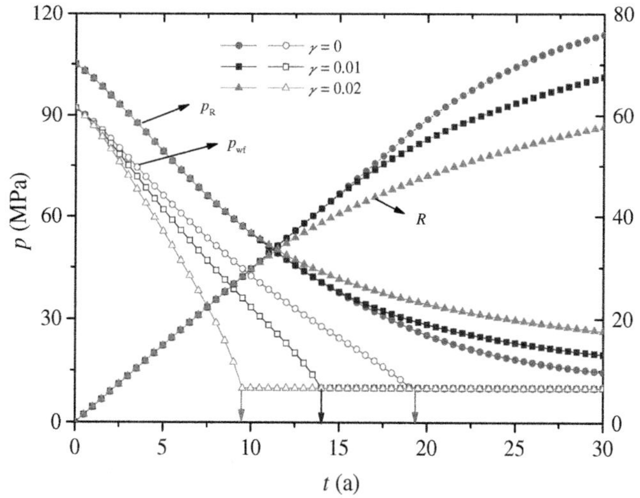

Fig. 5.3 Influence of the stress sensitivity of Well X on long-term performance (production allocation $45 \times 10^4 m^3$/d).

5.2.1.1.4 Simplification of method

If the influence of stress sensitivity is not considered, the preceding method can be simplified as the conventional production performance forecasting method combining the material balance equation and the turbulent flow deliverability equation. However, the biggest drawback of this method is that the indices tend to be optimistic.

5.2.1.2 *Performance forecasting method for low-permeability gas reservoirs*

For a low-permeability gas reservoir, the result of production performance forecasting combining the material balance equation and the turbulent flow deliverability equation is too optimistic. On one hand, the deliverability equation of the low-permeability gas reservoir is time-sensitive. On the other hand, the material balance equation is too simple, and cannot ideally represent the actual situation of the reservoir.

Accordingly, the pressure solution at any position of the rectangular boundary of the well and the material balance equation are combined to forecast the changes of bottomhole flow pressure and average reservoir pressure at the constant rate condition, so that the effectiveness of the deliverability equation is avoided. Based on the material balance equation and the deliverability equation, the production decline equation, under a constant-pressure condition, is built, and the performance of production under constant pressure is forecasted by using the Wattenbarger constant-pressure shape factor (Helmy and Wattenbarger, 1998).

5.2.1.2.1 Constant-rate production performance forecasting model

It is assumed that a well in a homogeneous rectangular gas reservoir produces at a stable rate q_g. Under the constant-rate condition, the change of bottom-hole flow pressure can be determined according to the superposition principle, and the average reservoir pressure can be obtained by the material balance equation.

5.2.1.2.2 Constant-pressure production performance forecasting model

Mathematical model The flow of real gas is nonlinear, so the pseudo-pressure is introduced

$$m(p) = 2\int_0^p \frac{p}{\mu Z}dp = 2\int_0^p \frac{1}{\mu c_g} d\left(\frac{p}{Z}\right) \tag{5.12}$$

The flow equation is linearized. When the production reaches a pseudo-steady state, the turbulent flow deliverability equation expressed in the pseudo-pressure form is

$$\Delta m(p) = Aq_g + Bq_g^2 \tag{5.13}$$

The coefficients A and B of the turbulent flow deliverability equation are, respectively

$$A = \frac{12.69T}{Kh} \ln\left(\frac{2.2458A}{C_A r_w^2}\right) \tag{5.14}$$

$$B = \frac{12.69T}{Kh} D \tag{5.15}$$

For a closed constant-volume gas reservoir, the material balance equation is

$$\frac{p}{Z} = \left(\frac{p}{Z}\right)_i \left(1 - \frac{G_p}{G}\right) \tag{5.16}$$

Eqs. (5.12)–(5.16) constitute the basic equations for forecasting the production performance of constant-volume gas reservoirs under constant pressure (Sun and Xu, 2011).

Model solution After differentiating Eqs. (5.12), (5.13), and (5.16) respectively, we have

$$dm = \frac{2}{\mu C_g} d\left(\frac{p}{Z}\right) \tag{5.17}$$

$$dm = \left(A + 2Bq_g\right) dq_g \tag{5.18}$$

$$d\left(\frac{p}{Z}\right) = -\left(\frac{p}{Z}\right)_i \frac{q_g}{G} dt \tag{5.19}$$

After combining the preceding three equations, we have

$$\left(\frac{A}{q_g} + 2B\right) dq_g = -2\left(\frac{p}{Z}\right)_i \frac{1}{G\mu c_g} dt \tag{5.20}$$

After integration, we have

$$\ln\left(\frac{q}{q_g}\right) + \frac{2Bq_g}{A}\left(\frac{q}{q_g} - 1\right) = -2\left(\frac{p}{\mu c_g Z}\right)_i \frac{1}{AG} t_a \tag{5.21}$$

Where

$$t_a = \int_0^t \frac{(\mu c_g)_i}{\mu c_g} dt \tag{5.22}$$

If $B=0$, Eq. (5.21) can be simplified as the Fraim model (Fraim, 1989). According to Eq. (5.21), the decline law of a closed gas reservoir is closely related to the initial reservoir, in addition to well-controlled reserves, geometrical shape, and reservoir pressure.

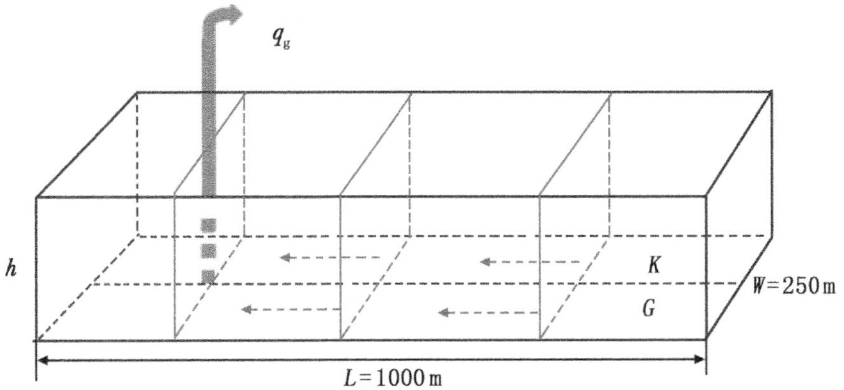

Fig. 5.4 Schematic diagram of gas reservoir parameters.

Decline law For a gas reservoir, the thickness is 30 m, the porosity is 0.2, the formation temperature is 80°C, the initial reservoir pressure is 30 MPa, the relative density of natural gas is 0.6, and the permeability is 10 mD. A well is deployed at 1/4 of the length of the rectangular gas reservoir, as shown in Fig. 5.4. The wellbore radius is 0.1 m, the non-Darcy coefficient D is $0.1\times(10^4\ \text{m}^3/\text{d})^{-1}$, the well-controlled reserves are $3.846\times10^8\ \text{m}^3$, and the absolute open flow potential pseudo-pressure is $97.4\times10^4\ \text{m}^3/\text{d}$. Thus, the deliverability equation is

$$\Delta m(p) = 341.583 q_g + 1.13048 q_g^2 \tag{5.23}$$

The forecasting results are shown in Fig. 5.5. Given the material balance time, the time and production of the Fraim model show an exponential relationship. The curve of the proposed model in this section is parallel to that of the Fraim model at the later stage, and the distance between the two curves is controlled by the deliverability equation coefficient, the initial flow rate, and other factors. Given the real time, the decline curve conforms to the exponential decline law at the later stage.

5.2.1.2.3 Example and discussion

Conventional index forecasting method For a gas reservoir, the thickness is 10 m, the porosity is 0.1, the permeability is 1.0 mD, and the well-controlled reserves are $0.641\times10^8\ \text{m}^3$; other parameters are the same as those in the preceding example. The single-well numerical simulation is conducted. The shape

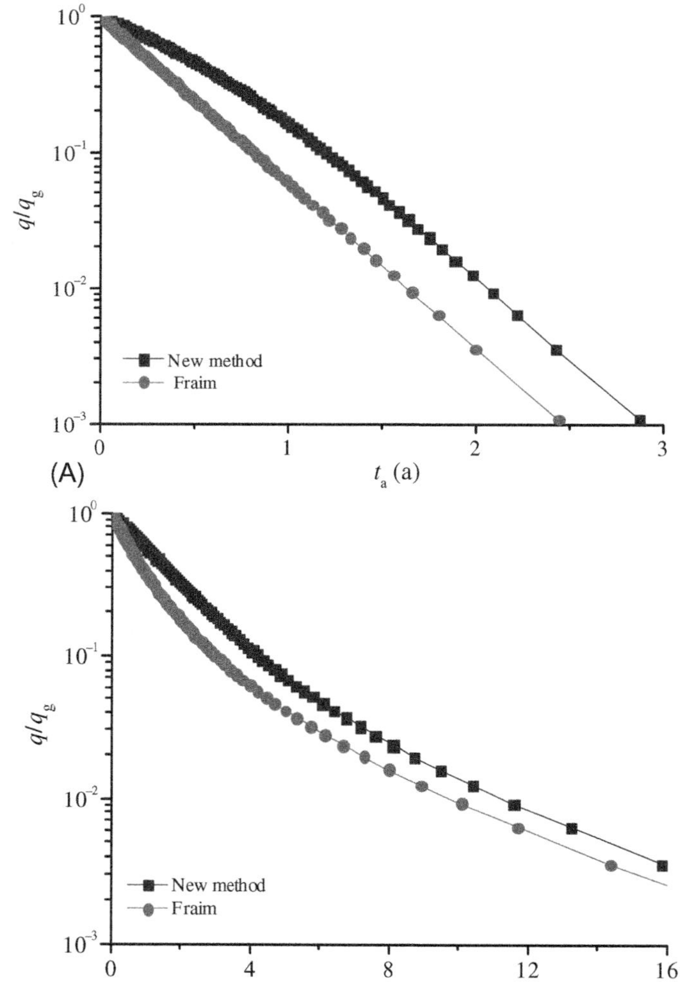

Fig. 5.5 Decline laws of two models under material balance time and real time. (A) Material balance time and (B) real time.

factors and forecasted absolute open flow potential are shown in Table 5.2. In a low-permeability reservoir, it takes a long time for the gas well to reach the pseudo-steady state, so the flow is unsteady during the early well testing period. The well-controlled geometry is different, but the absolute open flow potential calculated through simulation is basically the same. There is a significant difference between the simulation result and the results calculated by the pseudo-steady state Eqs. (5.14), (5.15), and the former is 1.8–2.5 times of the latter.

Table 5.2 Forecasting of absolute open flow potential

Reservoir shape, well location	Shape factor of constant rate, C_A	Shape factor of constant pressure, C_{ACP}	Dimensionless time to start of PSSF, t_{DA}	Time (d)	Absolute open flow potential (10^4 m^3/d)		
					Simulated (well test design)	Calculated (PSSF)	Simulated/calculated (%)
	30.88	29.34	0.1	14	13.06	7.0	187.8
	12.99	10.92	0.7	101	13.06	6.6	198.1
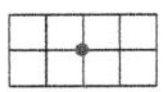	21.84	19.88	0.3	43	13.06	6.8	191.9
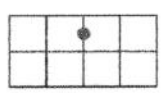	10.84	9.5	0.4	58	13.06	6.5	200.3
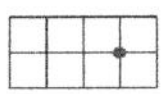	4.51	2.5	1.5	217	13.06	6.2	210.8
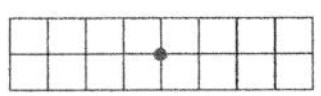	5.38	3.95	0.8	116	13.06	6.3	208.6
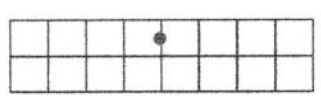	2.69	1.97	0.8	116	13.06	6.0	217.0
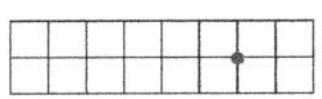	0.232	0.029	4.0	579	13.06	5.3	246.5

Now, the single-well numerical simulation method is used to analyze the recovery percent at the end of the stable production period and the stable production time. It is assumed that a well produces at a rate of 2×10^4 m^3/d, and the bottomhole flow pressure at the end of the stable production period is 3.0 MPa. Then, the stable production period determined by the turbulent flow deliverability equation using well testing design data is obviously longer, which is about 1.3–1.6 times of that calculated by the pseudo-steady state theory (Table 5.3).

Comparison between the proposed method and conventional method The length-width ratio of the well geometry is 1:4, as shown in Fig. 5.4. It is assumed that the well produces at a rate of 2.0×10^4 m^3/d, and the bottomhole flow pressure at the end of the stable production period is 3.0 MPa. The results of the proposed method and conventional method are shown in Fig. 5.6. It is known that the conventional method provides much longer stable production time and a higher recovery percent than the proposed method, showing optimistic indices.

5.2.1.2.4 Summary

The production performance of a single well is often forecasted under two conditions: constant rate and constant pressure. The forecasting at a constant rate can be further divided into unsteady state and pseudo-steady state. Due to the physical properties of reservoirs and the well-controlled geometry, it is hard to reach a pseudo-steady state during the gas well deliverability test; thus, the forecasting result may be biased when the conventional method is used.

In the case of material balance time, the time and flow rate of the Fraim model show an exponential relationship. The decline model proposed in this section is an improved Fraim model. The curve of the proposed model is parallel to that of the Fraim model at the later stage, and the distance between the two curves is controlled by the deliverability equation, the initial flow rate, and other factors. In the case of real time, the decline curve conforms to the exponential decline law at the later stage.

The proposed model in this section is combined with the unsteady-state well testing theory, avoiding the errors caused by the time-sensitive productivity equation, and errors caused by the shape factor. In this way, the forecasting results conform to the actual situations.

Table 5.3 Comparison of stable production time forecasted by conventional method and PSSF method

	Reservoir pressure		Degree of reserve recovery		Period of stable production		
Reservoir shape, well location	**PSSF theoretic (MPa)**	**Well test design (MPa)**	**PSSF theoretic**	**Well test design**	**PSSF theoretic (a)**	**Well test design (a)**	**Well test design/ PSSF theoretic**
	16.1	11.7	0.46	0.61	4.5	5.9	1.3
	16.6		0.45	0.61	4.3	5.9	1.4
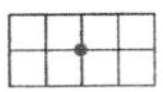	16.3		0.46	0.61	4.4	5.9	1.3
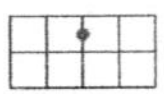	16.7		0.44	0.61	4.3	5.9	1.4
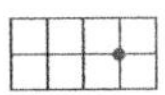	17.1		0.43	0.61	4.2	5.9	1.4
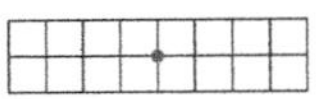	17.0		0.43	0.61	4.2	5.9	1.4
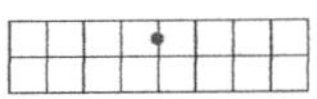	17.4		0.42	0.61	4.1	5.9	1.4
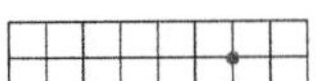	18.5		0.38	0.61	3.7	5.9	1.6

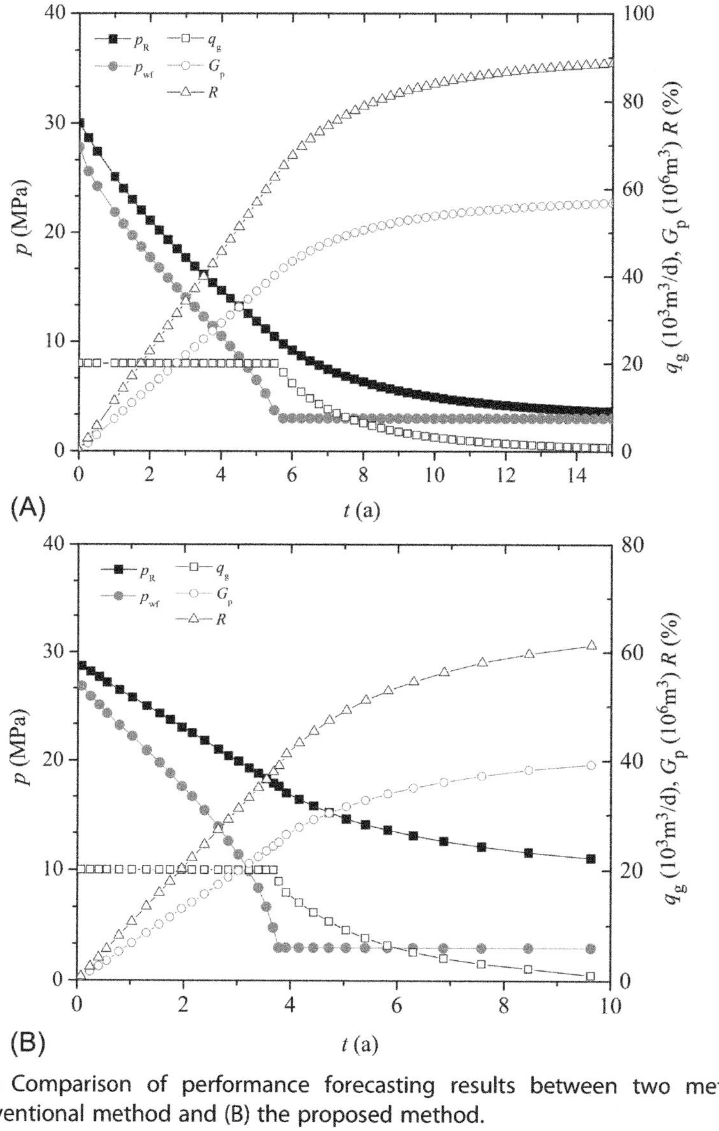

Fig. 5.6 Comparison of performance forecasting results between two methods. (A) Conventional method and (B) the proposed method.

5.2.1.3 Gas reservoir with recharge capacity

The traditional material balance method takes the entire gas reservoir as a container, neglecting the heterogeneity and the flow resistance of formations. Therefore, the results tend to be optimistic when the turbulent flow deliverability equation and the material balance equation are used to forecast the early performance of heterogeneous gas reservoirs. In this circumstance, the concept of the semi-permeable wall model (Gao and Sun, 2017) is proposed, and the material balance equation and theoretical model on performance forecasting are developed for heterogeneous gas reservoirs with recharge capacity according to the principle of mass conservation. It is assumed that a gas reservoir is divided by a low-permeability zone into two relatively homogeneous blocks that are interconnected. Imagining that the plane flow resistance is concentrated on the contact surface between the two blocks, and the flow resistance in blocks is zero, the wall surface can be considered semi-permeable. When fluids flow through the semi-permeable wall, the pressure jumps. The resistance of the semi-permeable wall to the flow can be calculated according to the principle of equivalent flow resistance (Fig. 5.7). The semi-permeable wall model can be used to ensure the pressure balance in each block at any time, so that each block can meet all conditions necessary for the traditional material balance method. This model considers the regional heterogeneity changes and flow resistance in the gas field, enabling the material balance model to be closer to practical cases. Also, it can easily and quickly forecast the performance trend of gas wells.

5.2.1.3.1 Material balance equation with recharge capacity

The traditional material balance method is too simply designed. It takes the gas field as a homogenous container, neglecting the heterogeneity of gas reservoirs, so its forecasting results are generally optimistic. An early performance forecasting method, which considers the regional heterogeneity change of gas

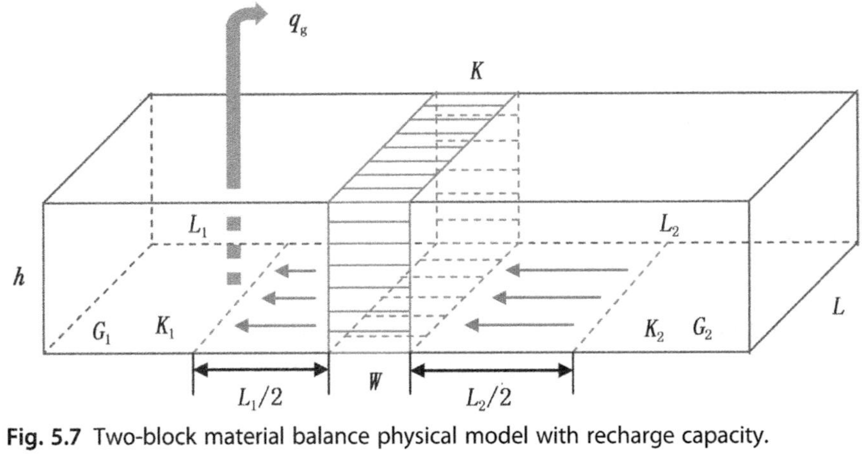

Fig. 5.7 Two-block material balance physical model with recharge capacity.

reservoirs and also easily and quickly forecasts the performance trend of gas reservoirs, is urgently needed in both production and research. Many researchers around the world worked on the material balance method of gas field partition and complementarity (Hower and Collins, 1989; Lord et al., 1992; Payne, 1996; Hagoort and Hoogstra, 1999; Gao et al., 2006; Sun, 2011) by incorporating this theory into the performance forecasting and parameter determination of real gas reservoirs, and they compared the results with those obtained by the numerical simulation method. It is indicated that the material balance method of gas field partition and complementarity is reliable and practical. This method is the development of the traditional material balance method, preliminarily addressing the regional heterogeneity.

Physical model According to the principle of equivalent flow resistance, the flow resistance recharged to the low-pressure block from the high-pressure block consists of three parts: resistance that flows to the low-permeability zone from the high-pressure block, resistance that flows in the low-permeability zone, and resistance that flows in the low-pressure block. Given the same viscosity, then

$$R = \frac{\mu}{h}\left(\frac{L_1}{2K_1} + \frac{L_2}{2K_2} + \frac{W}{K}\right) \tag{5.24}$$

The semi-permeable wall model can be used to ensure the pressure balance in each block at any time, so that each block can meet all conditions necessary for the traditional material balance method.

To make the study easier, the semi-permeable wall model is used to replace the real formation. (1) The reservoir in each block is a homogenous gas reservoir of equal thickness; the permeability is respectively K_1 and K_2, and the permeability in the transition area is K; and the formation thickness is h. (2) The reserves in the two blocks are G_1 and G_2, and the reserves in the low-permeability zone is ignored. (3) The production of a single well is q_g. (4) The gas flow in the reservoir meets the Darcy law, which is recharged to the low-pressure block from the high-pressure block. The recharge volume between blocks is in direct proportion to their pressure difference, and the recharge flow distance is $W+(L_1+L_2)/2$. (5) The influence of water is not considered.

Mathematical model and solution According to the material balance theory, the flow in the production area can be expressed as:

$$G_1\frac{d\rho_1}{dt}=-\rho_{sc}q_g+\frac{K^*Lh}{\mu}\rho\frac{\partial p}{\partial x} \tag{5.25}$$

The flow in the recharge area can be expressed as:

$$G_2\frac{d\rho_2}{dt}=-\frac{K^*Lh}{\mu}\rho\frac{\partial p}{\partial x} \tag{5.26}$$

Where, K^* is the average permeability within the recharge flow distance. According to the principle of equivalent flow resistance, it can be expressed as

$$K^*=\frac{W+\frac{L_1+L_2}{2}}{\frac{L_1}{2K_1}+\frac{W}{K}+\frac{L_2}{2K_2}} \tag{5.27}$$

The initial condition is:

$$\rho_2(t=0)=\rho_1(t=0)=\rho_0 \tag{5.28}$$

After solving Eqs. (5.25)–(5.28), we have:

$$\left(\frac{p}{Z}\right)_1=\left(\frac{p}{Z}\right)_i\left\{1-\frac{q_gt}{G_1+G_2}-\frac{q_g}{G_1\beta(1+\alpha)^2}\left[1-e^{-\beta(1+\alpha)t}\right]\right\} \tag{5.29}$$

$$\left(\frac{p}{Z}\right)_2=\left(\frac{p}{Z}\right)_i\left\{1-\frac{q_gt}{G_1+G_2}+\frac{q_g}{G_2\beta(1+\alpha)^2}\left[1-e^{-\beta(1+\alpha)t}\right]\right\} \tag{5.30}$$

Where

$$\alpha=\frac{G_1}{G_2},\beta=\frac{K^*Lh}{G_1\mu C_gW}$$

Under the condition of variable rate, according to the Duhamel's principle, we have

$$\left(\frac{p}{Z}\right)_{1,j}=\left(\frac{p}{Z}\right)_i\left\{\begin{array}{l}1-\frac{1}{G_1+G_2}\sum_{k=1}^{j}(q_k-q_{k-1})\times\left(t_j-t_{k-1}\right)\\-\frac{1}{G_1(1+\alpha)^2}\sum_{k=1}^{j}\frac{(q_k-q_{k-1})}{\beta_k}\left[1-e^{-\beta_k(1+\alpha)\left(t_j-t_k\right)}\right]\end{array}\right\} \tag{5.31}$$

$$\left(\frac{p}{Z}\right)_{2,j}=\left(\frac{p}{Z}\right)_{\mathrm{i}}\left\{\begin{array}{l}1-\dfrac{1}{G_1+G_2}\sum_{k=1}^{j}(q_k-q_{k-1})\times(t_j-t_{k-1}) \\ +\dfrac{1}{G_2(1+\alpha)^2}\sum_{k=1}^{j}\dfrac{(q_k-q_{k-1})}{\beta_k}\left[1-\mathrm{e}^{-\beta_k(1+\alpha)(t_j-t_k)}\right]\end{array}\right\} \tag{5.32}$$

Determination of performance forecasting parameters Assuming that there is only 1 well in the block, the following turbulent flow deliverability equation is obtained through deliverability testing.

$$\Delta p^2 = Aq + Bq^2 \tag{5.33}$$

The recharge volume between blocks is in direct proportion to the pressure difference between the two blocks, so we have

$$q_{\mathrm{c}} = \frac{\Delta p^2}{R} \tag{5.34}$$

Eqs. (5.29)–(5.32) can be combined to forecast the long-term production performance of a single well.

5.2.1.3.2 Example and discussion

Example For a gas reservoir, the thickness is 10 m, the porosity is 0.1, the formation temperature is 80°C, the initial reservoir pressure is 30 MPa, the relative density of natural gas is 0.6, the permeability of the production area is 10 mD, the permeability of the recharge area is 5 mD and the permeability of the intermediate low-permeability zone is 0.1 mD. The well is located in the middle part of the production area, as shown in Fig. 5.8. The absolute open flow potential is $59\times10^4\ \mathrm{m^3/d}$, thus the deliverability equation is

$$\Delta p^2 = 7.74344 q_{\mathrm{g}} + 0.127014 {q_{\mathrm{g}}}^2$$

Characteristics of *p/Z* curve It is assumed that this well produces at a rate of $10.0\times10^4\ \mathrm{m^3/d}$. The relationship between the apparent formation pressure and the cumulative gas production in the production area is shown in Fig. 5.9. The curve shows an obvious two-segment feature.

In the early stage of production, the pressure difference between the production area and the recharge area is very small, so the recharge volume can be neglected. At this moment, the first straight line segment appears on the

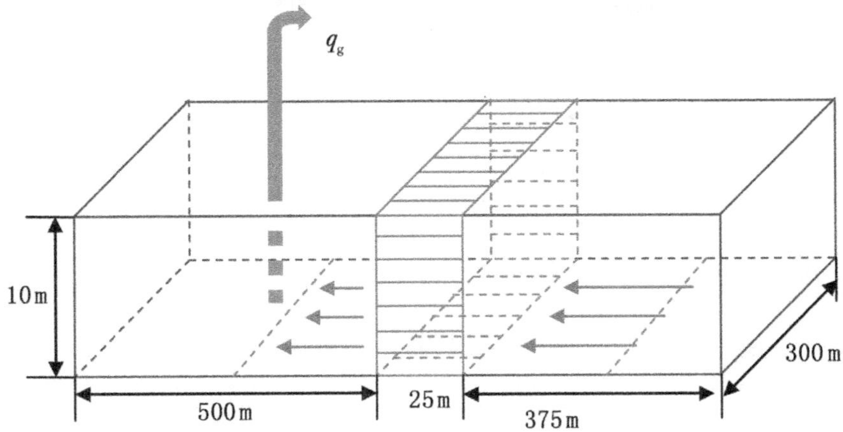

Fig. 5.8 Schematic diagram of gas reservoir parameters.

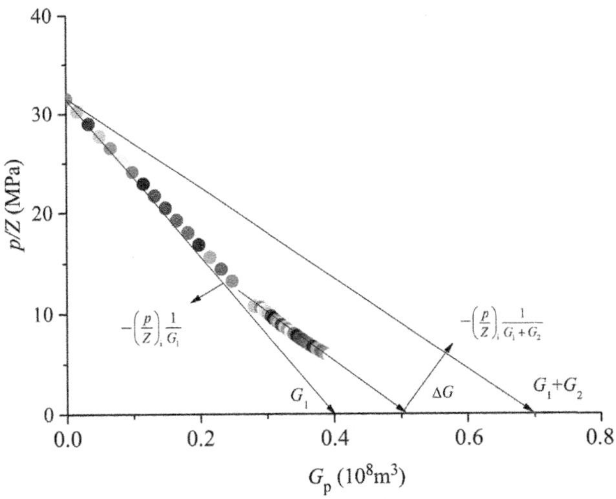

Fig. 5.9 *p/Z* curve of material balance of the two blocks with recharge capacity.

$\left(\frac{p}{Z}\right)_1 \sim G_p$ curve, with the slope of $-\left(\frac{p}{Z}\right)_i\frac{1}{G_1}$. Taking the derivative of Eq. (5.29), and assuming that the flow rate is a constant, then we have

$$\frac{d\left(\frac{p}{Z}\right)_1}{d(q_g t)} = -\left(\frac{p}{Z}\right)_i \frac{1}{G_1 + G_2}\left[1 + \frac{G_2}{G_1}e^{-\beta(1+\alpha)t}\right] \tag{5.35}$$

In the late stage of production, the exponential term of the preceding equation can be neglected. Next, the second straight line segment appears, with the slope of $-\left(\frac{p}{Z}\right)_i\frac{1}{G_1 + G_2}$. The middle is a transitional period, as shown in Fig. 5.9.

Due to the flow resistance in the low-permeability zone, the reserves calculated according to the p/Z curve by block are ΔG smaller than that calculated according to the traditional p/Z curve. According to Eq. (5.29), we have

$$\Delta G = \frac{\mu C_g W}{K^*(Lh)} \left(\frac{G_2^2}{G_1 + G_2} \right) q_g \tag{5.36}$$

According to Eq. (5.36), given certain reserves in the blocks, ΔG is not only sensitive to the characteristic parameters of the gas reservoir, but also related to the initial production allocation of the production area (Hower, 1989). On one hand, the larger the gas reservoir permeability and the contact area between blocks are, the smaller the low-permeability zone width between two blocks and the residual reserves will be. On the other hand, the larger the initial production allocation is, the larger the residual reserves will be. This is consistent with the experimental result (Hu et al., 2007).

Well performance forecasting It is assumed that this well produces at a rate of 10.0×10^4 m^3/d. The comparison between the traditional material balance method and the proposed method in this section is illustrated in Fig. 5.10. The proposed method corresponds to a stable production period of 0.85 year, while the traditional material balance method corresponds to a stable production period of 1.32 years, revealing a big difference.

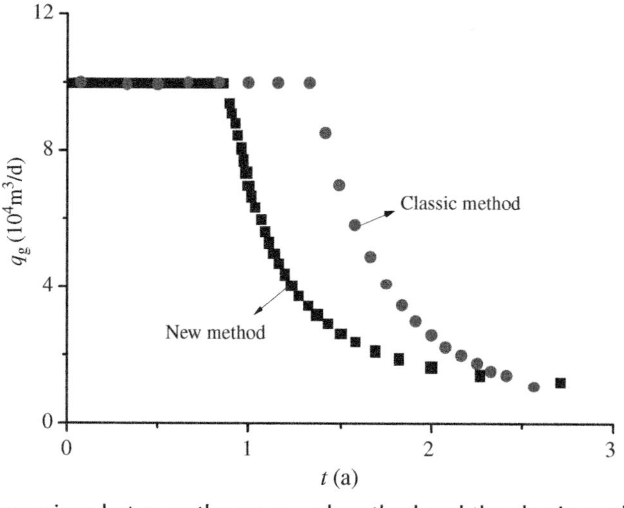

Fig. 5.10 Comparison between the proposed method and the classic method.

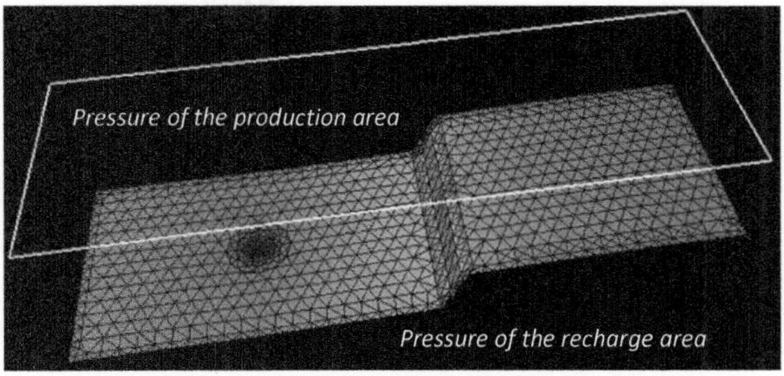

Fig. 5.11 Pressure field changes in numerical well testing simulation (at the end of the stable production period).

The proposed model is calculated by the numerical well testing method. If the flow rate is 10.0×10^4 m^3/d, the stable production period is 0.90 years. The result obtained by the proposed method is close to this value.

Essentially, the traditional material balance method takes the sweep range of a single well as a big homogeneous container, completely neglecting the heterogeneity of the reservoir, so the pressure balance declines during the production, ultimately leading to an optimistic index. According to Fig. 5.11, because of a low-permeability zone, there is a big pressure difference between the production area and the recharge area, which leads to the nonbalanced exploitation. The proposed method considers the heterogeneity of the reservoir and the flow resistance, so it conforms to the real conditions. This is also verified by the numerical simulation result.

It is assumed that this well produces at a rate of 10.0×10^4 m^3/d. The changes of bottomhole flow pressure and flow rate of the production area and the recharge area are shown in Fig. 5.12. As the production continues, the recharge area continues to play its role. Particularly, in the decline stage, the recharge area contributes an increasing proportion to the total production, so it cannot be neglected.

5.2.1.3.3 Summary

The proposed material balance equation, considering the regional heterogeneity and flow resistance, is the development of the material balance equation with recharge capacity. It is closer to real conditions and can provide more accurate results. This proposed model can be easily extended to the N blocks model, and can be used to simulate the production performance of multiple wells, or the whole block.

It is worth noting that the performance forecasting is based on the accurate deliverability equation of a single well. The traditional deliverability

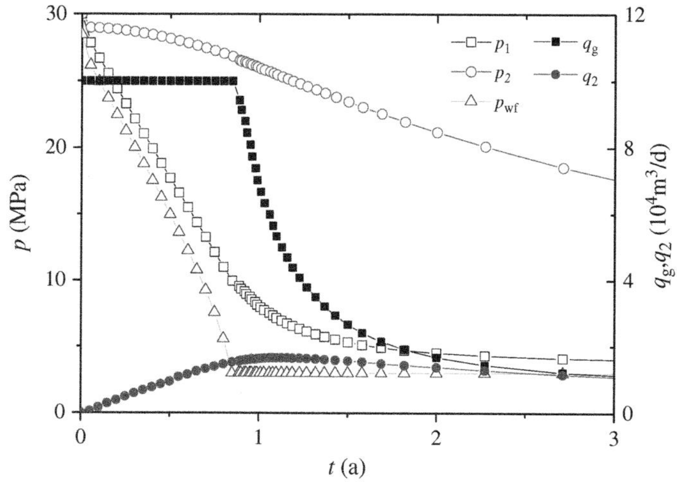

Fig. 5.12 Performance forecasting curve of the two-block material balance method with recharge capacity.

equation of gas wells can only be representative when the production reaches a pseudo-steady state. For low-permeability gas reservoirs, however, several months, years, or an even longer time is usually needed to reach the same state. Therefore, for low-permeability tight gas reservoirs, the pseudo-steady state is often hard to reach during the deliverability test. In this circumstance, deviation appears when the pseudo-steady state method is used for performance forecasting.

5.2.1.4 Fractured-vuggy condensate gas reservoir

5.2.1.4.1 Deliverability evaluation

Taking the Tazhong gas field as an example, the relationship between the absolute open flow potential of a gas well in a fractured vuggy carbonate gas reservoir, and the average flow rate in the first year, is shown in Fig. 5.13, and the production allocation coefficient is 3%–36%. Therefore, it is hard to use the deliverability test to determine the reasonable production of a single well. For example, the absolute open flow potential of Well G462 is 118×10^4 m^3/d. The production allocation is 20.0×10^4 m^3/d on a 1/5 basis, which is much larger than the average flow rate in the first year, that is, 7.5×10^4 m^3/d (Deng et al., 2015).

The relationship among the productivity index, production level, and RMS is shown in Fig. 5.14. The productivity index of gas wells is irrelevant to the distribution of high, medium, and low-productivity wells. Therefore, it is inadvisable to use the deliverability test to determine the reasonable

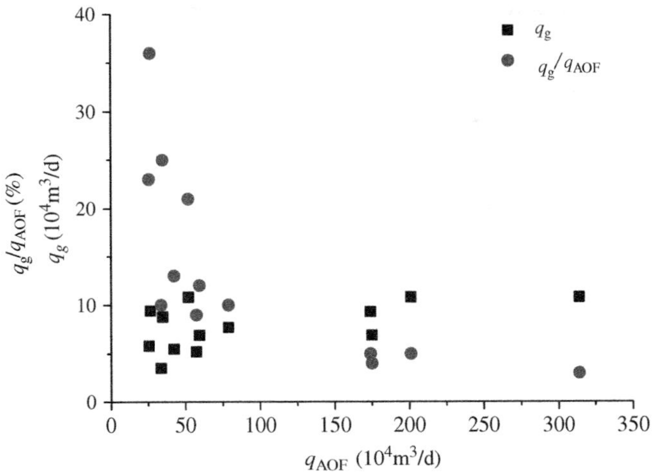

Fig. 5.13 Absolute open flow potential production in the first year and production allocation coefficient.

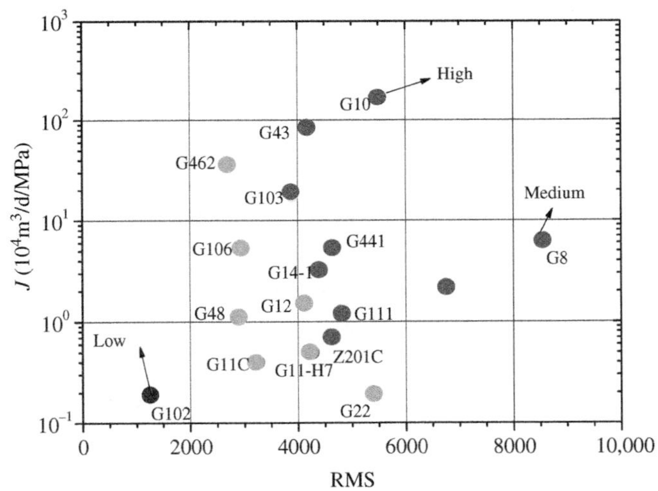

Fig. 5.14 Productivity index, production level, and RMS.

deliverability. In addition, for the carbonate gas wells in the Tazhong gas field, the initial productivity is high, and the production rate declines fast. Therefore, in the practical application, we must break through the conventional thought of the productivity evaluation of a sandstone gas reservoir, weaken the function of absolute open flow potential in the productivity evaluation, and evaluate the productivity by blocks, reservoir types, and well types. Meanwhile, the function of the dynamic reserves in the productivity evaluation should be highlighted, and the well testing dynamical description method, the production performance statistical method, and the empirical method should be combined to determine the productivity, as shown in Fig. 5.15. In the process of productivity evaluation, dynamic and static research results should be combined, as shown in Fig. 5.16.

Determination of the decline rate Once the producing well in fractured vuggy condensate gas reservoirs is put into production, the production begins to decline—rapidly in the early stage, and slowly in the late stage. The method to determine the decline rate in the forecasting period will directly influence the production performance forecasting result. If the producing well has reached the late stage of decline, the decline rate at the late stage can be used to forecast the subsequent production performance. If the producing well is put into production for a short time, the average decline rate in the block can be used to forecast the production rate.

The three oil wells in the Tazhong gas field have been producing for more than 2100 days. The decline curve is shown in Fig. 5.17A after the production of the three wells is averaged. The decline rate is 30% in the first year, 25% in the second to fourth years, and 20% after the fifth year. The seven gas wells have been producing for more than 1650 days (5 years), as shown in Fig. 5.17B. The decline rate is 25% in the first year, 20% in the second to fourth years, and 15% after the fifth year.

Determination of the production rate of condensate oil In the development process of condensate gas reservoirs, the decline rate can be used to forecast the condensate oil production at different stages. When the reservoir pressure is higher than the dewpoint pressure, the decline law at the wellhead is analyzed according to the converted condensate gas, and the condensate gas production rate is forecasted; then, the condensate oil production is converted according to the stable gas-oil ratio (GOR). When the reservoir pressure is lower than the dewpoint pressure, the decline law is analyzed according to the natural gas production rate at the wellhead (the average decline law of a well or a block), and the natural gas production rate is forecasted; then, the condensate oil production is determined according to the relationship between pressure and GOR.

When the average reservoir pressure is higher than the dewpoint pressure, the GOR is relatively stable. Accordingly, the condensate oil production is forecasted by the constant volume depletion test. When the average reservoir pressure is lower than the dewpoint pressure, the retrograde condensation appears in the formation. In this circumstance, the reservoir

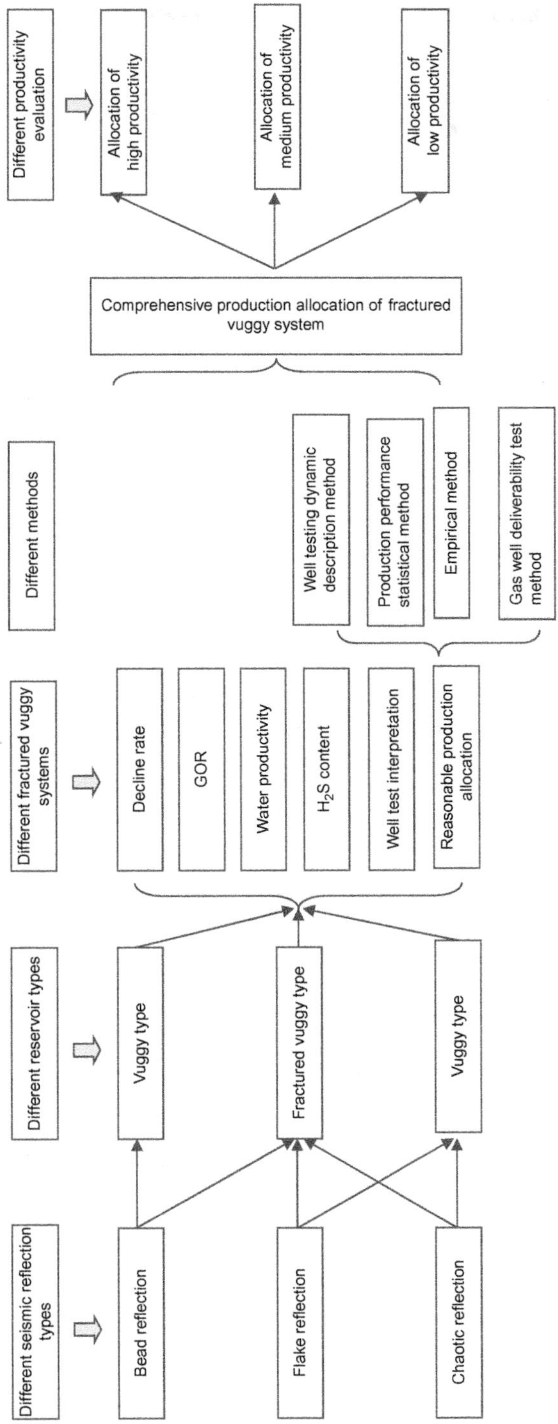

Fig. 5.15 Productivity evaluation process.

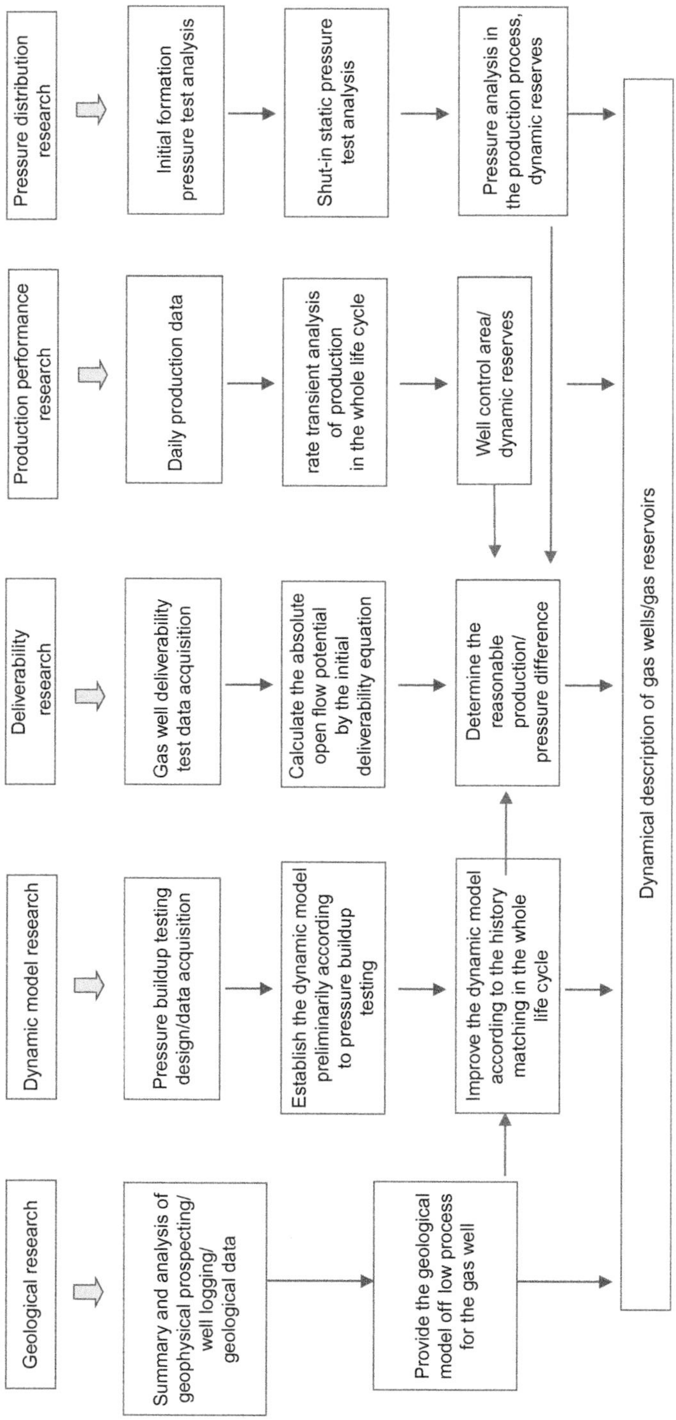

Fig. 5.16 Concept of optimized well production allocation.

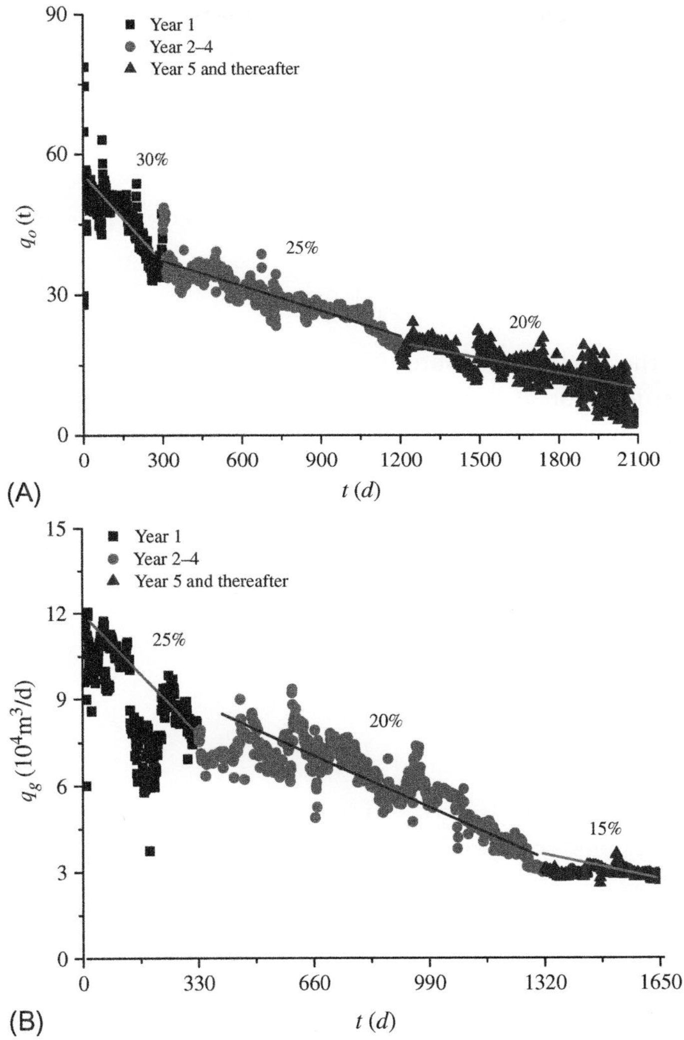

Fig. 5.17 Decline rate by years. (A) Decline rate of oil wells by years (7 years) and (B) decline rate of gas wells by years (5 years).

pressure is determined according to the material balance analysis, and then the condensate oil production is forecasted through the constant volume depletion test.

$$GOR = f(p, Z) \tag{5.37}$$

When the reservoir pressure is higher than the dewpoint pressure, the material balance equation of constant volume gas reservoirs is used directly.

When the reservoir pressure is lower than the dewpoint pressure, the single-phase volume factor and the deviation factor are considered as the two-phase volume factor and the two-phase deviation factor (Z_T), respectively, which can be determined by the laboratory test.

Determination of reservoir pressure When the reservoir pressure is higher than the dewpoint pressure, the average reservoir pressure is:

$$p = \frac{p_i Z}{Z_i}\left(1 - \frac{G_p}{G}\right) \tag{5.38}$$

When the reservoir pressure is lower than the dewpoint pressure, the two-phase deviation factor is used, and the average reservoir pressure is:

$$p = \frac{p_i Z_T}{Z_{Ti}}\left(1 - \frac{G_p}{G}\right) \tag{5.39}$$

Eqs. (5.37)–(5.39) can be combined into Eq. (5.40) to forecast the condensate oil production. According to the abandonment pressure of condensate gas reservoirs, the oil and gas production, reservoir pressure, and recovery percent of a well can be calculated.

$$q_o = f\left(p, Z, Z_T, G_p\right) \tag{5.40}$$

5.2.1.4.2 Analysis process

On the basis of dynamic reserves estimation and decline rate evaluation, the production change law is forecasted first. Within the given time period, the average pressure of the reservoir is determined according to the cumulative production and the oil-gas two-phase material balance equation, and the advanced production decline analysis method can be used to verify the average reservoir pressure. According to the results of the PVT test, the GOR change law under different pressures is determined. The wellhead oil and gas production is separated by GOR. The bottomhole flow pressure is calculated through the production performance analysis, or empirical analysis. The process is shown in Fig. 5.18.

For volatile oil wells, the analysis process that is similar to that used for condensate gas wells can be used. First, the crude oil production at the wellhead is forecasted. Then, according to the two-phase material balance equation, PVT test data, and GOR, the oil and gas production rate is separated, and the reservoir pressure is calculated. Finally, under the constraint of the quit flow pressure, the recoverable reserves and recovery percent are determined.

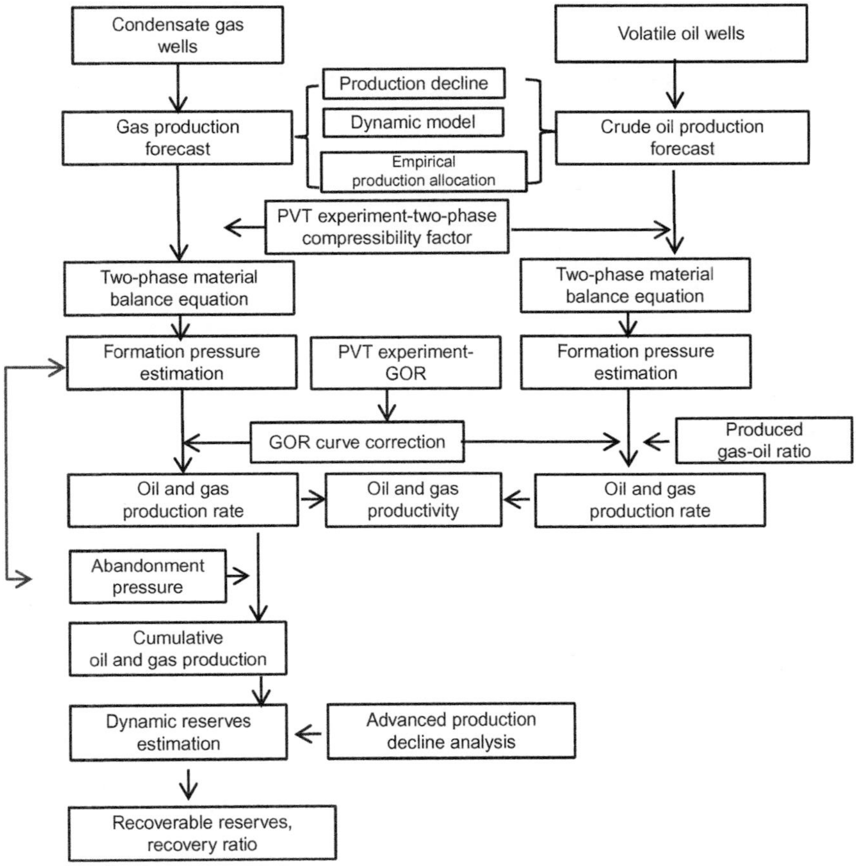

Fig. 5.18 Process of index forecasting for condensate gas reservoirs by the gas reservoir engineering method.

5.2.1.4.3 Example

For Well G43 in Zone II, the Tazhong gas field, the mid-depth of the reservoir is 5156m, and the formation temperature is 128.5°C. This well was put into production in March 2010. Up to December 2014, the cumulative effective production time was 1515 days, the cumulative gas production was 1.33×10^8 m^3, and the cumulative condensate oil production was 5.89×10^4 t. The production test curve is shown in Fig. 5.19. The GOR is steady in the early stage and gradually rises in the late stage.

According to the PVT test data, the dewpoint pressure is 56.99 MPa, the difference between the reservoir pressure and the dew point pressure is 6.34 MPa, and the condensate oil content is 451.63 g/m^3, so it belongs to

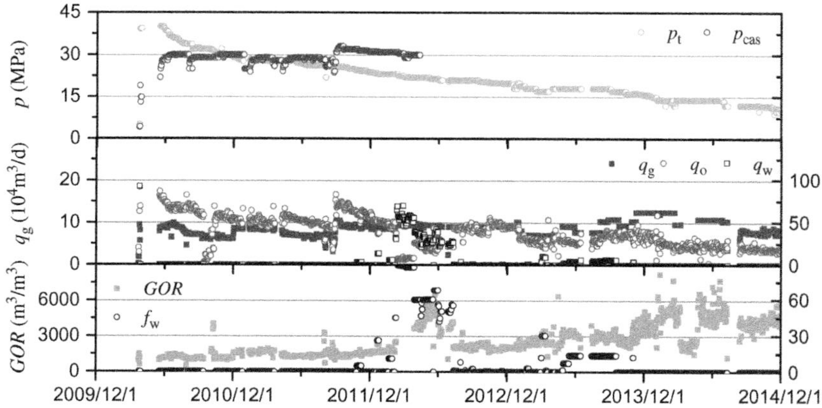

Fig. 5.19 Production test curve of Well G43.

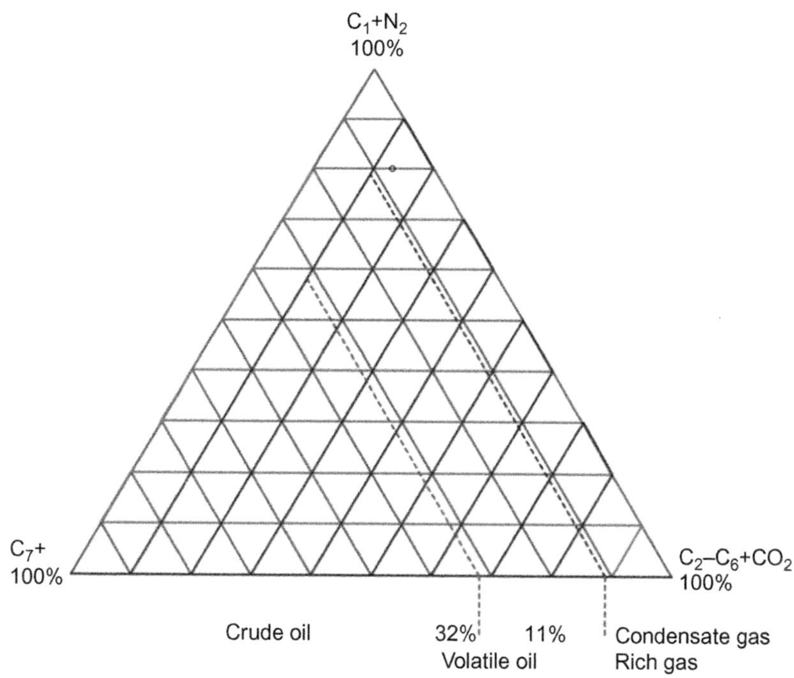

Fig. 5.20 Triangular chart of hydrocarbon fluid types.

a gas reservoir with a high content of condensate oil. Figs. 5.20 and 5.21 are the triangular chart and the phase-state diagram of hydrocarbon fluid types, respectively.

The dynamic reserves of this well (condensate gas) calculated by the advanced production decline analysis method are $3.12 \times 10^8\ m^3$; the

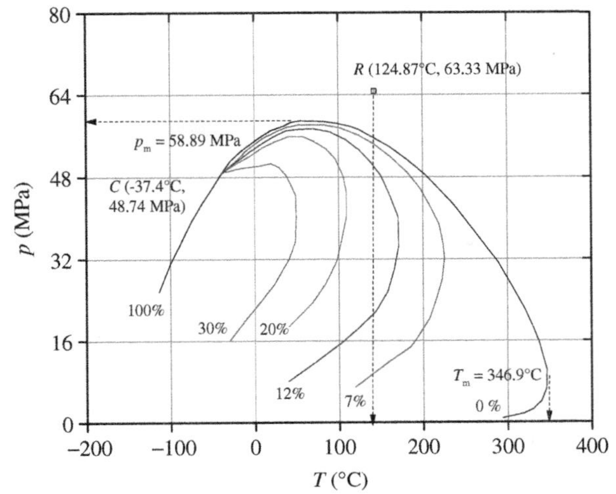

Fig. 5.21 Phase-state diagram of hydrocarbon fluid types.

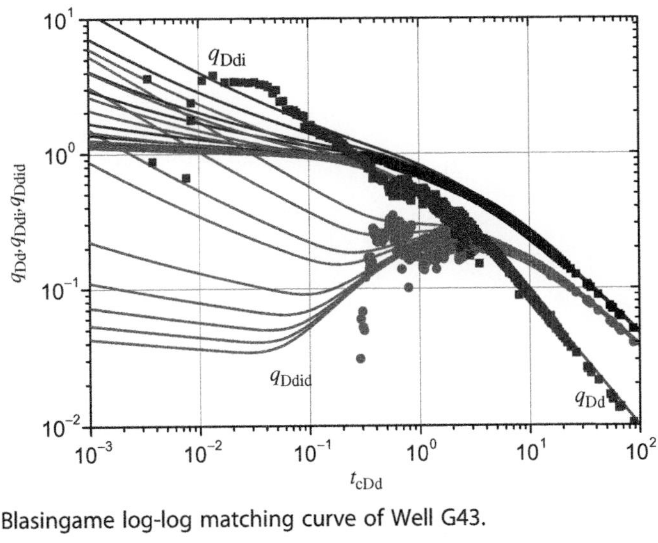

Fig. 5.22 Blasingame log-log matching curve of Well G43.

converted dynamic reserves of natural gas is 2.9×10^8 m^3, and dynamic reserves of condensate oil are 20.5×10^4 t. The Blasingame log-log matching curve and the history matching curve, respectively, are shown in Figs. 5.22 and 5.23.

The decline rate of this well in the production stage is 20%. On this basis, natural gas production in the following 8 years is forecasted, as shown in Fig. 5.24.

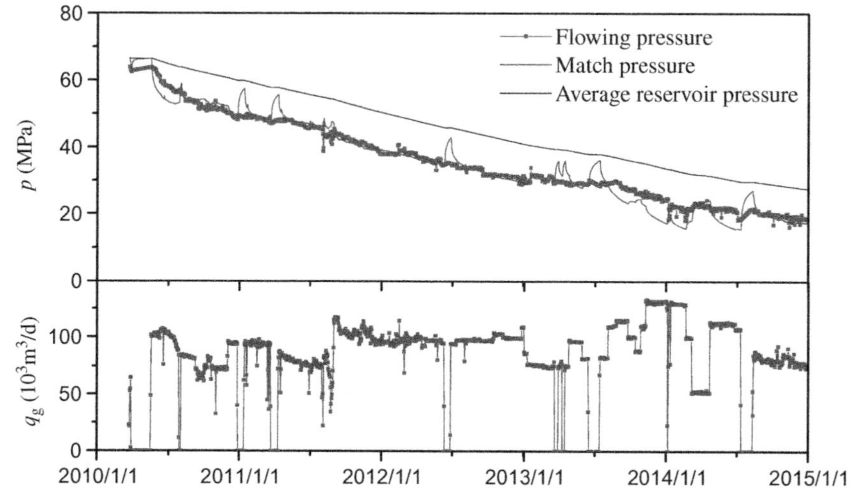

Fig. 5.23 History matching curve of Well G43.

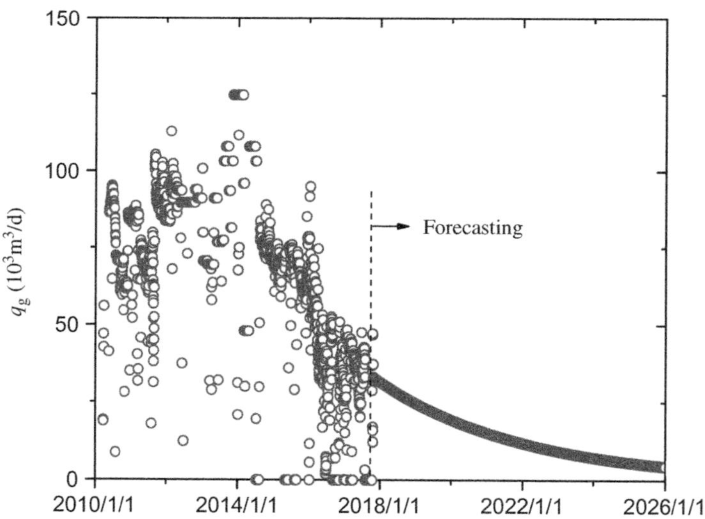

Fig. 5.24 Forecasting curve of natural gas production of Well G43.

According to the PVT constant volume depletion test, the relationships between pressure and GOR and between pressure and the two-phase deviation factor are determined, as shown in Fig. 5.25. The deviation factor under the measured reservoir pressure is 1.469 and the deviation factor under the dewpoint pressure is 1.454.

Substituting relevant parameters into the model helps to forecast the condensate oil production rate, as shown in Fig. 5.26. The GOR change is

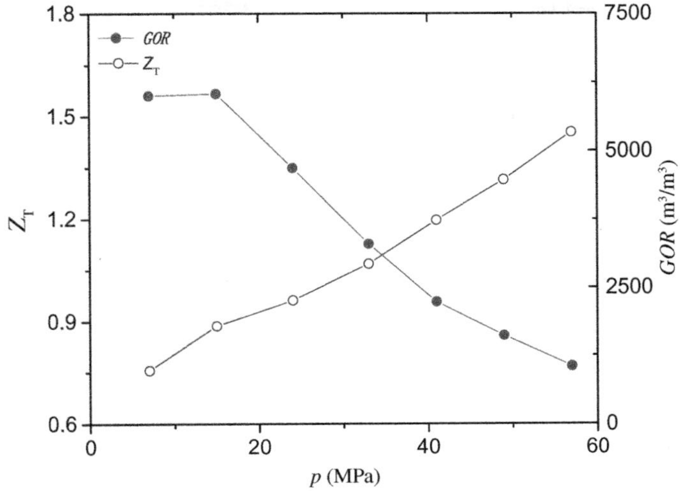

Fig. 5.25 GOR and two-phase deviation factor versus pressure of Well G43.

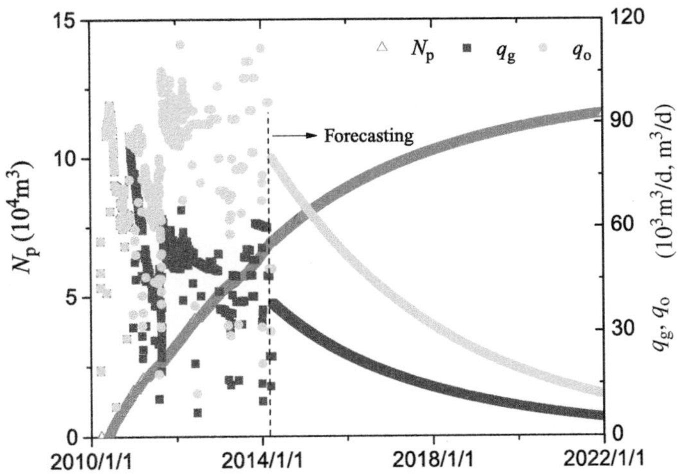

Fig. 5.26 Forecasting curve of the production of Well G43.

shown in Fig. 5.27, revealing basically the same trend for actual production data and calculated production data. The GOR is relatively stable in the early stage, and then rises gradually; and it declines gradually in the late stage. Essentially, the reservoir pressure is higher than the dewpoint pressure in the early stage, and gas is the single phase in the reservoir, so the GOR is relatively stable. When the reservoir pressure is lower than the dewpoint pressure, retrograde condensation occurs, and some condensate oil is

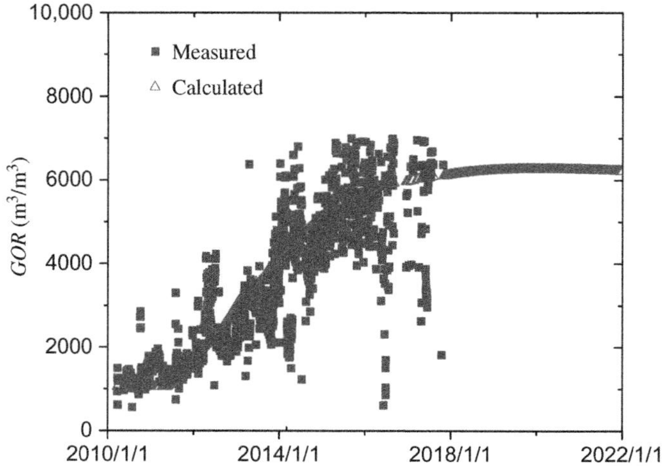

Fig. 5.27 Forecasting curve of GOR change of Well G43.

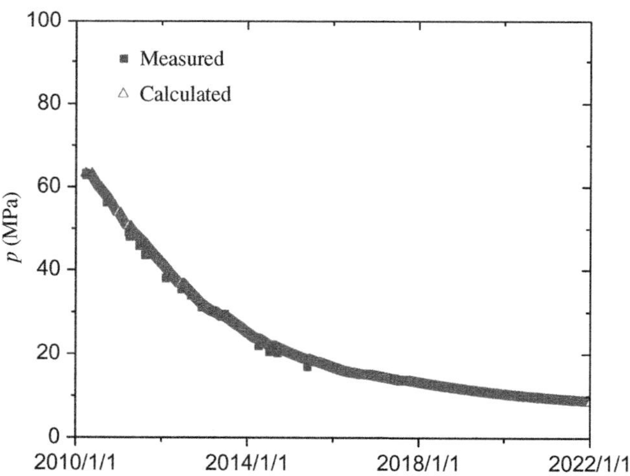

Fig. 5.28 Forecasting curve of reservoir pressure of Well G43.

accumulated in the formation, so the GOR rises gradually. In the late stage, evaporation occurs as the reservoir pressure continues to decline, so the GOR shows a declining trend.

The forecasted reservoir pressure is basically the same as the measured static pressure, as shown in Fig. 5.28. The abandonment pressure estimated according to the empirical equation is about 17.6 MPa; the forecasted cumulative gas production and cumulative oil production at the end of the period are 1.68×10^8 and 9.8×10^4 m^3, respectively. The present residual

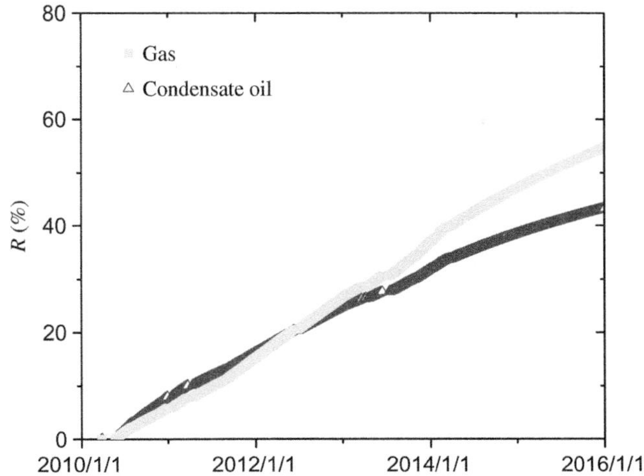

Fig. 5.29 Forecasting curve of recovery percent of Well G43.

recoverable reserves of natural gas and condensate oil are 0.35×10^8 and 2.44×10^4 m^3, respectively. The forecasted recovery percent of natural gas and condensate oil at the end of the period is 57.4% and 42.4%, respectively. The results are shown in Fig. 5.29.

5.2.2 Numerical well testing method

Based on the geological cognition on structures and faults, the numerical model of a reservoir built on numerical well testing analysis can be used to forecast the oil and gas production and the reservoir pressure.

5.2.2.1 Analysis process

First, the parameters such as formation coefficient, skin factor, boundary, and dynamic reserves are determined preliminarily according to the results of analytical well testing. Then, the numerical analysis is conducted depending on the structure, fault, fracture, seismic carving attributes, and well logging results, and a relatively reasonable single-well numerical model is obtained according to the analysis of the log-log curve and the verification by history matching. Next, the natural gas production is forecasted according to the production decline analysis results matched with the PVT test data, and separating the oil and gas production. Finally, the recoverable reserves of oil and gas are forecasted under the constraint of abandonment pressure. The process is shown in Fig. 5.30.

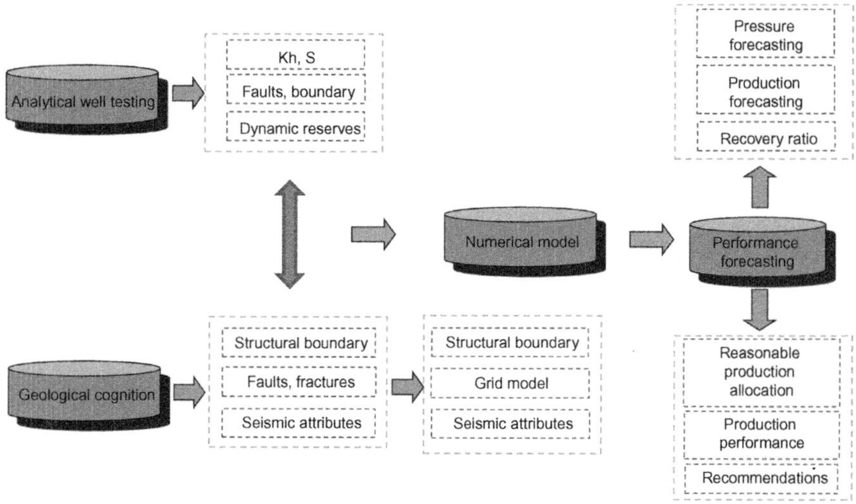

Fig. 5.30 Process of numerical well testing method.

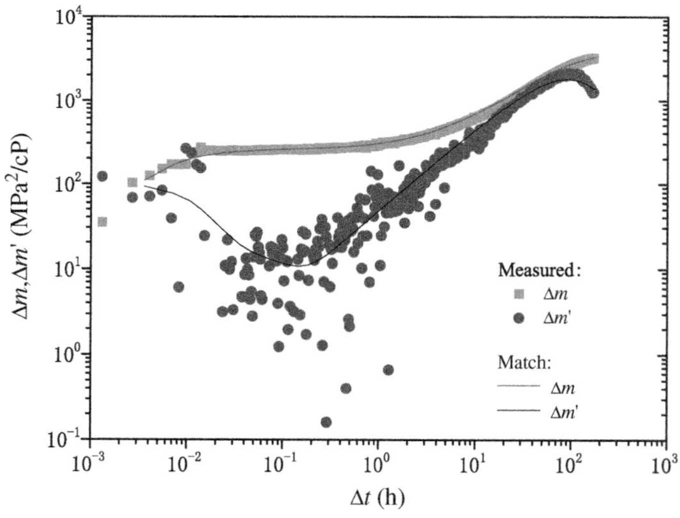

Fig. 5.31 Log-log matching curve of Well G14-1.

5.2.2.2 Example

The data of the pressure build-up test performed in March 2011 on Well G14-1 in the Tazhong gas field is interpreted with the variable wellbore storage + radial composite model. The log-log matching chart and the pressure history matching chart are shown in Figs. 5.31 and 5.32, respectively. According to the cognition on structures and faults and the results of

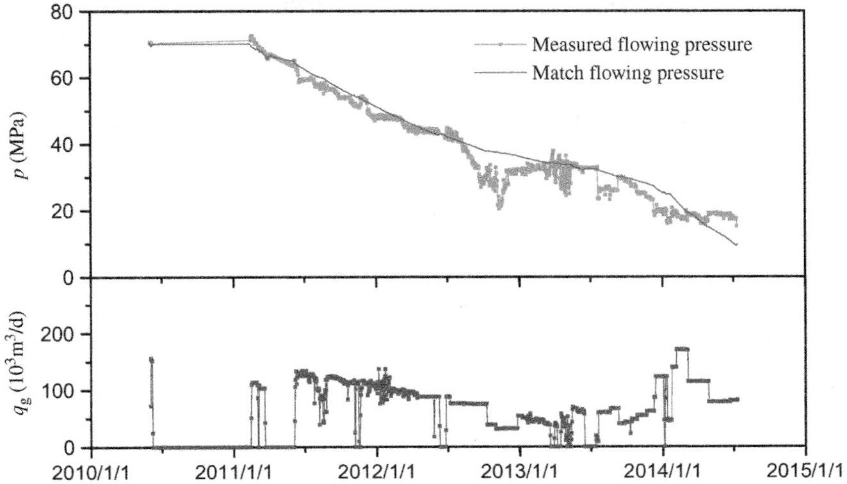

Fig. 5.32 Pressure history matching curve of Well G14-1.

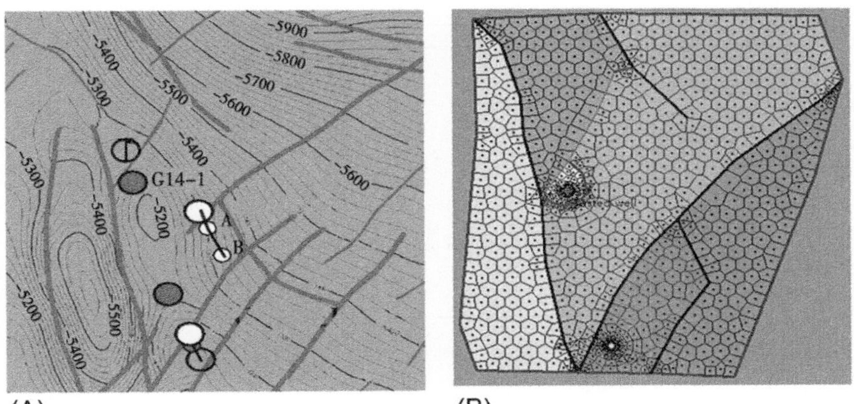

Fig. 5.33 Geological structure map and gridding composite numerical model of Well G14-1. (A) Geological structure map and (B) plane block-division numerical model.

analytical well testing, the plane block-division numerical model is built, as shown in Fig. 5.33.

The numerical well testing log-log matching is shown in Fig. 5.34. Compared with the analytic well testing method, this method is slightly improved, and can be used to forecast the pressure. Given the abandonment pressure of 17.6 MPa, the recoverable reserves and recovery percent at the end of the period are forecasted to be 1.1×10^8 m^3 and 62.9%, respectively, as shown in Fig. 5.35.

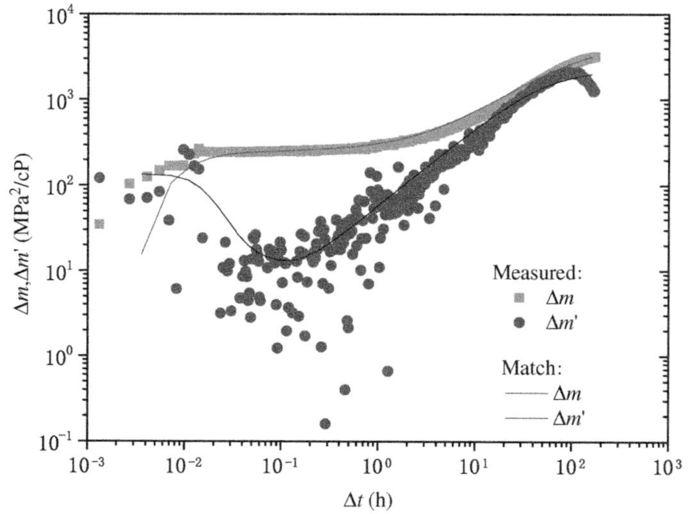

Fig. 5.34 Log-log matching curve of the Well G14-1 by numerical well testing.

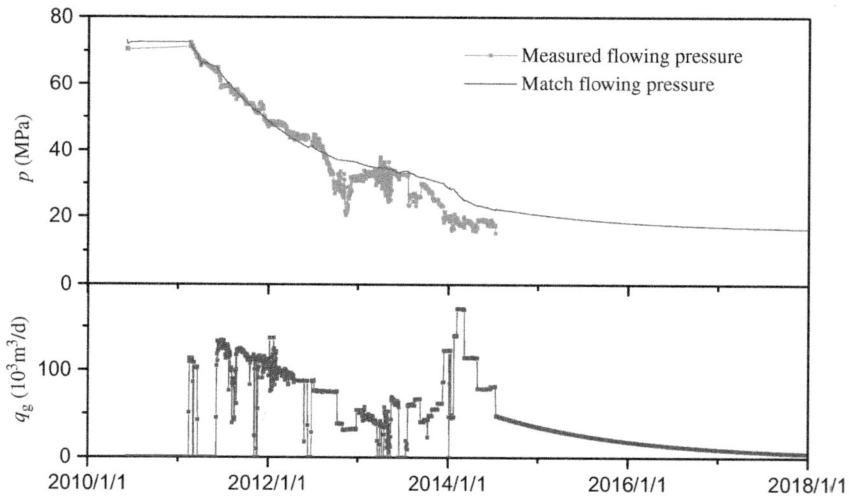

Fig. 5.35 Performance forecasting curve of Well G14-1 by numerical well testing method.

5.2.3 Advanced production decline analysis method

The advanced production decline analysis method has become a new technique for well performance forecasting. It is used to forecast the oil and gas production, reservoir pressure, recoverable reserves, and other development indices by using the dynamic model based on the advanced production decline analysis, and the production decline analysis results. Currently, the analytical model is mainly used for the whole history matching according to the superposition principle. However, this analytical model is insufficient for field application in gas wells due to significant production fluctuation, multiple flow stages, and long cycles of history matching. In this circumstance, taking the fractured vertical gas well as an example, a mathematical model of unsteady flow is established, and the mixed finite element method is adopted to obtain the numerical solutions under three production modes, that is, constant rate, constant pressure, and variable rate and variable pressure. The advanced production decline analysis curve is plotted according to the constant-rate solution. The whole history matching is conducted according to the variable-rate and variable-pressure solution. The production performance of gas wells is forecasted according to the solution of constant rate, and then, constant pressure.

5.2.3.1 Unsteady flow model of fractured vertical gas wells

5.2.3.1.1 Physical model

There is a fractured vertical gas well with finite conductivity in the homogeneous bounded reservoir. The physical model is built with some assumptions. First, gas flows in the reservoir in the form of single-phase seepage subject to the Darcy law. Second, the flow of gas in fractures is one-dimensional. Third, considering the damage of fracturing fluid to the reservoir, the rectangular low-permeability zone is adopted, and the fracture skin factor is used to measure the damage to the reservoir, as shown in Fig. 5.36. Fourth, the fracture volume is very small compared with the well-controlled volume, the fracture permeability is much larger than the reservoir

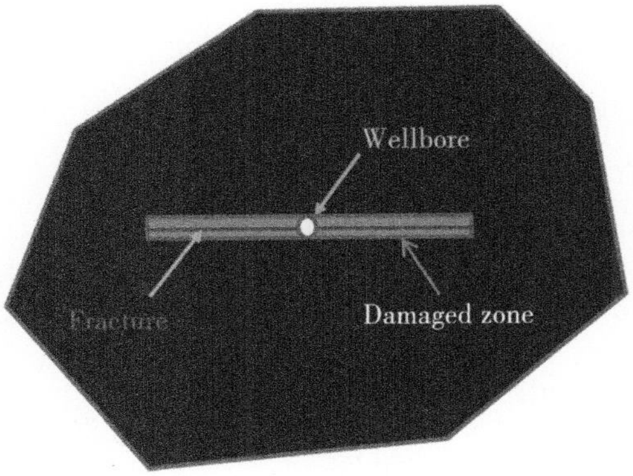

Fig. 5.36 Schematic diagram of physical model.

permeability, and the gas volume expansion in fractures resulting from fracture pressure decrease has little influence on the whole flow, so the derivative term of pseudo-pressure to time in the fracture control equation can be ignored. Fifth, the influence of the wellbore storage effect on the production is not considered. Moreover, the permeability, porosity, and rock compressibility do not change with the pressure. The rock compressibility is calculated by the Newman empirical equation (Newman, 1973), the gas compressibility and deviation factor are calculated by the DPR method (Dranchuk et al., 1973), and the gas viscosity is calculated by the Lee method (Lee et al., 1966).

5.2.3.1.2 Mathematical model

Reservoir:

$$\frac{K_{\mathrm{L}}}{\mu(p)}\left(\frac{\partial^2 m}{\partial x^2}+\frac{\partial^2 m}{\partial y^2}\right)-q_{\mathrm{f}}=\alpha_1\phi c_{\mathrm{t}}(p)\frac{\partial m}{\partial t} \tag{5.41}$$

Fracture:

$$\frac{K_{\mathrm{f}}}{\mu(p)}\frac{\partial^2 m}{\partial l^2}+q_{\mathrm{f}}=0 \tag{5.42}$$

Initial condition:

$$m(x,y,0)=m_{\mathrm{i}} \tag{5.43}$$

Inner boundary condition (production mode):

$$\text{Constant}-\text{rate}:\ \left.\frac{\partial m}{\partial l}\right|_{\Gamma_{\mathrm{in}}}=\frac{\alpha_2 q_{\mathrm{sc}}T}{K_{\mathrm{f}}hw_{\mathrm{f}}} \tag{5.44}$$

$$\text{Constant}-\text{pressure}:\ m|_{\Gamma_{\mathrm{in}}}=m_{\mathrm{w}} \tag{5.45}$$

Variable-rate and variable-pressure:

$$\begin{cases}\left.\dfrac{\partial m}{\partial l}\right|_{\Gamma_{\mathrm{in}}}=\dfrac{\alpha_2 q_{\mathrm{sc}}(t)T}{w_{\mathrm{f}}K_{\mathrm{f}}h} & \text{Obtain the pressure according to the variable rate}\\ m|_{\Gamma_{\mathrm{in}}}=m_{\mathrm{w}}(t) & \text{Obtain the production according to the variable pressure}\end{cases} \tag{5.46}$$

Closed outer boundary:

$$\left.\frac{\partial m}{\partial n}\right|_{\Gamma_{\mathrm{out}}}=0 \tag{5.47}$$

5.2.3.1.3 Model solution

The model is solved by the finite element method with hybrid elements (Wan et al., 2016), as shown in Eq. (5.48). The whole calculation area is divided into two parts: reservoir area of two-dimensional flow, and fracture area of one-dimensional flow.

$$\iint_{\Omega} FEQ d\Omega = \iint_{\Omega_m} FEQ d\Omega_m + w_f \cdot \int_{\bar{\Omega}_f} FEQ d\bar{\Omega}_f \tag{5.48}$$

Where, FEQ represents the fluid flow equation, Ω represents the whole flow area, Ω_m represents the reservoir flow area, and $\bar{\Omega}_f$ represents the fracture flow area.

The two-dimensional finite element equation of the reservoir area is obtained by using the Galerkin method of weighted residuals to disperse the governing equations of reservoir and fracture:

$$\begin{aligned}
& AK_L\left[b_ib_i + c_ic_i + \frac{\alpha_1\phi\mu(p)c_t(p)}{6\Delta t K_L}\right]m_i^{n+1} \\
& + AK_L\left[b_ib_j + c_ic_j + \frac{\alpha_1\phi\mu(p)c_t(p)}{12\Delta t K_L}\right]m_j^{n+1} \\
& + AK_L\left[b_ib_k + c_ic_k + \frac{\alpha_1\phi\mu(p)c_t(p)}{12\Delta t K_L}\right]m_k^{n+1} \\
& - \frac{K_L L}{3}\frac{\partial m_i^{n+1}}{\partial n} - \frac{K_L L}{6}\frac{\partial m_{j(k)}^{n+1}}{\partial n} = \frac{\alpha_1\phi\mu(p)c_t(p)}{6\Delta t}m_i^n \\
& + \frac{\alpha_1\phi\mu(p)c_t(p)}{12\Delta t}m_j^n + \frac{\alpha_1\phi\mu(p)c_t(p)}{12\Delta t}m_k^n
\end{aligned} \tag{5.49}$$

The one-dimensional finite element equation of the fracture area:

$$\frac{K_f w_f}{L}m_i^{n+1} - \frac{K_f w_f}{L}m_j^{n+1} + K_f w_f\frac{\partial m_i}{\partial l} = 0 \tag{5.50}$$

The finite element stiffness matrix of the reservoir area and the stiffness matrix of the fracture area are established according to Eqs. (5.49), (5.50), and then the two equations are combined into a system stiffness matrix. According to the fluid crossflow relationship, the boundary term at the junction between fracture and reservoir is eliminated by superposing the reservoir and fracture units, corresponding to the last two terms at the left side of Eq. (5.49) and the last term at the left side of Eq. (5.50). Specific combination methods are available in relevant literature (Wan et al., 2016). The gas viscosity and compressibility are the functions of pressure, so, in actual computation, at each time step, the viscosity and compressibility are calculated according to real reservoir pressure, and they are changed constantly in the governing equation. The SuperLU linear equation system solver is used to solve the combined linear equation set to obtain the solution of the model.

5.2.3.2 Numerical advanced production decline curve and comparative analysis

5.2.3.2.1 Comparison of calculation results

For a reservoir, the permeability is 1 mD, the initial reservoir pressure is 30 MPa, the reservoir temperature is 100°C, the net pay thickness is 8 m, and the porosity is 10%. The circular radius is 500 m, and the relative density of natural gas is 0.58. The fracture half-length is 100 m, the fracture permeability is 100 D, the fracture width is 0.01 m, and the fracture skin factor is 0.1. The width of the fracture damage zone is 0.2 m, and the permeability of the damage zone obtained according to the fracture skin factor is 0.03 mD. The numerical model and Topaze software are used to determine the Blasingame curve of this well, and the calculation results are compared. As shown in Fig. 5.37, the calculation result of the numerical model is exactly the same as that of the Topaze software, thus verifying the correctness of the numerical model and the calculation method.

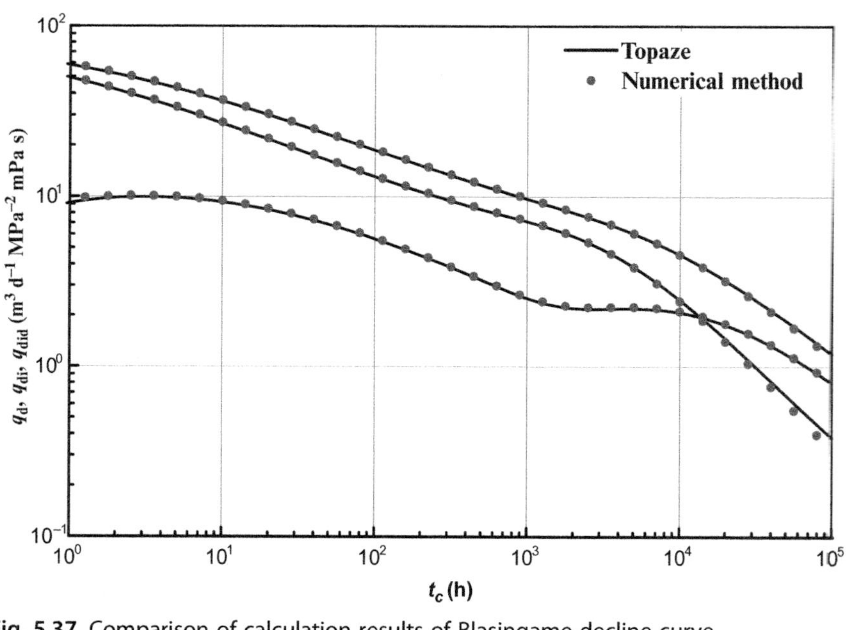

Fig. 5.37 Comparison of calculation results of Blasingame decline curve.

Table 5.4 Comparison of calculating speeds between the numerical model and the Topaze software

Item	Numerical model	Topaze software
Number of time points	3164	273
Number of grid elements	1418	/
Computation time of type curve	0.34 s	16.2 s
History match time	14.52 s	
Computation time of average single step	4.59 ms	59.34 ms

5.2.3.2.2 Comparison of calculating speeds

On the basis of the production data of a gas well, the calculating speeds of the numerical model and the Topaze software are compared. This well was put into production on December 6, 2002. Up to August 6, 2011, the total production time was 74,673 h, and the number of data points was 3164. The numerical model and the Topaze software are adopted for matching analysis, and their calculating times are recorded. The computer used is the ThinkPad L440, with an Intel Core i7-4712MG CPU @2.3 GHz processor, and 4 GB of memory. The comparison of the calculating speeds is shown in Table 5.4.

For the purpose of whole history matching, the Topaze software completes the calculation, not according to the number of time points input, but the number of data points diluted. In this example, the number of points calculated is only 273. In contrast, the numerical model completes the calculation step by step completely according to the number of input points, thus ensuring the data integrity. In this example, 3164 time steps are calculated with the numerical model; the total time needed is 14.52 s, and the average single-step time is 4.59 ms, while the single-step calculating time of Topaze is 59.34 ms. Obviously, the numerical model is superior in calculating speed, meeting the actual matching needs. For well examples with relatively long production times and large data size, the numerical model is superior in respect to calculating speed, and it can greatly improve the efficiency of whole history matching.

5.2.3.3 Simulation example

The production performance of a gas well is forecasted by using the constant rate first, and the constant pressure second. It is assumed that the reservoir permeability is 0.03 mD, the initial reservoir pressure is 40 MPa, the reservoir temperature is 100°C, the net pay thickness is 15 m, the porosity is 10%, the circular radius is 500 m, the fracture half-length is 100 m, the fracture permeability is 100 D, the fracture width is 0.01 m, the fracture skin factor is 0.1, the relative density of natural gas is 0.58, and the inertia-turbulence coefficient is 1e −7. The initial flow rate is 3×10^4 m^3/d, and the late constant bottomhole flow pressure is 4 MPa. Accordingly, the theoretical turbulent flow deliverability equation of the pseudo-steady state is obtained as follows:

$$m(p_i) - m(p_{wf}) = 1.741 q_{sc} + 1.07 \times 10^{-7} q_{sc}^2 \tag{5.51}$$

The comparison of results between the conventional forecasting method (based on the turbulent flow deliverability equation and the material balance equation) and the numerical model is shown in Fig. 5.38.

The result of the conventional forecasting method tends to be optimistic, with both the stable production period and the cumulative production higher than the numerical model, as shown in Fig. 5.38A. For the conventional method, the bottomhole flow pressure declines slowly, and the time to reach the constant pressure production is long. Compared with the bottomhole flow pressure, the difference of calculated average reservoir pressure is relatively small, as shown in Fig. 5.38B. In this example, the reservoir permeability is very low, and roughly 2 years will be needed to reach the pseudo-steady state. However, for the conventional method, the flow is assumed in a pseudo-steady state during the initial production, thus causing a big forecasting error.

5.2.3.4 Literature example

The numerical model is used to analyze the production decline curve analysis and forecast the production performance of a fractured well given in SPE literature (Pratikno et al., 2003). The curve matching of the production type is illustrated in Fig. 5.39, and the whole history pressure-production matching and forecasting are shown in Fig. 5.40 and 5.41. The interpretation results of the numerical model, Topaze software, and literature are compared, as shown in Table 5.5. It is found that the results of Blasingame type curve matching of the three are ideal, but the interpretation result of the numerical model is very close to that of the Topaze software, but different from that of the literature. Essentially, the interpretation result of the literature is ambiguous, because it is obtained only by the matching of production decline type curves, but not verified by the whole history curve matching. The whole history pressure and the production data are calculated according to the interpretation result of the literature, as shown in Fig. 5.40 and Fig. 5.41. The comparison of matching effects indicates that the

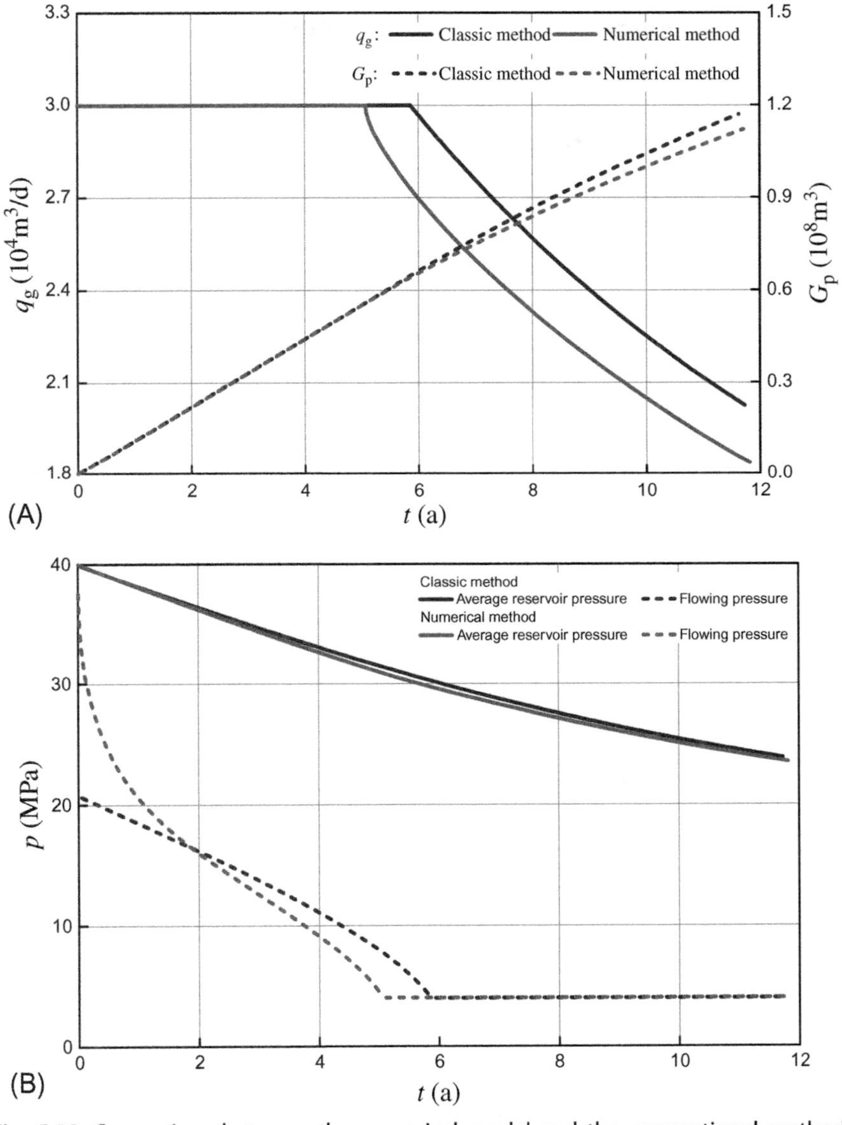

Fig. 5.38 Comparison between the numerical model and the conventional method. (A) Comparison of production forecasting curves and (B) comparison of pressure forecasting curves.

numerical model is obviously better than the literature, showing a large error in the interpretation result of the literature. Thus, the verification by whole history curve matching is important for reducing the ambiguity of interpretation result and improving the accuracy.

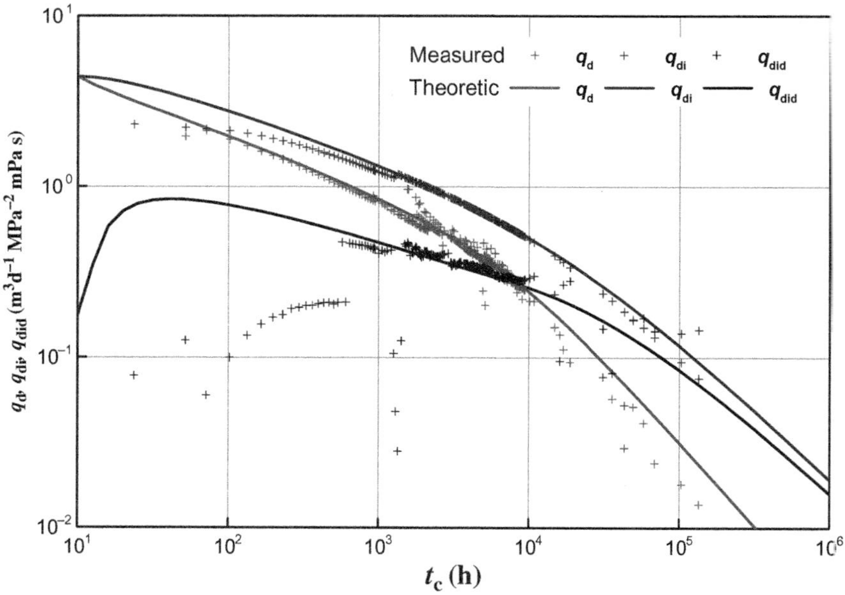

Fig. 5.39 Blasingame curve matching by the numerical method.

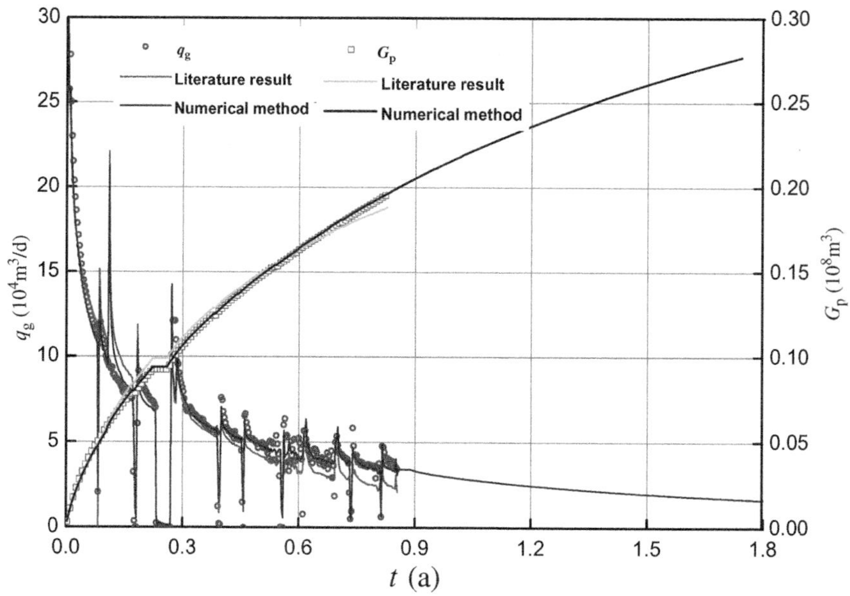

Fig. 5.40 History production matching and performance forecasting.

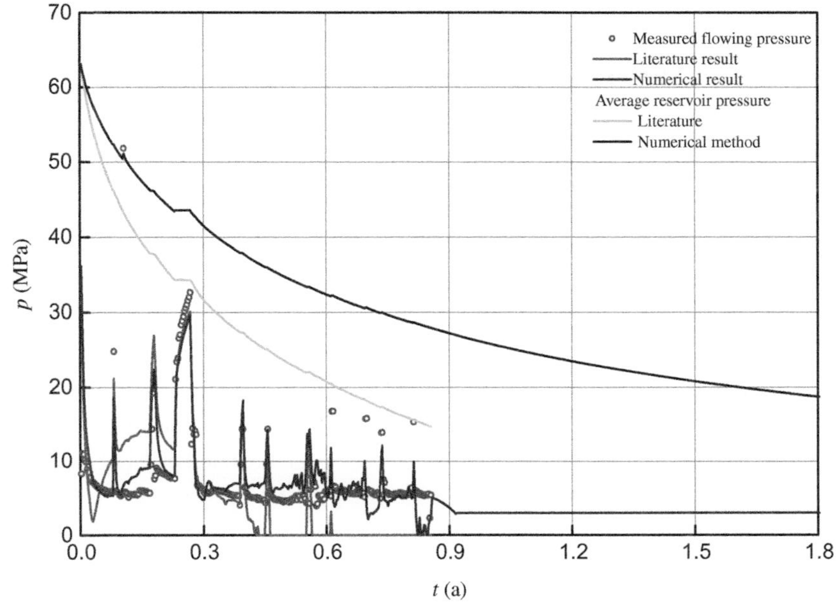

Fig. 5.41 History pressure matching and performance forecasting.

Table 5.5 Interpretation results of production decline analysis

Parameters	Reference	Topaze	Numerical model
Reservoir permeability (mD)	0.015	0.0056	0.0055
Fracture half-length (m)	42.1	88.4	90.5
Fracture conductivity (mD m)	3.16	4.7	5.1
Well-controlled radius (m)	84.3	103	107
Well-controlled reserves (10^8 m^3)	0.287	0.459	0.466

The production performance of this well is forecasted by the advanced production decline analysis method after production data matching. It is completed at the current rate, and then at the constant bottomhole flow pressure, which is assumed to be 3 MPa. Figs. 5.40 and 5.41 show the forecasted production and pressure changes in 1 year. It can be seen that the bottomhole flow pressure reaches the predetermined pressure rapidly. In the next year, the annual cumulative production is 868×10^4 m^3, and the average reservoir pressure is reduced to 18.4 MPa. If the abandonment production of this well is assumed at 3000 m^3/d for forecasting the production in the whole lifecycle, the total production time of this well is 6.3 years, the ultimate cumulative production is 0.404×10^8 m^3, and the ultimate recovery is 86.7%.

5.2.3.5 Example

For Well G162-1H in the Tazhong gas field, the mid-depth of the reservoir is 6250.06 m and the formation temperature is 144.9°C. This well was put into production on November 10, 2009. It produced with a 6 mm nozzle in the early stage, and then with a 5 mm nozzle after 3 months of production. The initial oil production was 81.5 t/d, and the daily gas production was 3.01×10^4 m^3. After oil was produced without water for 54 days, water breakthrough occurred suddenly. In the late stage, water was produced intermittently. Up to December 2014, the oil cumulative production was 7.4×10^4 t, the gas cumulative production was 0.56×10^8 m^3, and the cumulative water production was 0.88×10^4 m^3. The producing test curve is shown in Fig. 5.42. The data of only six bottomhole flow pressure gradient tests are available.

The advanced production decline analysis method is used, and the curve matching is shown in Fig. 5.43 and 5.44. The dynamic reserves of oil are 65.8×10^4 t.

Under the constraint of the quit flow pressure of 23.7 MPa, the dynamic model based on the advanced production decline analysis method is combined with the production decline analysis to forecast the key development indices. The forecasted recoverable reserves and recovery percent at the end of the forecasting period are 9.3×10^4 t and 14.13%, respectively, as shown in Fig. 5.45.

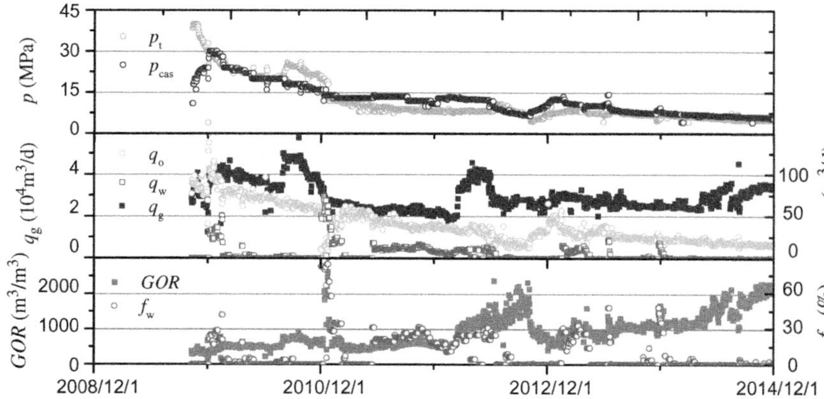

Fig. 5.42 Producing test curve of Well G162-1H.

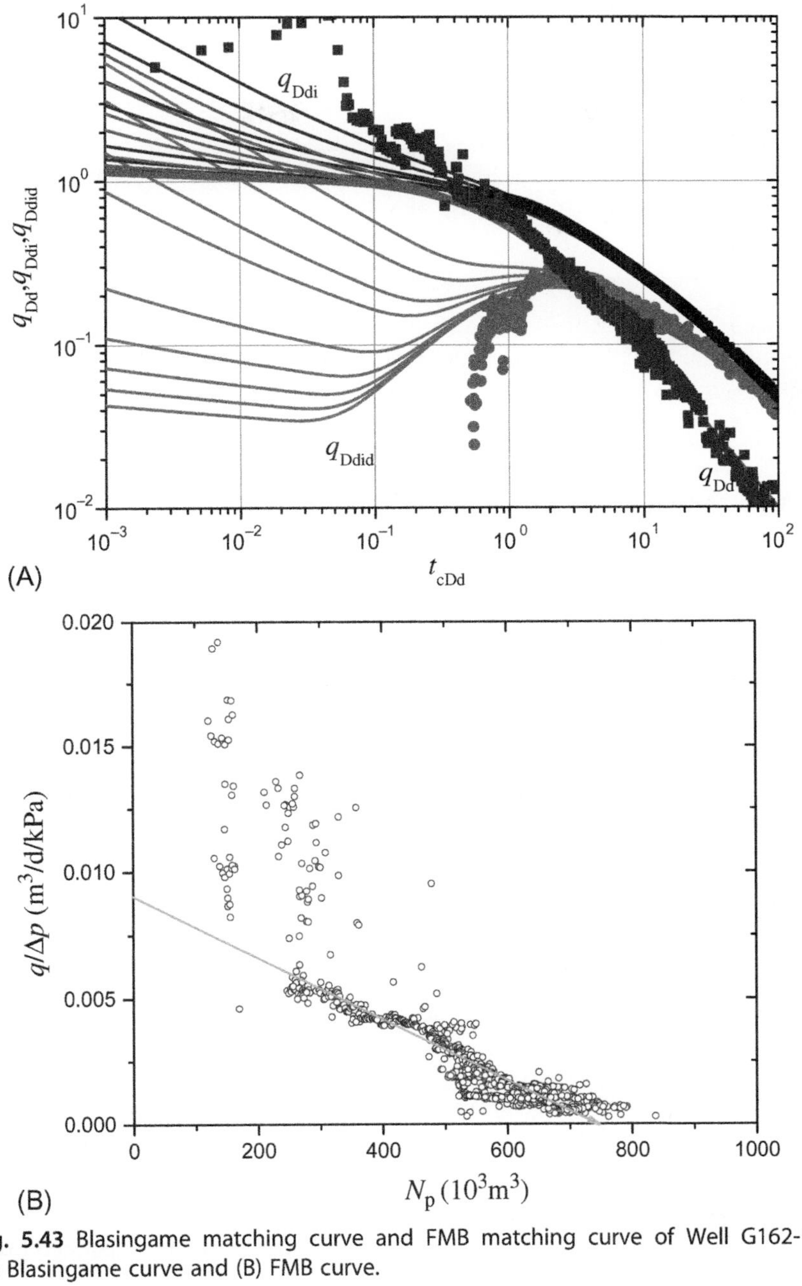

Fig. 5.43 Blasingame matching curve and FMB matching curve of Well G162-1H. (A) Blasingame curve and (B) FMB curve.

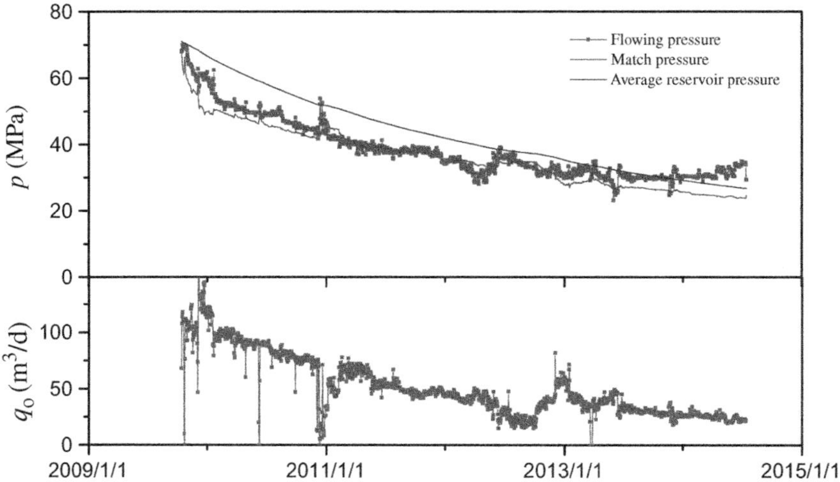

Fig. 5.44 Production history matching curve of Well G162-1H.

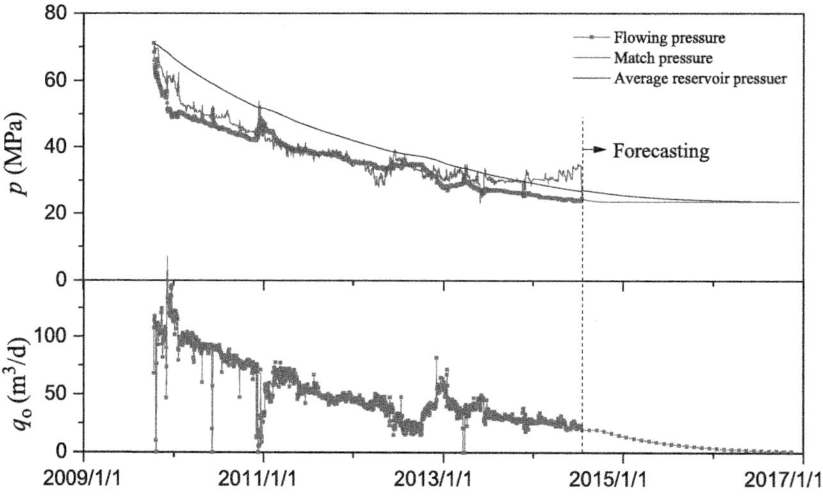

Fig. 5.45 Performance forecasting curve of the advanced production decline analysis method for Well G162-1H.

5.2.3.6 Summary

The numerical calculating method is a potential direction for advanced production decline analysis. It can realize the advanced production decline analysis of complex reservoir boundaries and complex well types, and can greatly improve the history matching efficiency of advanced production decline analysis. The production performance model, based on the history matching, can be used in three cases: constant-rate, constant-pressure, and variable-pressure and variable-rate. The numerical model can meet the actual needs in respect to both accuracy and speed. It is more accurate than the traditional method based on the pseudo-steady state deliverability equation and the material balance equation.

Example analysis suggests that history matching is an effective measure to reduce the ambiguity of the result of advanced production decline analysis. A complete advanced production decline analysis should include four steps, that is, data evaluation, type curve matching analysis, whole history matching, and production performance forecasting; but the last two steps are often ignored.

5.2.4 Numerical simulation method of a single well

The single-well numerical simulation method is mainly applicable to single-well hydrocarbon reservoirs of clear geological cognition. For this method, a three-dimensional numerical model is established according to vertical seismic attributes and the two-dimensional numerical well testing model, and then the history matching, rapid simulation analysis, and forecasting are conducted.

5.2.4.1 Analysis process

On the basis of the single-well and two-dimensional numerical well testing analysis and seismic attributes, a three-dimensional numerical model is established, and it is verified by matching the production history and pressure history, producing water trends and dynamic reserves; then a three-dimensional dynamic model is built. The oil and gas production and reservoir pressure are forecasted according to the production decline trend. Under the given constraints, the recoverable reserves of oil and gas are forecasted. The process is shown in Fig. 5.46.

5.2.4.2 Example

For Well G151, the mid-depth (vertical depth) of the reservoir is 6153.78 m and the reservoir temperature is 140°C. This well was put into production on April 20, 2011. During the initial production with a 4 mm nozzle, the daily gas production was $13.5 \times 10^4\ m^3$, the daily oil production was 45.56 t, and no water was produced; the average GOR is $2823\ m^3/t$.

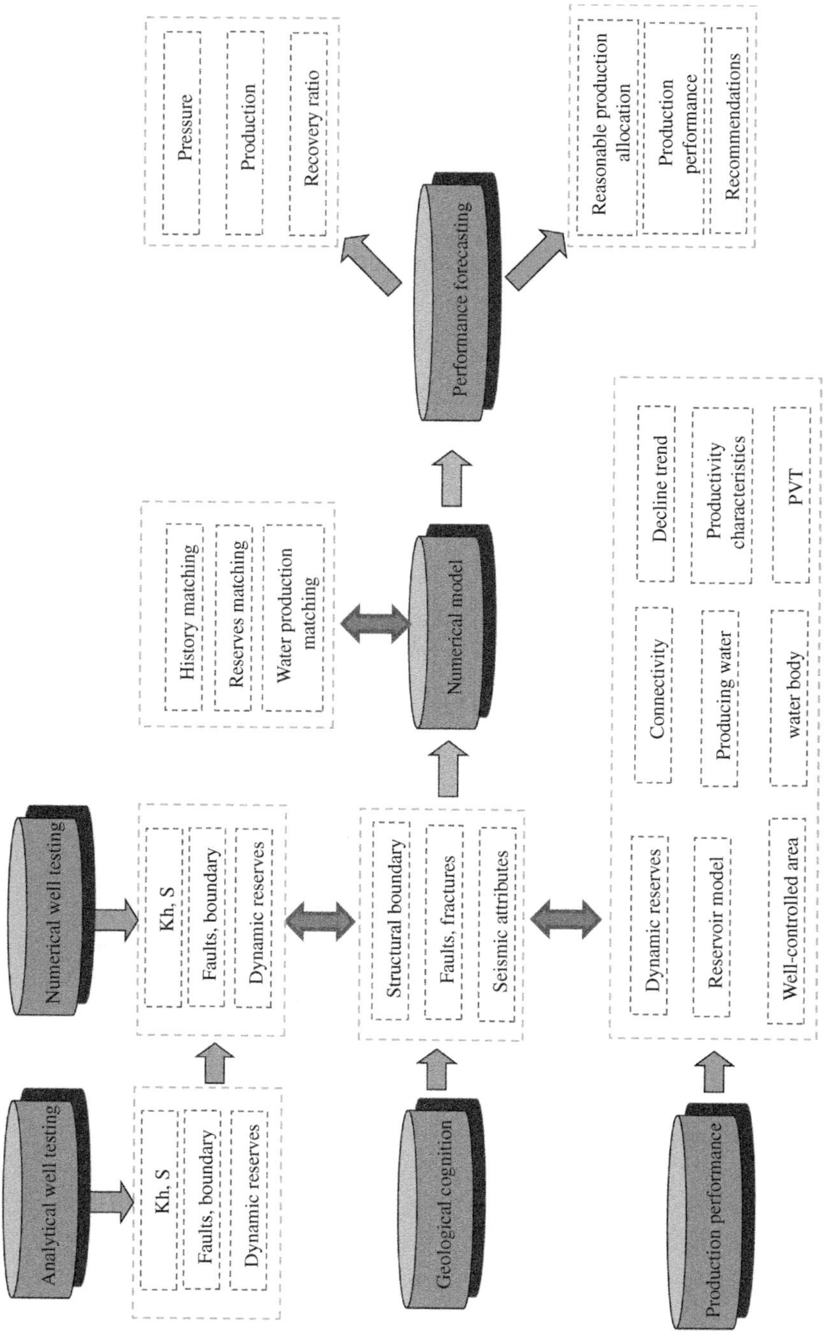

Fig. 5.46 Process of forecasting by single-well numerical simulation method.

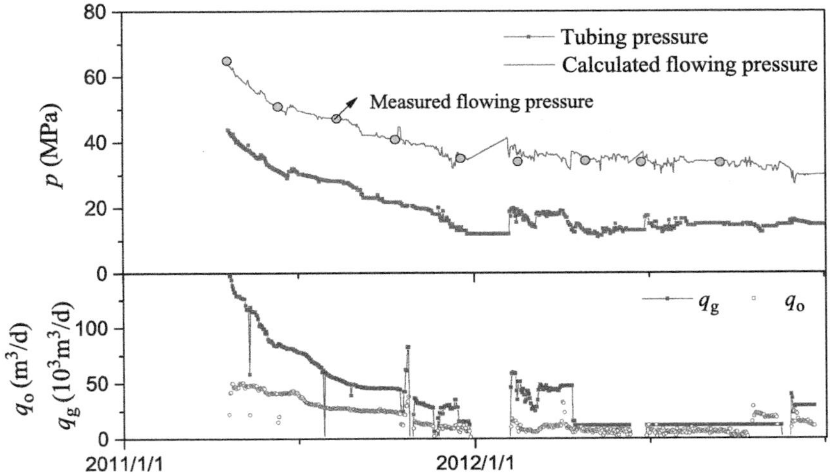

Fig. 5.47 Converted whole pressure history curve of Well G151.

The pressure buildup test was performed on October 2011. The pressure history is converted under the constraint of the measured bottomhole flow pressure gradient, as shown in Fig. 5.47.

First, analytical well testing is used. The log-log matching curve and whole pressure history matching curve are shown in Fig. 5.48. It can be seen that the matching effect is moderate.

The seismic carving model across this well is shown in Fig. 5.49. From a geological perspective, it is believed that there are complicated faults/fractures around this well. A numerical model based on the seismic attributes is used for the two-dimensional and three-dimensional numerical well testing analysis. The plane and three-dimensional initial models are shown in Fig. 5.50.

The log-log curve matching and pressure history matching results of the three-dimensional numerical well testing are shown in Figs. 5.51 and 5.52, respectively, suggesting better results than the analytical well testing method. Fig. 5.53 shows the changes of the pressure field at different times.

According to the forecasting results of natural gas production decline and under the constraint of the abandonment pressure of 17 MPa, the pressure forecasting results are shown in Fig. 5.54. The forecasted recovery percent of natural gas and condensate oil at the end of the period is 63.9% and 39.9%, respectively.

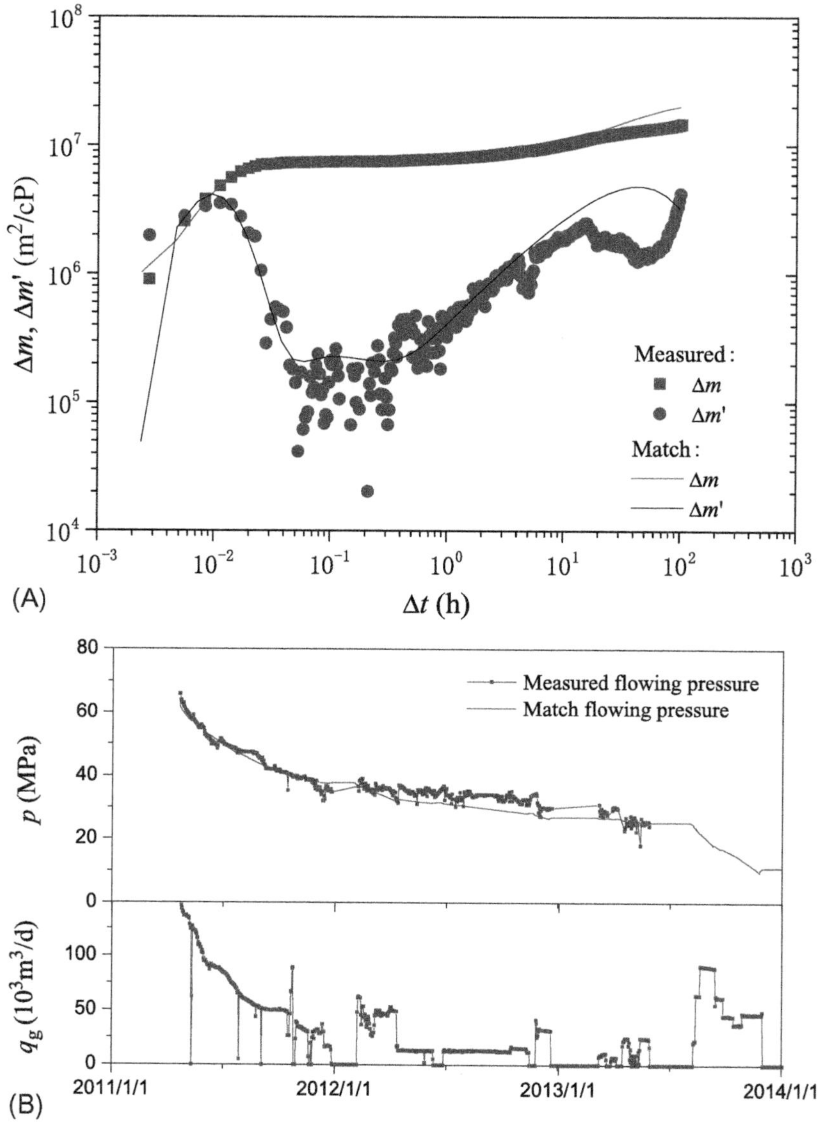

Fig. 5.48 Analytical well testing of Well G151. (A) Log-log matching curve and (B) history matching curve.

5.3 Performance forecasting method of units or connected clusters

In this section, the production performance forecasting methods of development units or connected clusters are introduced with examples. First, the interwell connectivity is analyzed to determine the connected clusters.

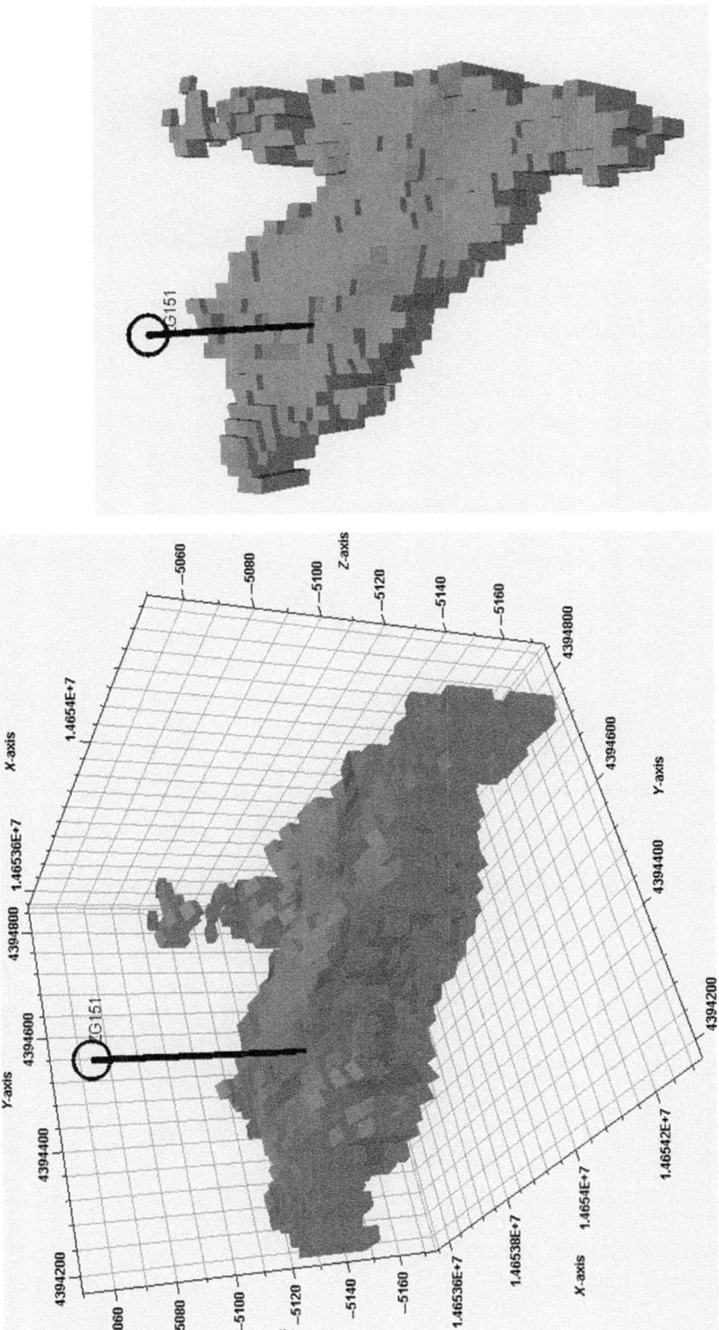

Fig. 5.49 Seismic carving model.

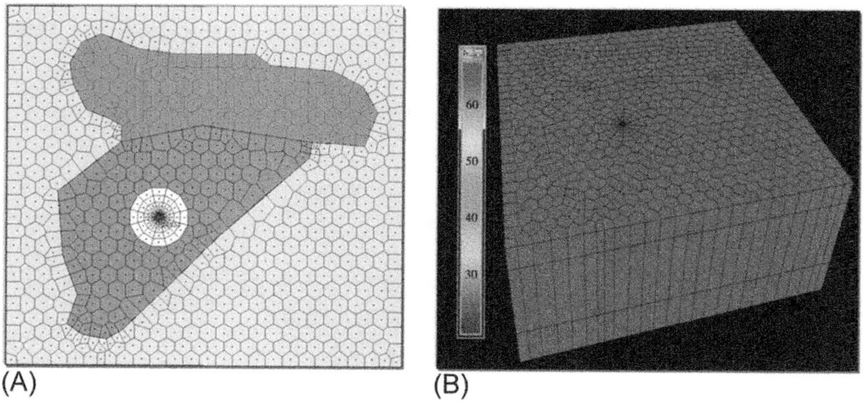

Fig. 5.50 Plane model and three-dimensional initial model. (A) Plane model and (B) three-dimensional model.

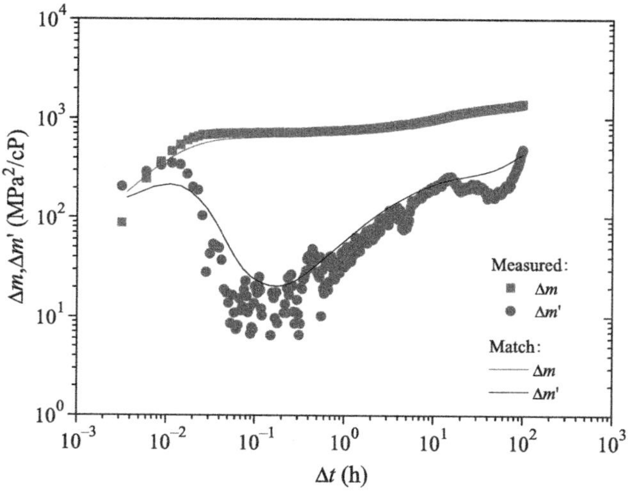

Fig. 5.51 Log-log curve of the Well G151—numerical well testing method.

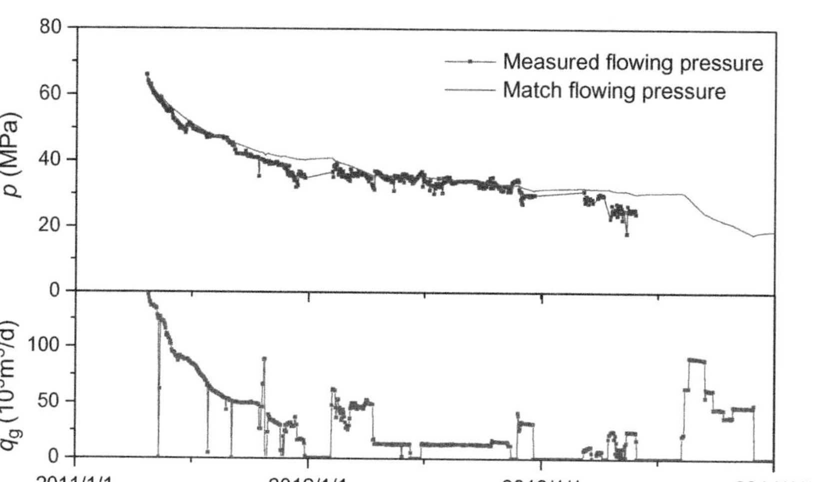

Fig. 5.52 Pressure history matching curve of the Well G151—numerical well testing method.

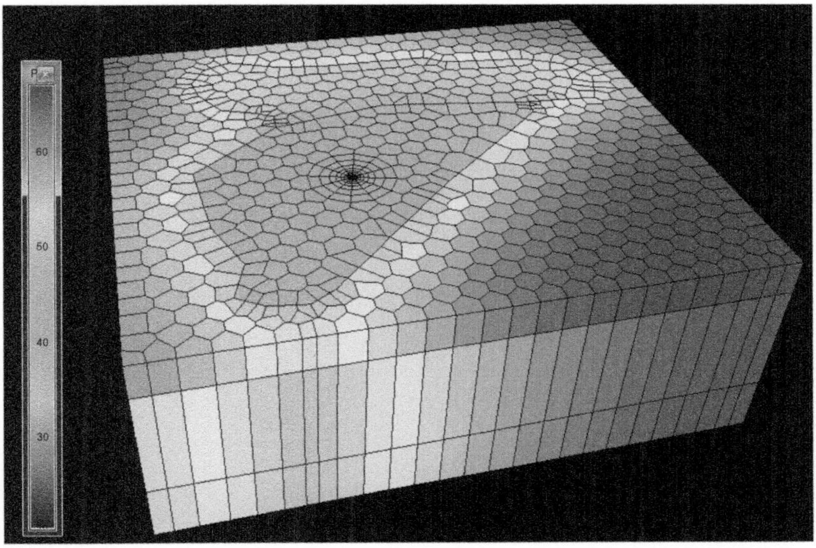

(A)

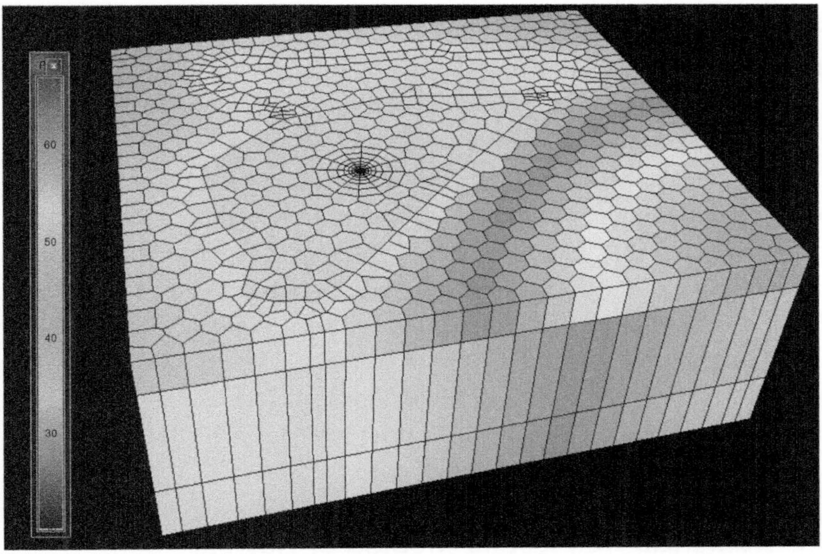

(B)

Fig. 5.53 Changes of the pressure field of Well G151. (A) September 9, 2011 and (B) October 18, 2012.

Then, on the basis of single-well numerical well testing and numerical simulation, the numerical well testing analysis of clusters is conducted with consideration of the interwell interference, and a three-dimensional numerical model is established to conduct the numerical simulation analysis and forecasting.

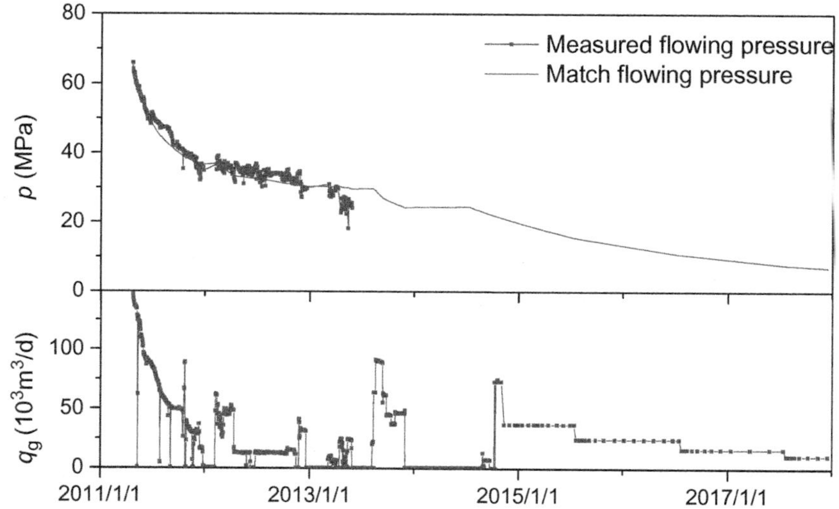

Fig. 5.54 Performance forecasting curve of Well G151—numerical simulation method.

5.3.1 Interwell connectivity analysis

Taking the carbonate reservoir as an example, on the basis of the well dynamic description and the dynamic and static characteristics, all wells in the suspected connected cluster are analyzed by several methods, including geological characteristics analogy, pressure analysis, fluid nature difference analysis, production performance analysis, interwell interference testing, and the well testing boundary control method. In practical application, depending on the availability of data, multiple factors can be combined to judge if the cluster is connected. The interwell connectivity analysis of the Tazhong gas field suggests that poor connectivity in the gas field, where some wells are only connected locally within a small range, causes the number of connected cluster and the number of wells in each cluster to increase gradually as the production time is extended.

5.3.2 Numerical well testing method of clusters

On the basis of the single-well numerical well testing analysis, the interwell interference, and natural gas/condensate oil production decline of the cluster, the key development indices of the cluster can be forecasted.

5.3.2.1 Example #1

For the T201C-G462 cluster (Fig. 5.55), the performance analysis suggests that the cluster is obviously connected. Well T201C was put into production in June 2010; during the initial production, the GOR was 3500 m^3/t,

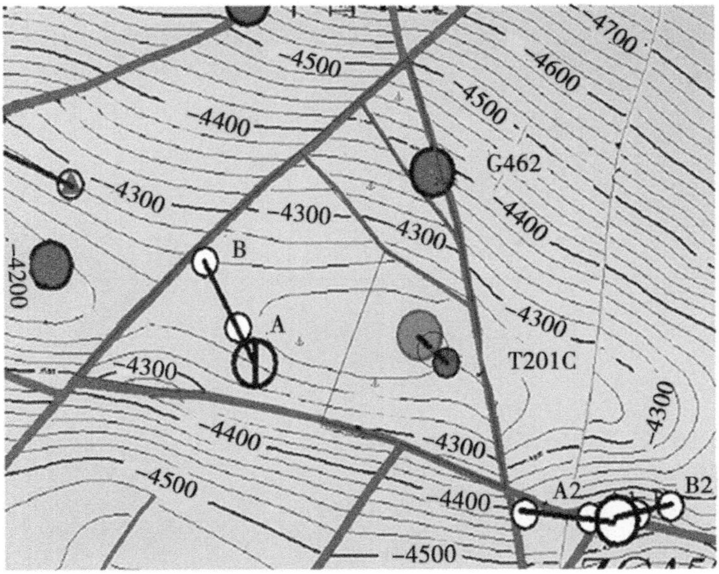

Fig. 5.55 Structural map of the cluster.

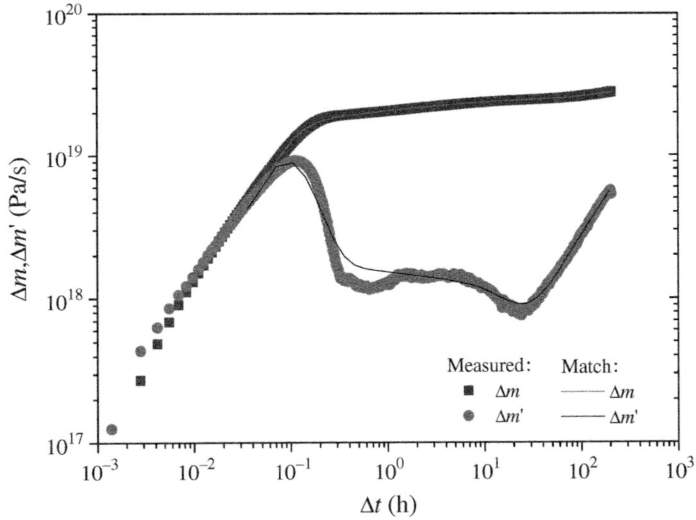

Fig. 5.56 Log-log matching curve of Well T201C.

and no water was produced; and the pressure buildup well test was performed on this well in March 2011. Well G462 was put into production in November 2010; during the initial production, the GOR was $1800\,m^3/t$, and no water was produced. The test data of Well T201C is interpreted by the variable wellbore storage + radial composite model, as shown in Figs. 5.56 and 5.57.

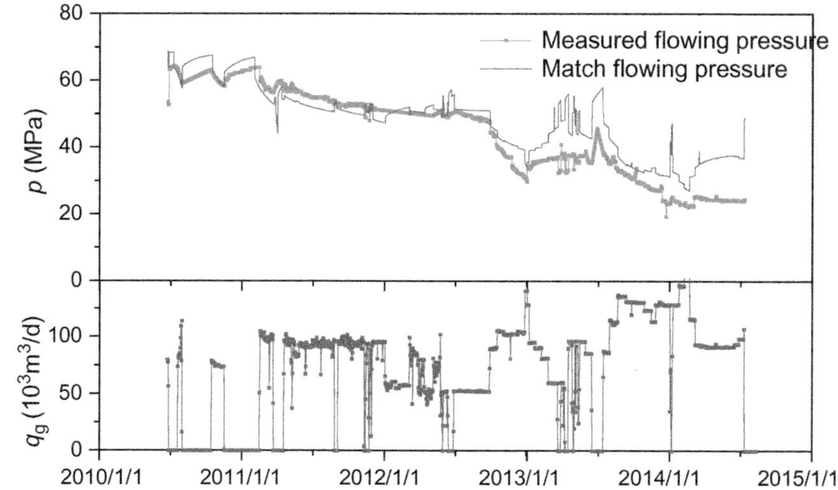

Fig. 5.57 History matching curve of Well T201C.

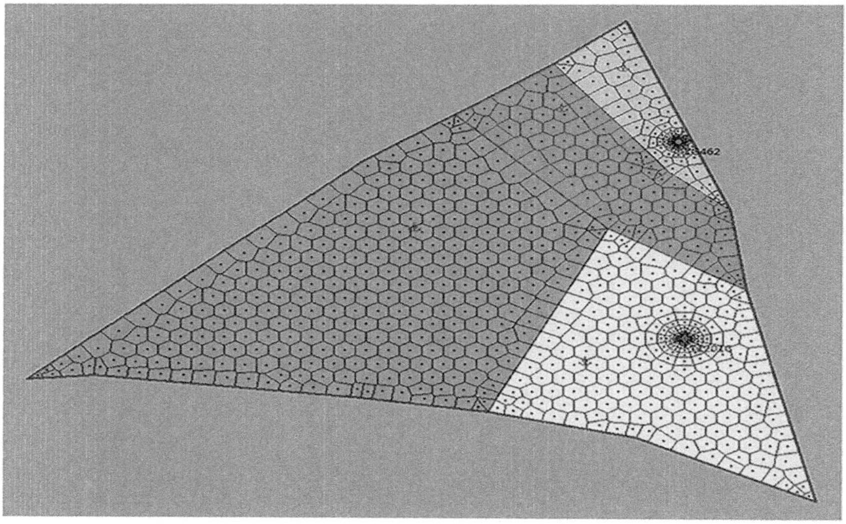

Fig. 5.58 Numerical model of the cluster.

The multiblock numerical well testing model is established according to the plane attributes of seismic, as shown in Fig. 5.58. The log-log curve matching and whole pressure history matching analysis results are shown in Figs. 5.59–5.61. The analysis results suggest that the two wells are connected, and the outer boundary fault plays the role of blocking, and the internal fault has a partially blocking function.

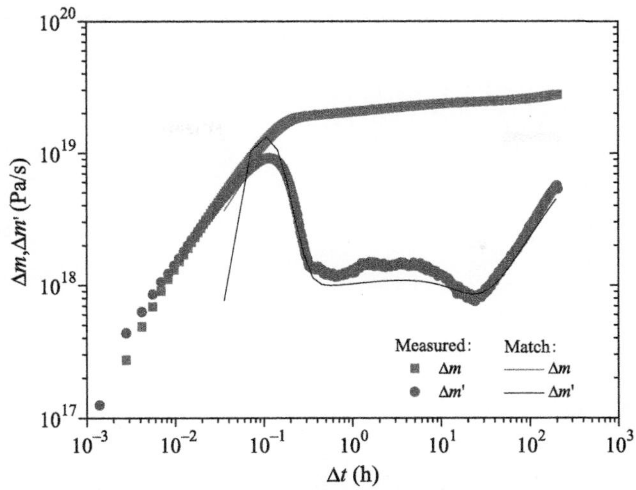

Fig. 5.59 Log-log matching curve of Well T201C—numerical well testing.

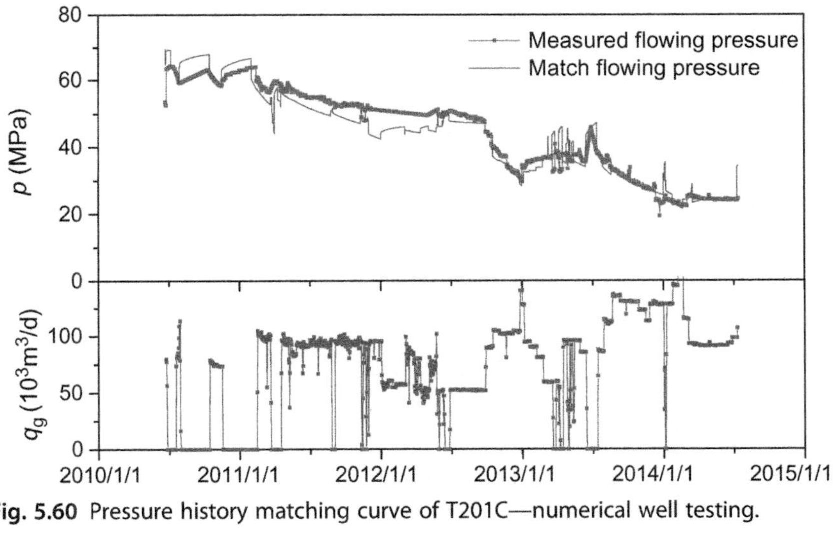

Fig. 5.60 Pressure history matching curve of T201C—numerical well testing.

The reservoir pressure and production forecasting results are shown in Fig. 5.62. The forecasted residual recoverable reserves of natural gas and condensate oil in the cluster at the end of the period are 1.48×10^8 m^3 and 2.64×10^4 t, respectively, and the recovery ratio percent of natural gas and condensate oil are 57.7% and 35.9%, respectively.

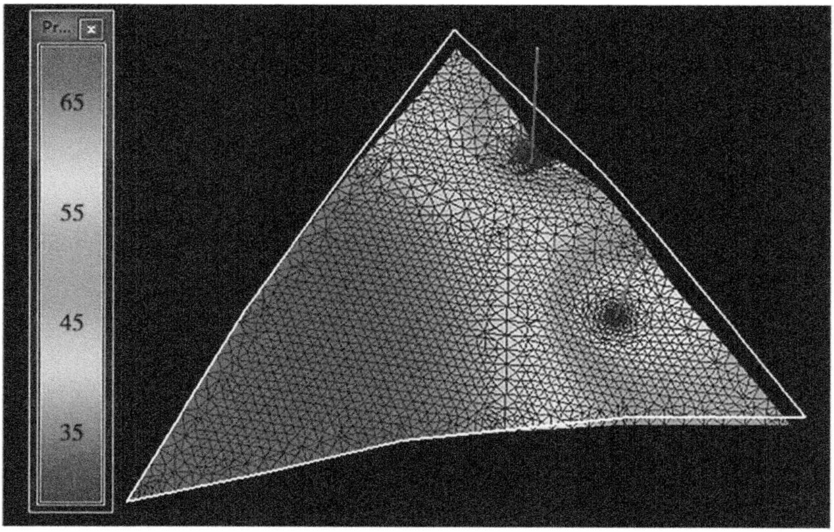

Fig. 5.61 Pressure sweep in the cluster.

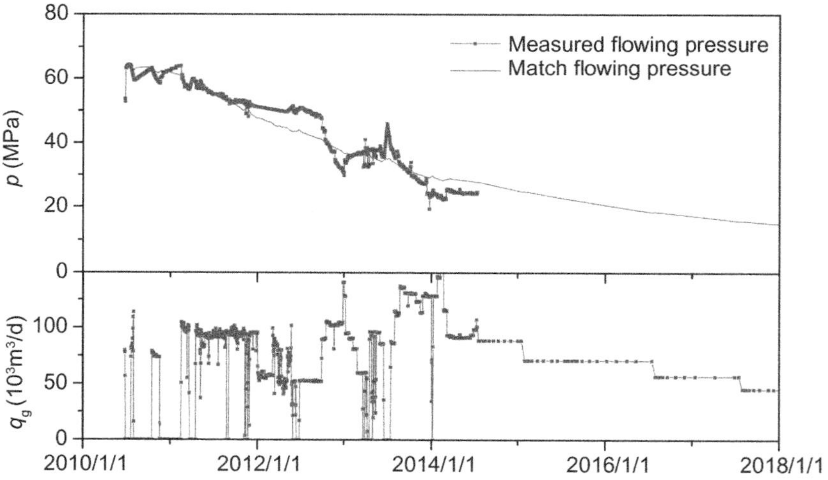

Fig. 5.62 Performance forecasting curve of the G462-T201C cluster.

5.3.2.2 Example #2

Well G15 is a discovery well in Tazhong III, a typical volatile oil reservoir. Wells G15-1H, G15-2, G15-H3, G15-4H, and G15-5H were put into production successively, as shown in Fig. 5.63. Analysis suggests that Wells G15, G15-4H, and G15-5H are connected, while the other three wells are not connected. The pressure buildup well test was performed on Well G15 in

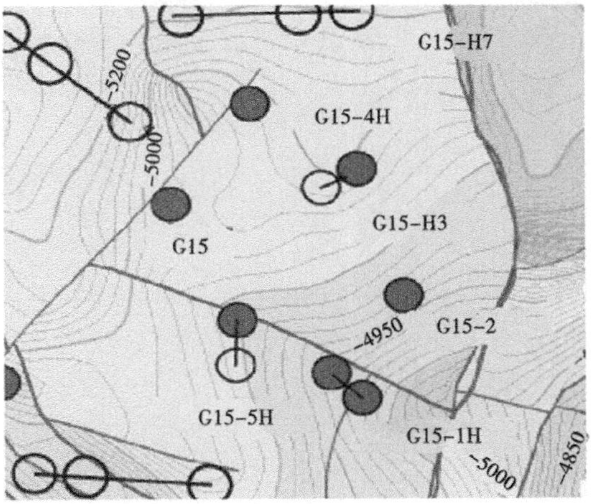

Fig. 5.63 Plane structural map of the G15 cluster.

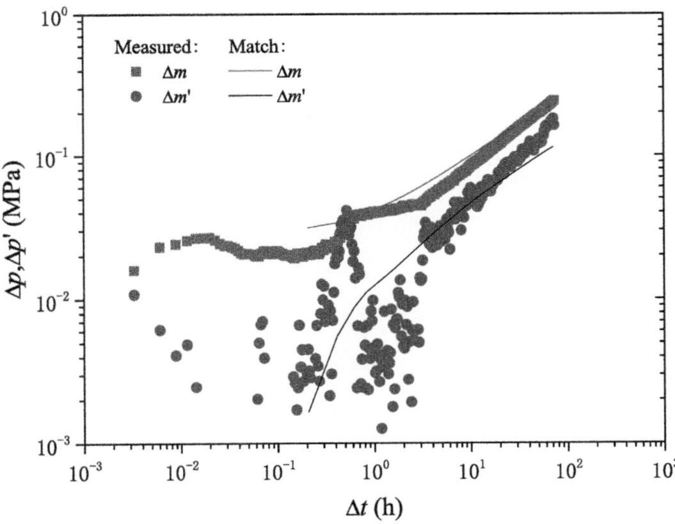

Fig. 5.64 Log-log matching curve of Well G15—analytical well testing.

October 2011. Considering the influence of the adjacent wells G15-4H and G15-5H, the numerical well testing analysis of the cluster is conducted.

The variable wellbore storage + radial composite + closed boundary model is used for the analytical well testing analysis. The log-log and pressure history matching curves are shown in Figs. 5.64 and 5.65. The results are taken as the initial parameters for the numerical well testing analysis.

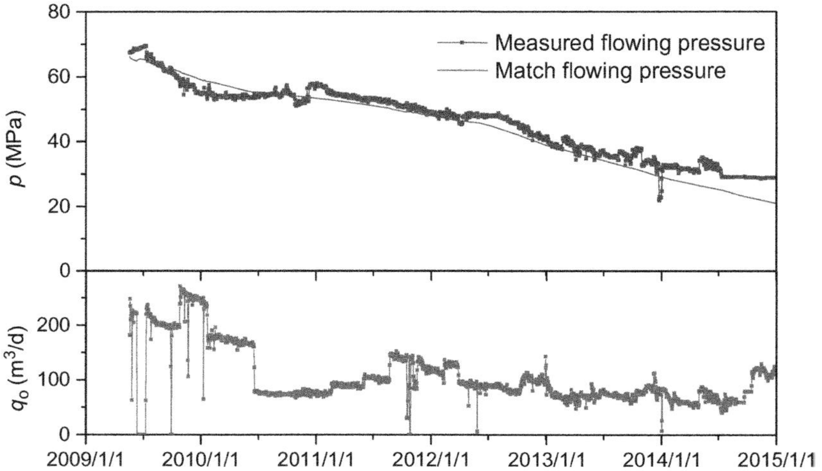

Fig. 5.65 Pressure history matching curve of Well G15—analytical well testing.

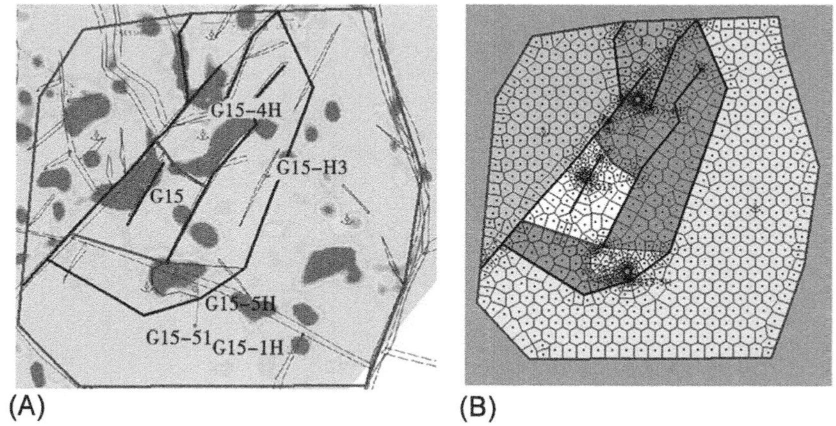

Fig. 5.66 Plane seismic attributes and gridding composite numerical model of the G15 cluster. (A) Plane seismic attributes and (B) plane composite numerical model.

The plane seismic attributes and plane numerical well testing model of the cluster are shown in Fig. 5.66A, and the numerical model of multiple wells and blocks is shown in Fig. 5.66B.

The log-log matching curve and whole pressure history matching curve of numerical well testing analysis of the cluster are shown in Fig. 5.67.

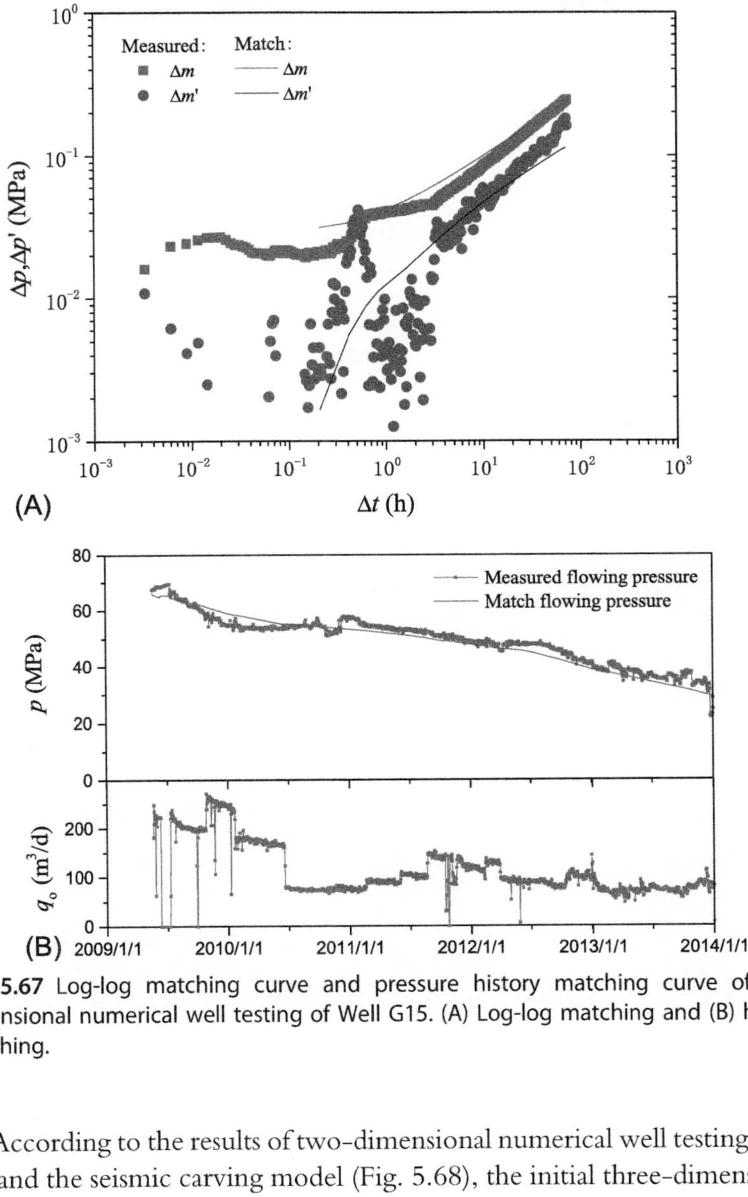

Fig. 5.67 Log-log matching curve and pressure history matching curve of two-dimensional numerical well testing of Well G15. (A) Log-log matching and (B) history matching.

According to the results of two-dimensional numerical well testing analysis and the seismic carving model (Fig. 5.68), the initial three-dimensional numerical model is established (Fig. 5.69). The statistical results of vertical porosity of the seismic carving model are shown in Fig. 5.70.

The pressure decline trend of the G15 cluster forecasted by the three-dimensional numerical well testing model is shown in Fig. 5.71. Under the constraint of the quit flow pressure, the forecasted cumulative

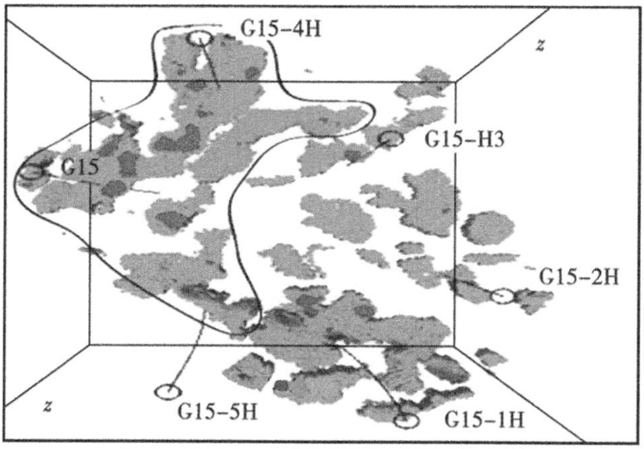

Fig. 5.68 Seismic carving model of the G15 cluster.

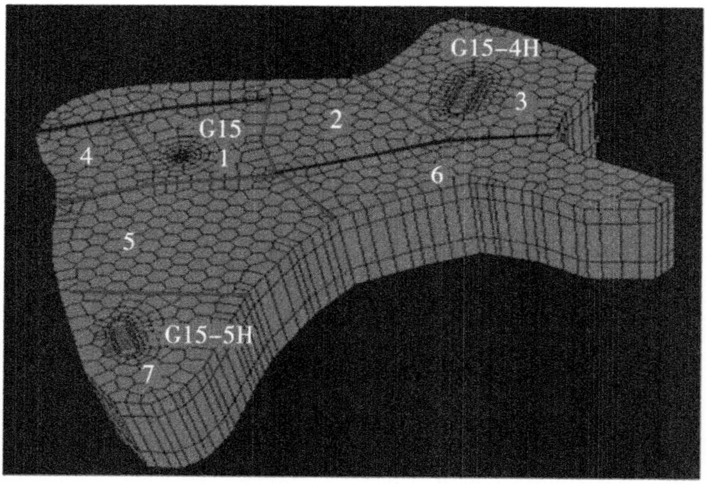

Fig. 5.69 Initial three-dimensional model of the G15 cluster (seven blocks and three layers).

recoverable reserves and recovery percent at the end of the period are 32.97×10^4 t and 8.2%, respectively.

5.3.3 Numerical simulation method of clusters

Some examples are taken to illustrate the forecasting of the cluster development indices based on the three-dimensional numerical well testing analysis and the production decline trend.

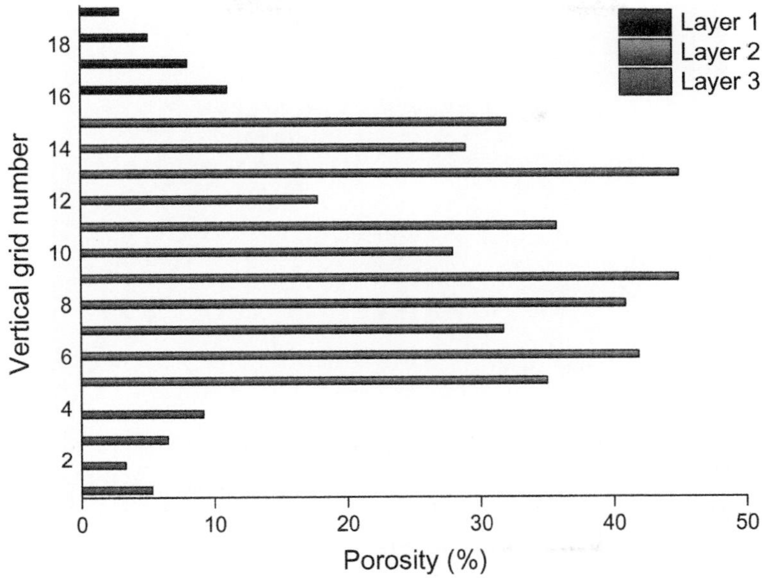

Fig. 5.70 Statistics of vertical porosity of seismic carving model.

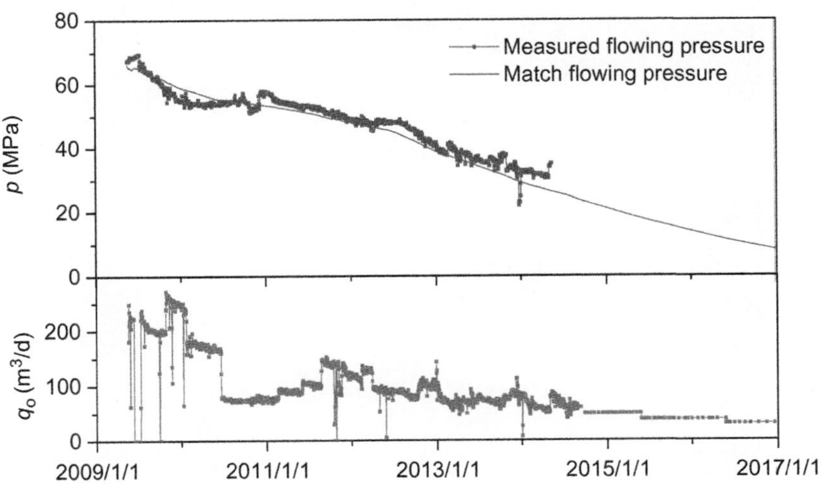

Fig. 5.71 The pressure forecasting curve of the G15, G15-4H, and G15-5H cluster—numerical well testing method.

5.3.3.1 Example #1

As introduced previously, the numerical well testing analysis is conducted for the G462-T201C cluster and the plane numerical model is established,

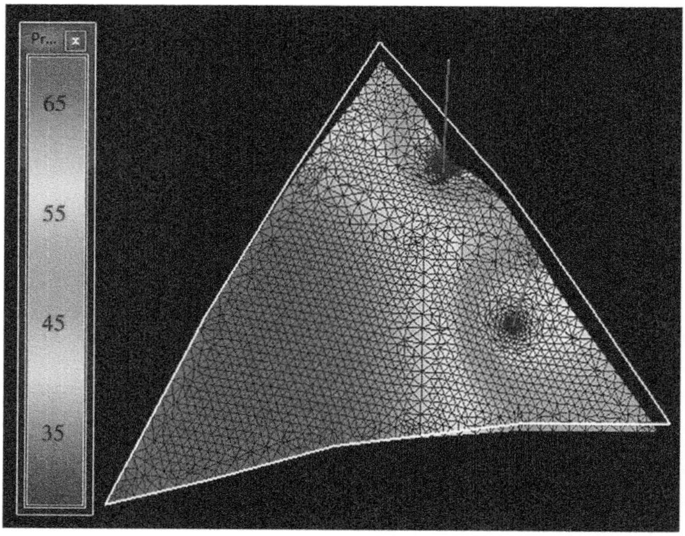

Fig. 5.72 Plane numerical model of the G462-T201C cluster.

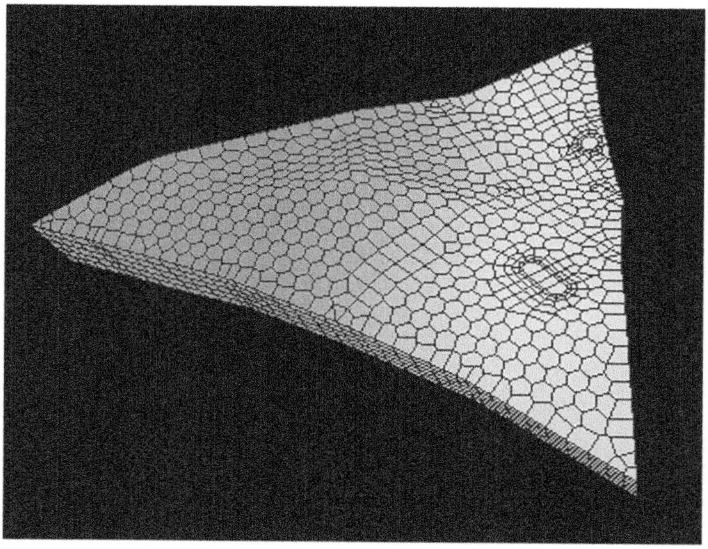

Fig. 5.73 Three-dimensional numerical model of the G462-T201C cluster.

as shown in Fig. 5.72. The established three-dimensional numerical model is shown in Fig. 5.73. This model is used for numerical simulation analysis, and the history matching curve of the cluster is shown in Fig. 5.74.

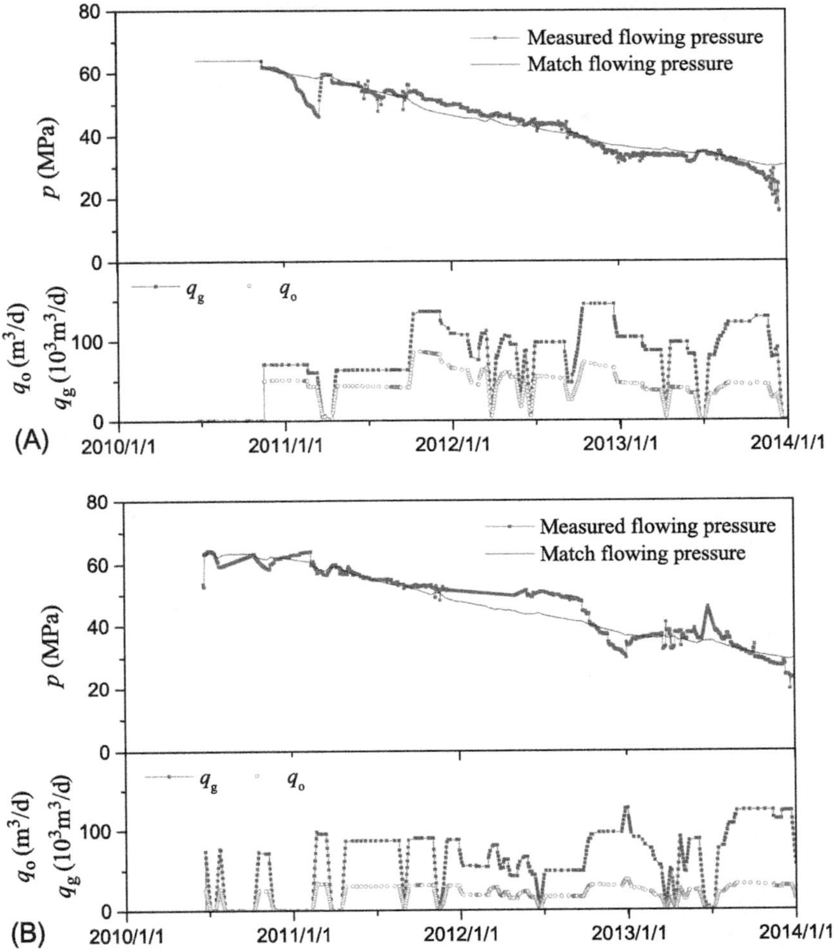

Fig. 5.74 History matching curve of the G462-T201C cluster. (A) Well G462 and (B) Well T201C.

Under the constraint of the abandonment pressure of 18 MPa, and according to the natural gas production decline forecasting results, the trends of condensate oil production and reservoir pressure change are shown in Fig. 5.75. The forecasted recoverable reserves of natural gas and condensate oil in the cluster at the end of the period are 3.33×10^8 m^3 and 9.06×10^4 t, respectively, and the recovery percent of natural gas and condensate oil are 56.2% and 33.8%, respectively, as shown in Fig. 5.76.

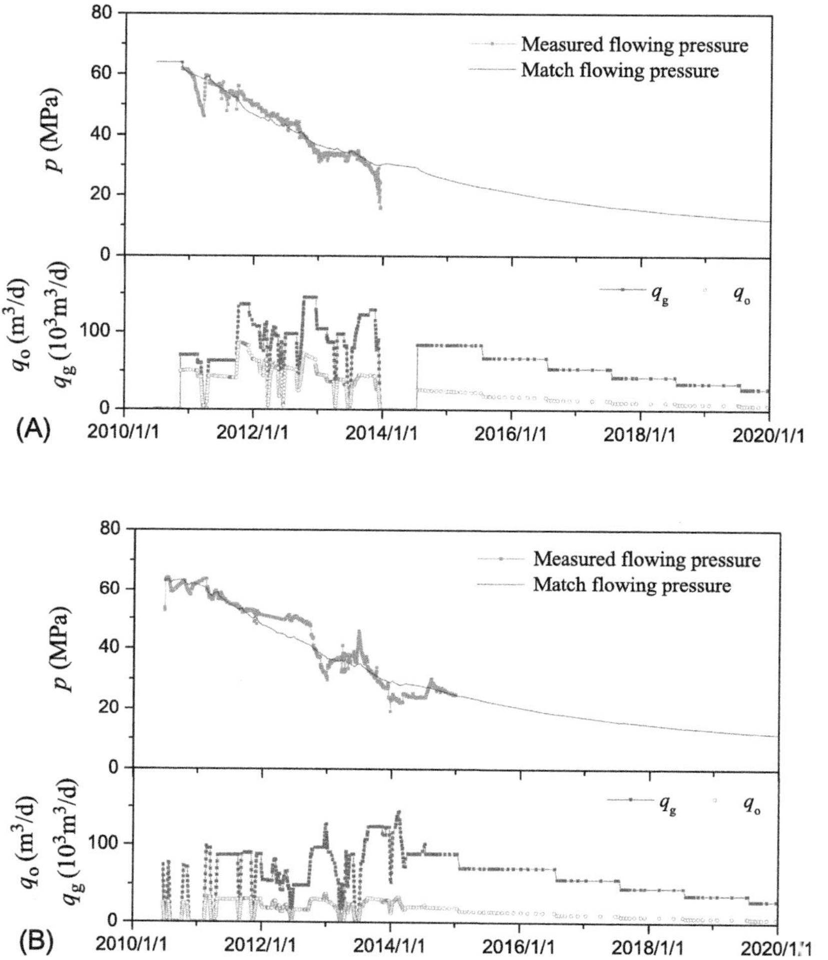

Fig. 5.75 Production forecasting curve of the G462-T201C cluster. (A) Well G462 and (B) Well T201C.

5.3.3.2 Example #2

The three-dimensional numerical model of the G15-G15-4H-G15-5H cluster is shown in Fig. 5.77, revealing an efficient history matching effect (Fig. 5.78). Therefore, this model is relatively reliable.

This model is used for numerical simulation analysis, and the key development indices of the cluster are forecasted according to the production decline and abandonment conditions. Under the constraint of the quit flow

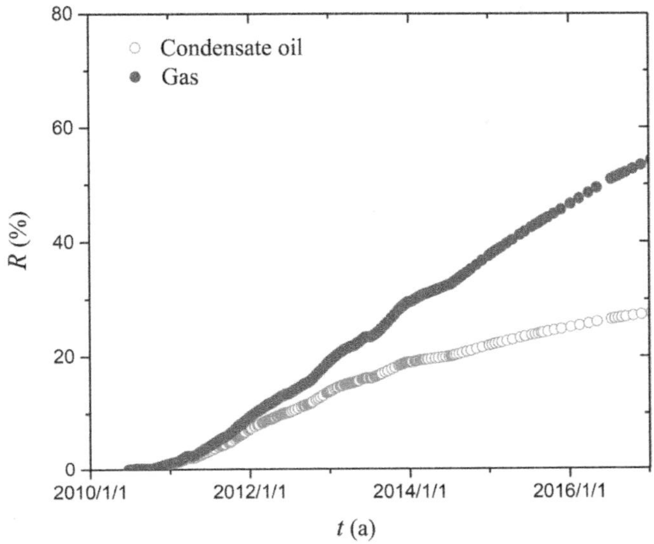

Fig. 5.76 Forecasting curve of the recovery percent of the G462-T201C cluster.

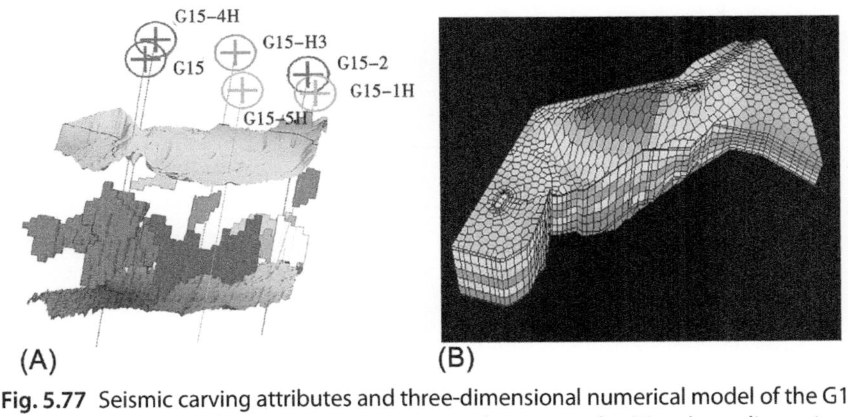

Fig. 5.77 Seismic carving attributes and three-dimensional numerical model of the G15 cluster. (A) Vertical seismic carving attributes and (B) three-dimensional numerical model.

pressure of 20 MPa, the forecasting results of production and reservoir pressure are shown in Fig. 5.79. The forecasted recoverable reserves and residual recoverable reserves of crude oil at the end of the period are 33.83×10^4 and 5.35×10^4 t, respectively, and the recovery percent is 10.43%, as shown in Fig. 5.80.

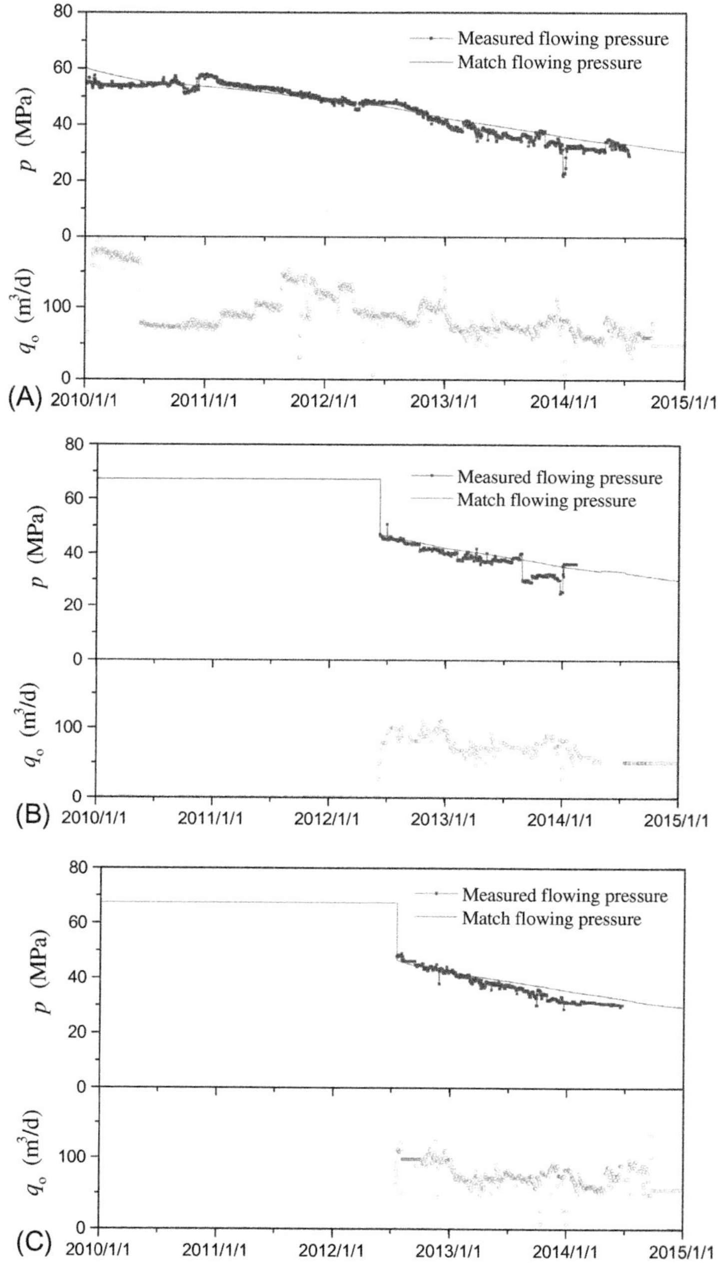

Fig. 5.78 History matching curve of the G15 cluster. (A) Well G15, (B) Well G15-4H, and (C) Well G15-5H.

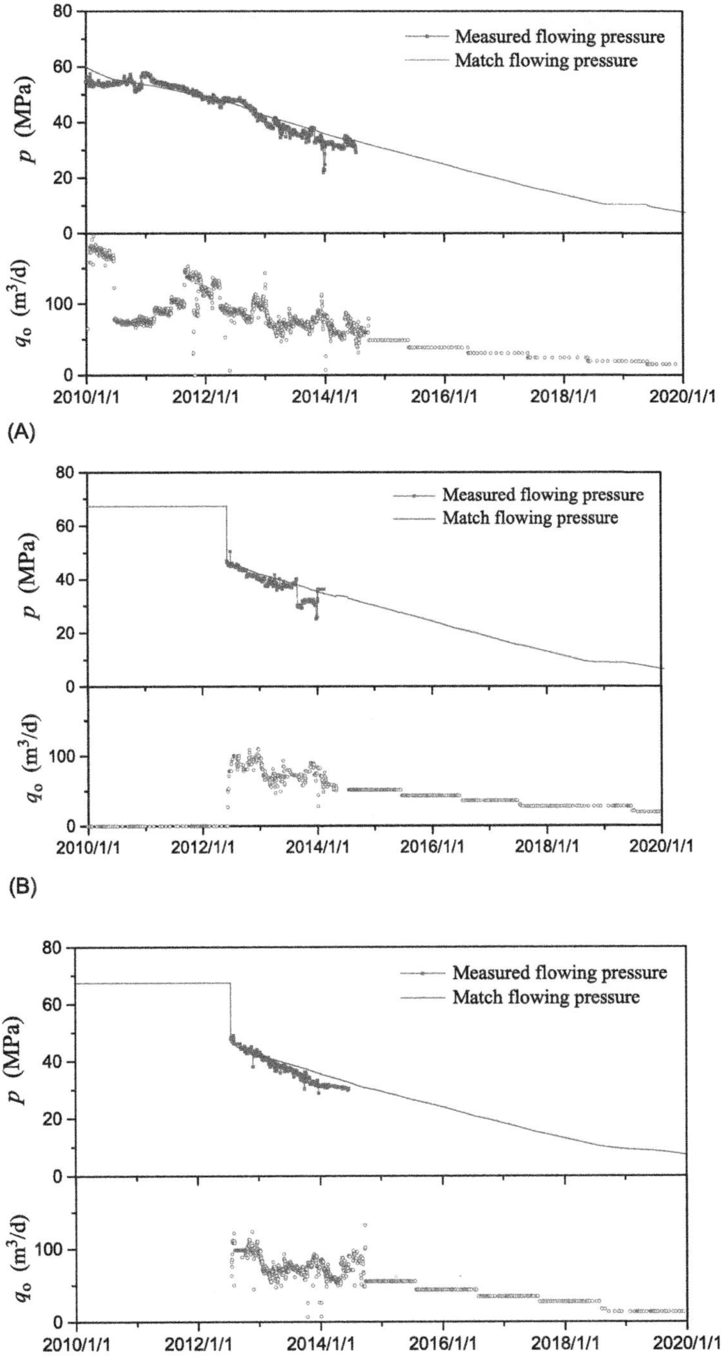

Fig. 5.79 Performance forecasting curve of the G15 cluster. (A) Well G15, (B) Well G15-4H, and (C) Well G15-5H.

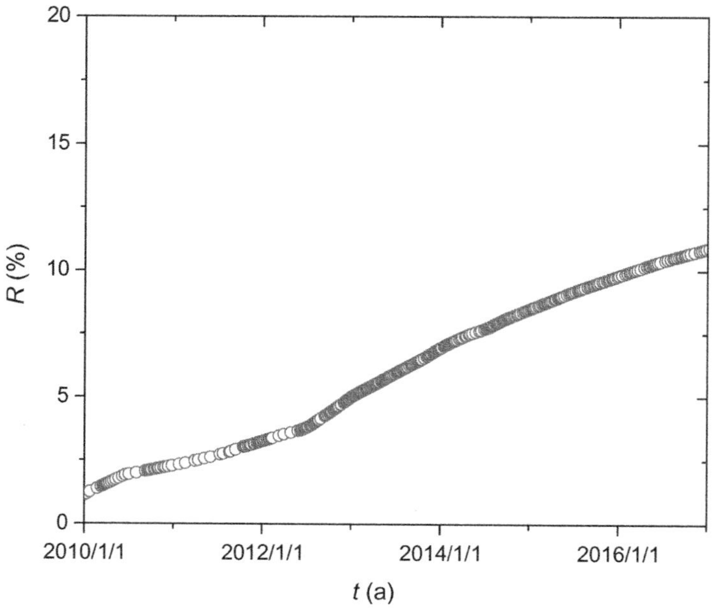

Fig. 5.80 Forecasting curve of the recovery percent of the G15 cluster.

5.4 Summary

Due to the complex properties of the fractured vuggy carbonate reservoirs and limited seismic data quality, only the gas reservoir engineering method can be used to forecast performance. In this chapter, three methods are proposed to forecast the indices of a well or unit, including the gas reservoir engineering method, based on the material balance theory and PVT test data, the numerical well testing method, based on well testing data and geological cognition; and the advanced production decline analysis method, based on daily production data.

When the material balance equation and the turbulent flow deliverability equation are combined for production performance forecasting, due to the physical properties of reservoirs and the well-controlled geometry, it is hard to reach a pseudo-steady state during the gas well deliverability test; thus, the forecasting result may be biased.

For gas wells in fractured vuggy carbonate condensate reservoirs, the initial productivity is high, and declines fast. Therefore, it is necessary to weaken the function of absolute open flow potential in the productivity evaluation and highlight the function of single-well dynamic reserves. On the basis of the dynamic reserves estimation and the decline rate evaluation, the two-phase material balance equation of oil and gas and the PVT test data should be combined to forecast the key development indices.

The reservoir dynamical model based on the numerical well testing method, or the advanced production decline analysis method, can be used to effectively forecast the key development indices, such as oil and gas production and reservoir pressure. Moreover, it can be extended to development units or connected clusters.

NOMENCLATURE

a_s, b_s, c_s	regression coefficients of different space variable conductivity relations, dimensionless; Eq. (4.54)
A	drainage area of well (m^2)
A	coefficient of laminar flow term in LIT deliverability equation ($MPa^2/(10^4\ m^3/d)$)
b	Fetkovich/Arps decline curve exponent (dimensionless)
b	gas slippage factor (MPa); Eq. (4.59)
B	formation volume factor (m^3/m^3)
B	water invasion coefficient ($10^4\ m^3/MPa$); Eq. (3.5)
B	coefficient of turbulent flow term in LIT deliverability equation ($MPa^2/(10^4\ m^3/d)^2$)
B_g, B_o, B_w	formation volume factors of gas, oil, and water, respectively (m^3/m^3)
B_{gi}	formation volume factor of gas evaluated at p_i (m^3/m^3)
B_{oi}	formation volume factor of oil evaluated at p_i (m^3/m^3)
B_{2g}	formation volume factors of gas-liquid two phase at pressure, p (m^3/m^3)
Const	constant
C	wellbore storage constant (m^3/MPa)
C	time coefficients of time variable conductivity (a); Eq. (4.57)
C_A	Dietz shape factor for single well drainage area (dimensionless)
c_e	effective compressibility of rock and fluid, $c_e = \dfrac{c_w S_{wi} + c_f}{1 - S_{wi}}$ (MPa^{-1})
c_g	gas compressibility (MPa^{-1})
c_o	oil compressibility (MPa^{-1})
c_f	effective rock compressibility (MPa^{-1})
c_w	water compressibility (MPa^{-1})
c_t	total compressibility (MPa^{-1})
c_{ti}	total compressibility evaluated at p_i (MPa^{-1})
C_D	dimensionless wellbore storage coefficient
$C_D e^{2S}$	shape parameter group of log-log pressure and pressure derivative curves
D	non-Darcy flow coefficient ($(10^4\ m^3/d)^{-1}$ or $(m^3/d)^{-1}$)
D_i	initial decline rate, d^{-1}
D	continuous decline rate, Figs. 3.62–3.64
e	=2.71828
E_i	exponential integration function
E_{go}	gas/oil ratio (m^3/m^3 or m^3/t)
$\overline{E}_{og}$	equivalent gas condensate coefficient (dimensionless)
EUR	cumulative production ($10^3\ m^3$)
f_w	correction coefficient of water content, dimensionless; Eq. (2.68)
f_w	water content (%)
f_{match}	parameter group of curve match, dimensionless; Eq. (4.40)
F_c	fracture conductivity varies across space and time (mD.m)
F_{ci}	initial fracture conductivity (mD.m)

F_{ci0}	initial fracture conductivity at the gap (mD.m)
F_{cl}	fracture conductivity varies across space (mD.m)
F_{c0}	fracture conductivity at the gap (mD.m)
F_{cr}	residual fracture conductivity (mD.m)
F_{ct}	fracture conductivity varies across time (mD.m)
F_{CD}	conductivity of fracture, $F_{CD}=K_{fD}\cdot W_{fD}$
g	gravity acceleration (m/s^2)
G	original gas in place (OGIP) (10^8 m^3)
G_e	gas equivalent of oil (m^3/m^3)
GOR	gas/oil ratio of gas condensate well (m^3/m^3)
G_p	cumulative gas production (10^8 m^3)
G_{pt}	cumulative gas condensate production (10^8 m^3)
G_{pseudo}	pseudo original gas in place (10^8 m^3)
G_{real}	real original gas in place (10^8 m^3)
G_t	original gas condensate in place (10^8 m^3)
h	formation thickness (m)
$I_0(x)$	modified Bessel function of the first kind (zero order)
$I_1(x)$	modified Bessel function of the first kind (first order)
J_o, J_w, J_L	oil, water, and liquid productivity index, respectively (m^3/(MPa.d))
$K_0(x)$	modified Bessel function of the second kind (zero order)
$K_1(x)$	modified Bessel function of the second kind (first order)
K	formation permeability (mD)
K_a	absolute permeability (mD)
K_f	fracture permeability (mD)
K_{fD}	dimensionless fracture permeability, $K_{fD}=K_f/K$
K_g	apparent permeability (mD)
K_H	horizontal formation permeability (mD)
K_o	effective permeability to oil (mD)
K_r	radial formation permeability (mD)
K_{rg}	relative permeability to gas (fraction)
K_{ri}	initial reservoir permeability (mD)
K_{ro}	relative permeability to oil (mD)
K_V	vertical formation permeability (mD)
K_∞	absolute permeability (mD)
Kh	formation flow capacity (mD.m)
Kh/μ	formation flow coefficient ((mD.m)/(mPa.s))
lg	common logarithm
ln	natural logarithm
L_D	dimensionless distance from the gap, $L_D=L/x_f$
m	real gas pseudo-pressure (MPa^2/mPa.s)
Δm	real gas pseudo-pressure difference (MPa^2/mPa.s)
$\Delta m'$	derivative of real gas pseudo-pressure (MPa^2/mPa.s)
$\frac{\partial m}{\partial n}$	derivative along the exterior normal direction
M	mobility ratio (dimensionless)
M_o	mean molecular weight of condensate oil (dimensionless)
N	original oil in place (10^4 m^3 or 10^4 t)

N_p	cumulative oil production (10^4 m^3 or 10^4 t)
N_p	normalized cumulative production, $N_p=\int_0^t \frac{q}{p_i-p_{wf}}d\tau$ (m^3/MPa)
N_{pDd}	dimensionless cumulative production, $N_{pDd}=\int_0^{t_{Dd}} q_{Dd}(\tau)d\tau$
N_o	cumulative condensate oil production (10^4 m^3)
p	pressure (MPa)
$\overline{p}$	average pressure (MPa)
p_0	reference pressure (MPa)
p_c	critical pressure of gas (MPa)
p_{cas}	casing pressure (MPa)
p_{cr}	critical pressure of liquid phase to flow (MPa)
p_{dew}	dewpoint pressure (MPa)
p_D	dimensionless pressure
p_D'	derivative of dimensionless pressure p_D with respect to t_D/C_D
$(p_D)_M$	matched value of dimensionless pressure, i.e., the value of dimensionless pressure at the match point on type curve
p_e	pressure at external boundary (MPa)
p_i	initial pressure (MPa)
p_m	pressure (MPa)
p_{pc}	pseudo critical pressure of gas (MPa)
p_r	reduced pressure of gas (dimensionless)
p_{pr}	pseudo reduced pressure of gas (dimensionless)
p_{sc}	standard pressure condition, $p_{sc}=0.101325$ MPa
p_t	tubing pressure (MPa)
p_w	bottomhole pressure (MPa)
p_{wf}	flowing bottomhole pressure (MPa)
p_{ws}	bottom hole shut in pressure (MPa)
p_{Di}	dimensionless pressure integral, $p_{Di}=\frac{1}{t_{DA}}\int_0^{t_{DA}} p_D dt_{DA}$
p_{Did}	dimensionless pressure integral derivative, $p_{Did}=\frac{dp_{Di}}{d\ \ln t_{DA}}=t_{DA}\frac{dp_{Di}}{dt_{DA}}$
Δp	pressure difference (MPa)
$\Delta p'$	pressure derivative (MPa)
Δp_{skin}	additional pressure drop across altered zone (MPa)
$\Delta p_{ws}(\Delta t)$	bottomhole shut in pressure at time Δt (MPa)
q	daily flow rate (m^3/d or t/d)
q_{ab}	abandonment rate (m^3/d)
q_{AOF}	absolute open flow potential (AOFP) of a gas well (10^4 m^3/d)
q_f	forecast abandonment rate (m^3/d)
q_f	standard volume of fluid entering the fractured from unit volume reservoir in unit time, $1/d$; Eq. (4.45)
q_g	gas flow rate (10^4 m^3/d)
q_{go}	gaseous hydrocarbon flow rate (10^4 m^3/d)
q_h	well flow rate (m^3/d)
q_i	initial flow rate (m^3/d)
q_L	liquid flow rate (m^3/d)
q_n, q_N	flow rate of nth, Nth time quantum (m^3/d)
q_o	oil flow rate (m^3/d or t/d)
q_{og}	equivalent gas condensate flow rate (10^4 m^3/d)

q_{sc} gas production under standard condition (10^4 m^3/d)
q_t equivalent gas condensate flow rate (10^4 m^3/d)
q_{tgo} equivalent oil and gas mass rate, $q_{tgo} = q_g \rho_{gsc} + q_o \rho_{osc}$ (kg/d)
q_{tgow} equivalent gas flow rate (10^4 m^3/d)
q_w water flow rate (m^3/d)
q_{Dd} Fetkovich dimensionless decline flow rate function, $q_{Dd} = \frac{q}{q_i} = q_D\left(\ln r_{eD} - \frac{1}{2}\right)$
q_{Dd} Blasingame dimensionless decline flow rate function for oil well, $q_{Dd} = \frac{q}{p_i - p_{wf}} b_{pss}$
q_{Dd} Blasingame dimensionless decline flow rate function for gas well, $q_{Dd} = \frac{b_{a,pss}}{\Delta p_p}$
q_D Agarwal-Gardner dimensionless flow rate, $q_D = \frac{1}{p_D} = \frac{q\mu B}{2\pi Kh(p_i - p_{wf})}$
q_{Ddi} dimensionless decline flow rate integral function, $q_{Ddi} = \frac{N_{pDd}}{t_{Dd}} = \frac{1}{t_{Dd}} \int_0^{t_{Dd}} q_{Dd}(\tau) d\tau$
q_{Ddid} dimensionless decline flow rate integral derivative function, $q_{Ddid} = -t_{Dd} \frac{d}{dt_{Dd}} \left[\frac{1}{t_{Dd}} \int_0^{t_{Dd}} q_{Dd}(\tau) d\tau \right]$
Q cumulative gas production (10^3 m^3)
Q dimensionless water influx
Q_f forecast cumulative production (10^3 m^3)
r radial distance from a well (m)
r_d drainage radius (m)
r_D dimensionless radius, $r_D = r/r_w$
r_e external drainage radius (m)
r_i radius of investigation (m)
r_j distance between tested well and adjacent well (m)
r_M inner radius of oil bank in radial composite reservoir (m)
r_w wellbore radius (m)
r_{wa} apparent wellbore radius (m)
R universal gas law constant ((MPa.m^3)/(kmol.K))
R_s gas/oil ratio (m^3/m^3)
S skin factor (dimensionless)
S_a apparent skin factor (dimensionless)
S_f fracture skin factor of a fractured well (dimensionless)
S_g, S_o, S_w gas, oil and water saturations, respectively (%)
S_{gi}, S_{oi}, S_{wi} initial gas, oil and water saturations, respectively (fraction)
S_p permeability stress sensitivity coefficient (dimensionless)
t time (h)
t_a pseudo-time (h)
t_{ab} Abandonment time (h)
t_c material balance pseudo-time for oil well, $t_c = N_p/q$ (d)
t_{ca} material balance pseudo-time for gas well, $t_{ca} = \frac{(\mu c_t)_i}{q} \int_0^t \frac{q}{\mu c_t} dt$ (d)
t_{cD} dimensionless material balance pseudo-time for oil well, $t_{cD} = \frac{K}{\phi \mu c_t A} t_c$
t_{cDd} Blasingame dimensionless material balance pseudo-time for oil well, $t_{cDd} = \frac{m}{b_{pss}} t_c = t_{cD} \frac{1}{\frac{1}{2}\left(r_{eD}^2 - 1\right)\left(\ln r_{eD} - \frac{1}{2}\right)}$
t_{caDd} Blasingame dimensionless material balance pseudo-time for gas well, $t_{caDd} = \frac{m_a}{b_{a,pss}} t_{ca}$

.. Fetkovich decline curve dimensionless time, $t_{Dd} = t_D \cdot \frac{1}{\frac{1}{2}\left(r_{eD}^2 - 1\right)\left(\ln r_{eD} - \frac{1}{2}\right)}$

t_D dimensionless time within term of wellbore radius, $t_D = \frac{3.6 \times 10^{-3} Kt}{\mu \phi c_t r_w^2}$

t_{DA} dimensionless time with drainage, $t_{DA} = \frac{3.6 \times 10^{-3} Kt}{\mu \phi c_t A}$

t_{De} dimensionless time with r_e, $t_{De} = \frac{3.6 \times 10^{-3} Kt}{\mu \phi c_t r_e^2}$

t_{Dxf} dimensionless time with x_f, $t_{Dx_f} = \frac{3.6 \times 10^{-3} Kt}{\mu \phi c_t x_f^2}$

$(t_D)_{pss}$ dimensionless time when pseudo-steady state flow begins

t_f time at start of forecast date (h)

t_M match time point of field data graph (h)

t_{ps}, t_{pss} time when pseudo-steady state flow begins (h)

t_{ss} time when steady state flow begins (h)

t_{pD} dimensionless production time

Δt_D dimensionless shut-in time

T temperature (K)

T_c critical temperature (K)

T_f temperature of reservoir (K)

T_r reduced temperature (dimensionless)

T_{pr} pseudo reduced temperature (dimensionless)

T_{pc} pseudo critical temperature (K)

T_{sc} standard temperature (K), 293.15 K

V volume (m^3)

V_f fissures volume per unit reservoir volume (dimensionless)

V_m matrix volume per unit reservoir volume (dimensionless)

V_p reservoir pore volume (m^3)

V_u volume per unit length of wellbore (m^3/m)

V_w wellbore volume (m^3)

W width of fracture (m)

W_e cumulative water encroachment (10^4 m^3)

W_{fD} dimensionless width of fracture, $W_{fD} = W/x_f$

W_g mass flow rate of gaseous hydrocarbon (kg/d)

W_p cumulative water production (10^4 m^3)

W_w mass flow rate of water (kg/d)

x coordinate of reservoir control equation

x_f half fracture length (m)

y coordinate of reservoir control equation

z Laplace variable (dimensionless)

Z real gas deviation factor (dimensionless)

$\overline{Z}$ real gas deviation factor at average test pressure (dimensionless)

Z_i real gas deviation factor at original pressure (dimensionless)

Z_{ws} real gas deviation factor at p_{ws} and T_R (dimensionless)

$\frac{1}{DER}$ inverse normalized pressure derivative for oil well, $\frac{1}{DER}=\frac{1}{t_c\frac{\partial\left(\frac{\Delta p}{q}\right)}{\partial t_c}}$ (m^3/d/MPa)

$\frac{1}{DER}$ inverse normalized pressure derivative for gas well, $\frac{1}{DER}=\frac{1}{\frac{\partial\left(\frac{\Delta p_p}{q}\right)}{\partial \ln t_{ca}}}=\frac{1}{t_{ca}\frac{\partial\left(\frac{\Delta p_p}{q}\right)}{\partial t_{ca}}}$ (m^3/d/MPa)

$\frac{1}{DERI}$ inverse normalized pressure integral derivative for oil well, $\frac{1}{DERI}=\frac{1}{t_c\frac{d\left(\frac{\Delta p}{q}\right)_i}{dt_c}}$ (m^3/d/MPa)

$\frac{1}{DERI}$ inverse normalized pressure integral derivative for gas well, $\frac{1}{DERI}=\frac{1}{t_{ca}\frac{d\left(\frac{\Delta p_p}{q}\right)_i}{dt_{ca}}}$ (m^3/d/MPa)

$\frac{1}{p_{di}}$ inverse normalized pressure integral for oil well, $\frac{1}{p_{di}}=\frac{1}{\left(\frac{\Delta p}{q}\right)_i}$ (m^3/d/MPa)

$\frac{1}{p_{did}}$ inverse normalized pressure integral derivative for oil well, $\frac{1}{p_{did}}=\frac{1}{\left(\frac{\Delta p}{q}\right)_{id}}$ (m^3/d/MPa)

$\frac{1}{p_{di}}$ inverse normalized pressure integral for gas well, $\frac{1}{p_{pi}}=\frac{1}{\left(\frac{\Delta p_p}{q}\right)_i}$ (m^3/d/MPa)

$\frac{1}{p_{did}}$ inverse normalized pressure integral derivative for gas well, $\frac{1}{p_{pid}}=\frac{1}{\left(\frac{\Delta p_p}{q}\right)_{id}}$ (m^3/d/MPa)

$\frac{\Delta p}{q}$ normalized pressure for oil well, $\frac{\Delta p}{q}=\frac{p_i-p_{wf}}{q}$ (MPa/m^3/d)

$\frac{\Delta p_p}{q}$ normalized pressure for gas well, $\frac{\Delta p_p}{q}=\frac{p_{p_i}-p_{p_{wf}}}{q}$ (MPa/m^3/d)

$\left(\frac{\Delta p}{q}\right)_i$ normalized pressure integral for oil well, $\left(\frac{\Delta p}{q}\right)_i=\frac{1}{t_c}\int_0^{t_c}\frac{p_i-p_{wf}}{q}d\tau$ (MPa/m^3/d)

$\left(\frac{\Delta p_p}{q}\right)_i$ normalized pressure integral for gas well, $\left(\frac{\Delta p_p}{q}\right)_i=\frac{1}{t_{ca}}\int_0^{t_{ca}}\frac{\Delta p_p}{q}d\tau$ (MPa/m^3/d)

$\left(\frac{\Delta p}{q}\right)_{id}$ normalized pressure integral derivative for oil well,

$$\left(\frac{\Delta p}{q}\right)_{id}=t_c\frac{d\left(\frac{\Delta p}{q}\right)_i}{dt_c}\ \text{(MPa/m}^3\text{/d)}$$

$\left(\frac{\Delta p_p}{q}\right)_{id}$ normalized pressure integral derivative for gas well,

$$\left(\frac{\Delta p_p}{q}\right)_{id}=t_{ca}\frac{d\left(\frac{\Delta p_p}{q}\right)_i}{dt_{ca}}\ \text{(MPa/m}^3\text{/d)}$$

$\frac{q}{\Delta p}$ normalized rate for oil well, $\frac{q}{\Delta p}=\frac{q}{p_i-p_{wf}}$ (m^3/d/MPa)

$\frac{q}{\Delta p_{\mathrm{p}}}$ normalized rate for gas well, $\frac{q}{\Delta p_{\mathrm{p}}} = \frac{q}{p_{p_{\mathrm{i}}} - p_{p_{\mathrm{wf}}}}$ (m^3/d/MPa)

$\left(\frac{q}{\Delta p}\right)_{\mathrm{i}}$ normalized rate integral for oil well, $\left(\frac{q}{\Delta p}\right)_{\mathrm{i}} = \frac{1}{t_{\mathrm{c}}}\int_0^{t_{\mathrm{c}}} \frac{q}{p_{\mathrm{i}} - p_{\mathrm{wf}}} d\tau$ (m^3/d/MPa)

$\left(\frac{q}{\Delta p_{\mathrm{p}}}\right)_{\mathrm{i}}$ normalized rate integral for gas well, $\left(\frac{q}{\Delta p_{\mathrm{p}}}\right)_{\mathrm{i}} = \frac{1}{t_{\mathrm{ca}}}\int_0^{t_{\mathrm{ca}}} \frac{q}{\Delta p_{\mathrm{p}}} d\tau$ (m^3/d/MPa)

$\left(\frac{q}{\Delta p}\right)_{\mathrm{id}}$ normalized rate integral derivative for oil well,

$$\left(\frac{q}{\Delta p}\right)_{\mathrm{id}} = -t_{\mathrm{c}} \frac{d\left(\frac{q}{\Delta p}\right)_{\mathrm{i}}}{dt_{\mathrm{c}}} \ (\mathrm{m^3/d/MPa})$$

$\left(\frac{q}{\Delta p_{\mathrm{p}}}\right)_{\mathrm{id}}$ normalized rate integral derivative for gas well,

$$\left(\frac{q}{\Delta p_{\mathrm{p}}}\right)_{\mathrm{id}} = -t_{\mathrm{ca}} \frac{d\left(\frac{q}{\Delta p_{\mathrm{p}}}\right)_{\mathrm{i}}}{dt_{\mathrm{ca}}} \ (\mathrm{m^3/d/MPa})$$

GREEK LETTER

α_1 dimension conversion coefficient, 277.8 (dimensionless)
α_2 dimension conversion coefficient, 0.004 (dimensionless)
α_3 dimension conversion coefficient, 0.008 (MPa/K)
β the ratio of the derivative of regularized production integral to the regularized production integral (dimensionless)
χ the regression coefficient of the time variable conductivity logarithmic relationship, dimensionless; Eq. (4.55)
ϕ porosity (%)
ϕc_{t} storability coefficient (MPa^{-1})
γ the regression coefficient of the time variable conductivity exponential relationship; Eq. (4.56)
γ relative density (dimensionless)
γ_{g} relative density of gas (dimensionless)
γ_{o} relative density of stock tank oil (dimensionless)
η diversion coefficient of variable fracture conductivity, dimensionless; Eq. (4.57)
η diffusivity coefficient, $\eta = K/\mu\phi C_{\mathrm{t}}$
μ viscosity (mPa .s)
$\overline{\mu}$ viscosity at average test pressure (mPa .s)
μ_{g} reservoir gas viscosity (mPa .s)
μ_{o} reservoir oil viscosity (mPa .s)
μ_{w} formation water viscosity (mPa .s)
$\rho_{\mathrm{g}}, \rho_{\mathrm{o}}, \rho_{\mathrm{w}}$ densities of gas, oil and water, respectively (kg/m^3)
$\rho_{\mathrm{gsc}}, \rho_{\mathrm{osc}}$ densities of gas, oil and water, respectively at standard state (kg/m^3)
σ overburden rock stress (MPa); Eq. (4.61)
σ parameter cluster; Eq. (3.7)
σ_{e} reservoir effective stress (MPa); Eq. (4.61)
σ_{ei} initial reservoir effective stress (MPa); Eq. (4.61)

ω	storativity ratio of dual-porosity reservoir (dimensionless)
ψ	pseudo-pressure (MPa); Eq. (2.54)
Γ_{in}	inner boundary
Γ_{out}	outer boundary
∇	divergence

SUBSCRIPTS

b	pressure build-up
d	pressure drawdown
D	dimensionless
e	external boundary
f	formation
f	fracture or forecast
g	gas
h	horizontal
i	initial
L	liquid
m	matrix
max	maximum
min	minimum
M	type curve match point
o	oil
R	reservoir
S	skin effected zone
t	total
v	vertical
w	well bottom
wf	wellbore and flowing
m	pseudo-pressure

References

Abbaszadeh, M., Cinco-Ley, H., 1995. Pressure-transient behavior in a reservoir with a finite-conductivity fault. SPEFE 10 (1), 26–32.

Abdassah, D., Ershaghi, I., 1986. Triple-porosity systems for representing naturally fractured reservoirs. SPEFE 1 (2), 113–127.

Acuna, J.A., Ershaghi, I., Yortsos, Y.C., 1995. Practical application of fractal pressure-transient analysis in naturally fractured reservoirs. SPEFE 10 (3), 173–179.

Agarwal, R.G., 1979. Real gas pseudotime-a new function for pressure buildup analysis of MHF gas wells. In: SPE8279.

Agarwal, R.G., AI-Hussainy, R., Ramey Jr., H.J., 1970. An investigation of wellbore storage and skin effect in unsteady liquid flow: I. Analytical treatment. SPE J. 10 (3), 279–290.

Agarwal, R.G., Gardner, D.C., Kleinsteiber, S.W., 1998. Analyzing well production data using combined type curve and decline curve concepts. In: SPE49222.

Ahmed, T., 2010. Reservoir Engineering Handbook. Gulf Professional Publishing.

Al-GhamDi, A., Ershaghi, I., 1996. Pressure transient analysis of dually fractured reservoirs. SPE J. 1 (1), 93–100.

Al-Hussainy, R., Ramey Jr., H.J., 1966a. Application of real gas flow theory to well testing and deliverability forecasting. JPT 18 (5), 637–642.

Al-Hussainy, R., Ramey Jr., J.J., Crawford, P.B., 1966b. The flow of real gases through porous media. JPT 18 (5), 624–636.

Al-Hussainy, R., Ramey Jr., H.J., Crawford, P.B., 1966c. The flow of real gases through porous media. JPT 18 (5), 624–636.

Ambastha, A.K., 1988. Pressure Transient Analysis for Composite Systems (Ph.D. Thesis). Stanford University.

Amini, S., Ilk, D., Blasingame, T.A., 2007. Evaluation of the elliptical flow period for hydraulically fractured wells in tight gas sands-theoretical aspects and practical considerations. In: SPE106308.

Anderson, D.M., Mattar, L., 2003. Material balance time during linear and radial flow. In: PETSOC2003-201.

Anderson, D., Mattar, L., 2004. Practical diagnostics using production data and flowing pressures. In: SPE89939.

Anderson, D.M., Mattar, L., 2007. An improved pseudo-time for gas reservoirs with significant transient flow. JCPT 46 (7), 49–54.

Anderson, D.M., Stotts, G.W.J., Mattar, L., et al., 2010. Production data analysis challenges, pitfalls, diagnostics. SPEREE 13 (3), 538–552.

Arévalo-Villagrán, J.A., Gutiérrez-Acosta, T., 2005. Analysis of long term behavior in tight gas reservoirs: case histories. In: SPE95117.

Arévalo-Villagrán, J.A., Wattenbarger, R.A., 2001. Production analysis of long term linear flow in tight gas reservoirs: case histories. In: SPE71516.

Arps, J.J., 1945. Analysis of decline curves. Trans. AIME 160, 228–247.

Aziz, K., Mattar, L., Ko, S., et al., 1976. Use of pressure, pressure-squared or pseudo-pressure in the analysis of transient pressure drawdown data from gas wells. JCPT 15 (2), 58–65.

Bahadori, A., 2012. Analysing gas well production data using a simplified decline curve analysis method. Chem. Eng. Res. Des. 90 (4), 541–547.

Bahadori, A., Vuthaluru, H.B., 2010. A simple decline curve analysis approach for evaluating gas reserves and predicting future production in gas reservoirs. In: SPE139057.

Billy, P.R., Fred, F.F., 1981. P/Z abnormally pressured gas reservoirs. In: SPE10125.

Bixel, H.C., Larkin, B.K., Van Poollen, H.K., 1963. Effect of linear discontinuities on pressure build-up and drawdown behavior. JPT 15 (8), 885–895.

Blasingame, T.A., Lee, W.J., 1986a. Properties of homogeneous reservoirs, naturally fractured reservoirs, and hydraulically fractured reservoirs from decline curve analysis. In: SPE15018.

Blasingame, T.A., Lee, W.J., 1986b. Variable rate reservoir limits testing. In: SPE15028.

Blasingame, T.A., Johnston, J.L., Lee, W.J., 1989. Type curve analysis using the pressure integral method. In: SPE18799.

Blasingame, T.A., McCray, T.L., Lee, W.J., 1991. Decline curve analysis for variable pressure drop/variable flow rate systems. In: SPE21513.

Bourdet, D., 1985. Pressure behavior of layered reservoir with cross flow. In: SPE13628.

Bourdet, D., 2003. Well Test Analysis: The Use of Advanced Interpretation Model-Handbook of Petroleum Exploration & Production. Elsevier Science B.V.

Bourdet, D., Whittle, T.M., Douglas, A.A., Pirard, Y.M., 1983. A new set of type curves simplifies well test analysis. World Oil 196 (6), 95–106.

Bourdet, D., Ayoub, J.A., Pirard, Y.M., 1989. Use of pressure derivative in well test interpretation. SPEFE 4 (2), 293–302.

Bourgeois, M.J., Daviau, F.H., Boutaud de la Combe, J.L., 1996. Pressure behavior in finite channel-levee complexes. SPEFE 11 (3), 177–184.

Brown, L.P., Hawkes, R.V., 2005. Rules of thumb in well testing: what works and doesn't work-and why. JCPT 44 (5), 1–16.

Callard, J.G., 1995. Reservoir performance history matching using rate cumulative type curves. In: SPE30793.

Camacho, R.G., Raghavan, R., Reynolds, A.C., 1987. Response of wells producing layered reservoirs: unequal fracture length. SPEFE 2 (1), 9–28.

Camacho-Velázquez, R., 2005. Pressure transient and decline curve behaviors in naturally fractured vuggy carbonate reservoirs. SPEREE 8 (2), 95–112.

Carter, R.D., 1966. Pressure behavior of a limited circular composite reservoir. SPE J. 6 (4), 328–334.

Carter, R.D., 1981. Characteristic behavior of finite radial and linear gas flow systems - constant terminal pressure case. In: SPE9887.

Carter, R.D., 1985. Type curves for finite radial and linear gas flow systems: constant terminal pressure case. SPE J. 25 (5), 719–728.

Chang, J., Yortsos, Y.C., 1990. Pressure transient analysis of fractal reservoirs. SPEFE 5 (1), 31–38.

Chatas, A.T., 1966. Unsteady spherical flow in petroleum reservoirs. SPE J. 6 (2), 102–114.

Chen, Y., Li, D., 2001. Modern Reservoir Engineering. Petroleum Industry Press, Beijing.

Chen, H.Y., Teufel, L.W., 2000. A new rate time type curve for analysis of tight gas linear and radial flows. In: SPE63094.

Chen, C.C., Chu, W.C., Sadighi, S., 1996. Pressure-transient testing of gas reservoirs with edge-water drive. SPEFE 11 (4), 251–256.

Chen, X., Wang, H., Liu, H., 2009. An overview of prediction methods for dynamic reserves of gas reservoirs. Spec. Oil Gas Reserv. 16 (2), 9–13 + 103.

Chen, F., Zhang, F., Deng, X., et al., 2015. A numerical well test model for wells drilled out of big size cavity of fractured carbonate reservoirs. Sci. Technol. Rev. 33 (9), 46–49.

Cinco-Ley, H., Samaniego, F., 1981. Transient pressure analysis: finite conductivity fracture versus damaged fracture case. In: SPE10179.

Cinco-Ley, H., Miller, F.G., Ramey Jr., H.J., 1975. Unsteady-state pressure distribution created by a directionally drilled well. JPT 27 (11), 1392–1400.

Cinco-Ley, H., Samaniego-V, E., Domingoez-A, N., 1976. Unsteady-state flow behavior for a well near a natural fracture. In: SPE6019.

Cinco-Ley, H., Samanieco-V, F., Dominguez, N., 1978. Transient pressure behavior for a well with a finite conductivity vertical fracture. SPE J. 18 (4), 253–264.

Cinco-Ley, H., Samaniego-V, F., Viturat, D., 1985. Pressure transient analysis for high-permeability reservoirs. In: SPE14314.
Clossman, P.J., 1975. An aquifer model for fissured reservoirs. SPE J. 15 (5), 385–398.
Coats, K.H., Rapoport, L.A., McCord, J.R., Drews, W.P., 1964. Determination of aquifer influence functions from field data. JPT 16 (12), 1417–1424.
Coleman, S.B., Clay, H.B., McCurdy, D.G., et al., 1991. A new look at predicting gas-well load-up. JPT 43 (3), 329–333.
Compiling Group of Handbook of Mathematics, 2010. Mathematics Handbook. Higher Education Press, Beijing.
Cook, D.C., 2001. Screening wells for tight gas from single point flow data. In: SPE71065.
Corbett, P.W.M., Mesmari, A., Stewart, G., 1996. A method for using the naturally-occurring negative geoskin in the description of fluvial reservoirs. In: SPE36882.
Crawford, G.E., Hagedorn, A.R., Pierce, A.E., 1976. Analysis of pressure buildup tests in a naturally fractured reservoir. JPT 28 (11), 1295–1300.
Daungkaew, S., Hollaender, F., Gringarten, A.C., 2000. Frequently asked questions in well test analysis. In: SPE63077.
Deng, X., Sun, H., Shi, Y., et al., 2010. Research on Reservoir Engineering of Tazhong No. 1 Gas Field. Tarim Oilfield of PetroChina, Langfang Petroleum Science and Technology Research Institute.
Deng, X., Sun, H., Shi, Y., et al., 2013a. Test Production Characteristics and Well Test Research in Tazhong No. 1 Gas Field Pilot Production Area. Tarim Oilfield of PetroChina, Langfang Petroleum Science and Technology Research Institute.
Deng, X., Liu, J., Pei, G., et al., 2013b. Research on multi-subject integrated division of fracture and cave units in carbonate reservoirs: a case of well lg-11 area in lungu oil field, Tarim basin. Mar. Origin Pet. Geol. 18 (2), 72–78.
Deng, X., Sun, H., Shi, Y., et al., 2014a. Calculation Method for Dynamic Reserves of Complex Carbonate Condensate Gas Reservoirs in Tazhong No. 1 Gas Field. Tarim Oilfield of PetroChina, Langfang Petroleum Science and Technology Research Institute.
Deng, X., Liu, X., Shiyin, L., et al., 2014b. A flow unit division method for horizontal wells in carbonate reservoirs. Chinese patent: CN201410131727.3.
Deng, Q., Nie, R., Jia, Y., et al., 2015. A new method of pressure buildup analysis for a well in a multiwell reservoir. In: SPE175866.
Deng, X., Sun, H., Li, S.Y., et al., 2015a. Research on Gas Reservoir Engineering Method in Tazhong No. 1 Gas Field. Tarim Oilfield of PetroChina. Langfang Petroleum Science and Technology Research Institute.
Deng, X., Zhang, Q., Liang, B., et al., 2015b. Reconstruction of karst palaeogeomorphology for the Ordovician Yingshan formation in the central Tarim basin. Carsol. Sin. 34 (2), 154–158.
Deng, X., Li, S., Jiang, J., et al., 2015c. A method for distinguishing fractured vuggy carbonate reservoir types. Chinese patent: CN201510034228.7.
Deng, X., Sun, H., Li, S.Y., et al., 2016. Research on Numerical Well Test Analysis in Tazhong No. 1 Gas Field. Tarim Oilfield of PetroChina, Langfang Petroleum Science and Technology Research Institute.
Dietz, D.N., 1965. Determination of average reservoir pressure from buildup surveys. JPT 17 (8), 955–959.
Doublet, L.E., Pandie, P.K., McCollum, T.J., et al., 1994. Decline curve analysis using type curves analysis of oil well production data using material balance time: application to field cases. In: SPE28688.
Dranchuk, P.M., Purvis, R.A., Robinson, D.B., 1973. Computer calculation of natural gas compressibility factors using the Standing and Katz correlation. In: Annual Technical Meeting, Edmonton.

Duong, A.N., 1989. A new approach for decline curve analysis. In: SPE18859.

Earlougher Jr., R.C., 1977. Advances in Well Test Analysis. Monograph Series, vol. 5. SPE, Richardson, TX.

Ebrahimi, M., 2010. Enhanced estimation of reservoir parameters using decline curve analysis. In: SPE133432.

Ehlig-Economides, C.A., Economides, M.J., 1985. Pressure transient analysis in an elongated linear flow system. SPE J. 25 (6), 839–847.

Ehlig-Economides, C.A., Ramey Jr., H.L., 1981. Transient rate decline analysis for wells produced at constant pressure. SPE J. 21 (1), 98–104.

Ehlig-Economides, C.A., Hegeman, P., Vik, S., 1994. Guidelines simplify well test interpretation. Oil Gas J. 92 (29), 33–40.

Elshahawi, H., Hite, R.H., Hows, M.P., 2008. Optimal value testing-moving from vision to reality. In: SPE114869.

Fair Jr., W.B., 1981. Pressure buildup analysis with wellbore phase redistribution. SPE J. 21 (2), 259–270.

Fair Jr., W.B., 1996. Generalization of wellbore effects in pressure transient analysis. SPEFE 11 (2), 114–119.

Fast Harmony, 2018. https://www.ihsenergy.ca/support/documentation_ca/Harmony_Enterprise/2018_1/Default.htm.

Feng, X., He, W., Xu, Q., 2002. Discussion on calculating dynamic reserves in the early stage of heterogeneous gas reservoir development. Nat. Gas Ind. 22 (S1), 87–90+5-4.

Fernando, R., 2004. Well test characterization of small- and large-scale secondary porosity in naturally fractured reservoirs. In: SPE90287.

Fetkovich, M.J., 1980. Decline curve analysis using type curves. JPT 32 (6), 1065–1077.

Fetkovich, M.J., Vienot, M.E., Bradley, M.D., et al., 1987. Decline curve analysis using type curves case histories. SPEFE 2 (4), 637–656.

Fetkovich, M.J., Fetkovich, E.J., Fetkovich, M.D., 1996. Useful concepts for decline curve forecasting, reserve estimation, and analysis. SPERE 11 (1), 13–22.

Fevang, I., Whitson, C.H., 1995. Modeling gas-condensate well deliverability. SPERE 11 (4), 221–230.

Fraim, M.J., 1989. Gas reservoir decline-curve analysis using type curves with real gas pseudo pressure and normalized time. SPEFE 12, 671–682.

Gao, C., 1993. The material balance equation with a supplying region and its use in gas reservoir. Pet. Explor. Dev. 20 (2), 53–61.

Gao, C., Zhang, M., Yang, L., 1997. A study of unbalance development for closed gas reservoir. Acta Pet. Sin. 18 (1), 70–76.

Gao, C., Lu, T., Gao, W., et al., 2006. Multi-region material balance method and its use in the performance prediction and optimal well pattern design of edge water reservoir. Pet. Explor. Dev. 33 (1), 103–106.

Gao, C., Sun, H., 2017. Well Test Analysis for Multilayered Reservoirs With Formation Crossflow. Elsevier.

Gentry, R.W., 1972. Decline curve analysis. JPT 24 (1), 38–41.

Gerami, S., Pooladi-Darvish, M., Hong, H., 2007. Decline curve analysis for naturally fractured gas reservoirs: a study on the applicability of pseudo-time and material balance pseudo-time. In: IPTC11278.

Govier, G.W., 1979. Theory and Practice of the Testing of Gas Wells, fourth ed. Guide G-3, Alberta Energy Resources and Conservation Board, Calgary.

Gringarten, A.C., 1984. Interpretation of tests in fissured and multilayered reservoirs with double-porosity behavior: theory and practice. JPT 36 (4), 549–564.

Gringarten, A.C., 2008. From straight lines to deconvolution: the evolution of the state of the art in well test analysis. SPEREE 11 (1), 41–62.

Gringarten, A.C., Ramey Jr., H.J., Raghavan, R., 1974. Unsteady state pressure distributions created by a well with a single infinite conductivity vertical fracture. SPE J. 14 (4), 347–360.
Gringarten, A.C., Bourdet, D.P., Landel, P.A., et al., 1979. A comparison between different skin and wellbore storage type-curves for early-time transient analysis. In: SPE8205.
Gross, H.E., 1938. Decline curve analysis. Oil Gas J. 37 (18), 55–56 + 58 + 61.
Guppy, K.H., Cinco-Ley, H., Ramey Jr., H.J., 1981. Transient flow behavior of a vertically fractured well producing at constant pressure. In: SPE9963.
Hagoort, J., 2003. Automatic decline curve analysis of wells in gas reservoirs. SPEREE 6 (6), 433–440.
Hagoort, J., Hoogstra, R., 1999. Numerical solution of the material balance equations of compartmented gas reservoirs. SPEREE 2 (4), 385–392.
Han, Y., 2006. Study on Gas Well Performance Analysis in Low-Permeability and Tight Reservoirs (Ph.D. Thesis). China University of Geosciences, Beijing.
Han, Y., Zhuang, H., Sun, H., 2006. Application of numerical well test analysis technique in gas reservoir characterization. Well Test. 15 (2), 9–11.
Hao, Y., Bian, X.Y., 1999. About dynamic reserves. Well Test. Prod. Technol. 20 (2), 8–11.
Hawkins Jr., M.F., 1956. A note on the skin effect. JPT 8 (12), 65–66.
Hazlett, W.G., Lee, W.J., Narahara, G.M., et al., 1986. Production data analysis type curves for the devonian shales. In: SPE15934.
Hegeman, P.S., Hallford, D.L., Joseph, J.A., 1993. Well test analysis with changing wellbore storage. SPEFE 8 (3), 201–207.
Helmy, M.W., Wattenbarger, R.A., 1998. New shape factors for wells produced at constant pressure. In: SPE 39970.
Hong, K.C., 1975. Productivity of perforated completions in formations with or without damage. JPT 27 (8), 1027–1038.
Horne, R.N., 2009. Computer-assisted interpretation. In: Transient Well Testing. Monograph Series, vol. 23. SPE, Richardson, TX, pp. 179–201.
Horner, D.R., 1951. Pressure build-up in wells. In: Proc., Third World Petroleum Congress, Leiden, The Netherlands, Section II, Preprint7, pp. 25–43.
Hower, T.L., Collins, R.E., 1989. Detecting compartmentalization in gas reservoirs through production performance. In: SPE19790.
Hu, Y., Zhu, H.-Y., Chen, J.-J., et al., 2007. Physical simulated study on the gas-supplying mechanism of high-low permeable serial-connection-formations. Nat. Gas Geosci. 18 (3), 469–472.
Ilk, D., Valkó, P.P., Blasingame, T.A., 2005. Deconvolution of variable-rate reservoir performance data using B-splines. In: SPE95571.
Ilk, D., Valkó, P.P., Blasingame, T.A., 2007. A deconvolution method based on cumulative production for continuously measured flowrate and pressure data. In: SPE111269.
Ilk, D., Rushing, J.A., Perego, A.D., et al., 2008. Exponential vs. hyperbolic decline in tight gas sands understanding the origin and implications for reserve estimate using Arps decline curves. In: SPE116731.
Issaka, M.B., Ambastha, A.K., 1998. Decline curve analysis for composite reservoirs. JCPT 37 (6), 17–24.
Jiang, L.S., Chen, Z.X., 1985. Well Test Analysis Theory. Petroleum Industry Press, Beijing.
Johnston, L., 1992. Variable Rate Analysis of Transient Well Test Data Using Semi-Analytical Methods (M.S. Thesis). Texas A&M University, College Station, TX.
Jones, S.C., 1987. Using the inertial coefficient, p, to characterize heterogeneity in reservoir rock. In: SPE16949.
Kazemi, H., 1974. Determining average reservoir pressure from pressure buildup tests. SPE J. 14 (1), 55–62.

Khanamiri, H.H., 2010. A non-iterative method of decline curve analysis. J. Pet. Sci. Eng. 73 (1–2), 59–66.
Klinkenberg, L.J., 1941. The permeability of porous media to liquids and gases. Socar Proc. 2 (2), 200–213.
Kuchuk, F.J., Habashy, T.M., 1997. Pressure behavior of laterally composite reservoirs. SPEFE 12 (1), 47–56.
Kumar, A., Ramey, H.J., 1974. Well test analysis for a well in a constant pressure square. SPE J. 14 (02), 107–116.
Kumar, V., Dunlap, J., Klimentos, T., et al., 2010. The use of wireline formation tester for optimization of conventional well test design and interpretation. In: SPE129031.
Lane, H.S., Lee, W.J., Watson, A.T., 1991. An algorithm for determining smooth, continuous pressure derivatives from well-test data. SPEFE 6 (4), 493–499.
Larsenf, L., Kviljo, K., 1990. Variable skin and cleanup effects in well-test data. SPEFE 5 (3), 272–276.
Lee, W.J., Holditch, S.A., 1982. Application of pseudotime to buildup test analysis of low-permeability gas wells with long-duration wellbore storage distortion. JPT 34 (12), 2877–2888.
Lee, W.J., Wattenbarger, R.A., 1996. Gas Reservoir Engineering. Henry L. Doherty Memorial Fund of AIME, Society of Petroleum Engineers.
Lee, A.L., Gonzalez, M., Eakin, B., 1966a. The viscosity of natural gases. J. Pet. Technol. 18 (8), 997–1000.
Lee, A.L., Gonzalez, M.H., Eakin, B.E., 1966b. The viscosity of natural gases. JPT 18 (8), 997–1000.
Lee, W.J., Kuo, T.B., Holditch, S.A., et al., 1984. Estimating formation permeability from single-point flow data. In: SPE12847.
Lee, J.W., Rollins, J.B., Spivey, J.P., 2003. Pressure Transient Testing [M]. Textbook Series, SPE, Richardson, TX.
Levitan, M.M., 2005. Practical application of pressure/rate deconvolution to analysis of real well tests. SPEREE 8 (2), 113–121.
Levitan, M.M., 2007. Deconvolution of multiwell test data. In: SPE10248.
Levitan, M.M., Crawford, G.E., 2002. General heterogeneous radial and linear models for well-test analysis. SPE J. 1 (2), 131–138.
Levitan, M.M., Crawford, G.E., Hardwick, A., 2006. Practical considerations for pressure-rate deconvolution of well-test data. SPE J. 11 (1), 35–47.
Li, M., Yang, Y., 1994. Discussing on dynamic reserve of low permeability gas reservoir calculated by pressure build-up data. Nat. Gas Ind. 14 (3), 36–38+99-100.
Li, Y., Luo, J., Guo, J., et al., 2006. Numerical simulation of post-frac performance in naturally fractured gas reservoir with medium deformation and long term fracture conductivity. Nat. Gas Ind. 26 (9), 103–105.
Li, C., Liu, Q., Zhang, R., 2006. The research of well test of horizontal well in triple medium reservoir. J. Southwest Pet. Inst. 28 (4), 32–35.
Li, Q., Guo, P., Huang, Q., 2008. Methods of gas well reserves estimation. J. Chongqing Univ. Sci. Technol. 10 (6), 34–36.
Li, H., Ren, D., Guo, P., 2016. et al. In: Handbook of Gas Reservoir Engineering. Petroleum Industry Press, Beijing.
Liao, X., 2003. Study on the Transient Flow of Gas Reservoir and Condensate Gas Reservoir. Exploration and Development Research Institute of Petrochina, Beijing.
Lin, J., Yang, H., 2007. Analysis of welltest data in a multiwell reservoir with water injection. In: SPE 110349.
Lin, J., Liu, W., Chen, Q., 1994. Production engineering characteristics of pressure buildup curve for a well in a multiwell system. Pet. Geol. Oilfield Dev. Daqing 13 (4), 56–59.

Liu, N., 2008. Practical Modern Well Testing Method, fifth ed. Petroleum Industry Press, Beijing.
Liu, Q., Chen, Y., Zhang, L., et al., 2006. Study for disturbing test model with the well opening and its interpreting method. Well Test. 15 (1), 10–12.
Liu, W., Liu, Y., Han, G., et al., 2017. An improved deconvolution algorithm using b-splines for well-test data analysis in petroleum engineering. J. Pet. Sci. Eng. 149, 306–314.
Liu, W., Liu, Y., Zhu, W., Sun, H., 2018. Improvement and application of ILK flow-rate deconvolution algorithm based on the second-order B-splines. Acta Pet. Sin. 39 (3), 327–334.
Londono, F.E., Blasingame, T.A., 2002. Simplified correlations for hydrocarbon gas viscosity and gas density-validation and correlation of behavior using a large scale database. In: SPE75721.
Lord, M.E., Collins, R.E., Kocberber, S., 1992. Compartmented simulation system for gas reservoir evaluation with application to fluvial deposits in the frio formation, south Texas. In: SPE24308.
Lu, X., 1999. Performance forecasting based on material balance method. Nat. Ctas Explor. Dev. 22 (3), 29–39.
Lu, D., 2009. Modern Well Test Theory and Its Application. Petroleum Industry Press, Beijing.
Luo, R., Feng, J., Tang, M., et al., 2008. Probe into evaluation methods for stress sensitivity of low permeability reservoirs. J. Southwest Pet. Univ. 30 (5), 161–164.
Marhaendrajana, T., Blasingame, T.A., 2001. Decline curve analysis using type curves-evaluation of well performance behavior in a multiwell reservoir system. In: SPE71517.
Marhaendrajana, T., Kaczorowski, N.J., Blasingame, T.A., 1999. Analysis and interpretation of well test performance at Arun field, Indonesia. In: SPE56487.
Mattar, L., 1996. Critical evaluation and processing of data before pressure-transient analysis. SPEFE 11 (2), 120–127.
Mattar, L., Anderson, D.M., 2003. A systematic and comprehensive methodology for advanced analysis of production data. In: SPE84472.
Mattar, L., McNeil, R., 1998. The flowing gas material balance. JCPT 37 (2), 52–55.
Mattar, L., Moghadam, S., 2009. Modified power law exponential decline for tight gas. In: PETSOC 2009-198.
Mattar, L., Santo, M., 1992. How wellbore dynamics affect pressure transient analysis. JCPT 31 (1), 32–40.
Mattar, L., Hawkes, R.V., Santo, M.S., et al., 1993. Prediction of long term deliverability in tight formation. In: SPE26178.
Mattar, L., Anderson, D., Stotts, G., 2006. Dynamic material balance oil or gas-in-place without shut-ins. JCPT 45 (11), 7–10.
Matthews, C.S., Russell, D.G., 1967. Pressure Buildup and Flow Tests in Wells [M]. Monograph Series, SPE of AIME, Storm Printing Corp., Dallas, TX.
McCray, T.L., 1990. Reservoir Analysis Using Production Decline Data and Adjusted Time (M.S. Thesis). Texas A&M University, College Station, TX.
McLeod Jr., H.O., 1983. The effect of perforating conditions on well performance. JPT 35 (1), 31–39.
Meunier, D.F., Kabir, C.S., Wittmann, M.J., 1987. Gas well test analysis: use of normalized pseudo variables. SPEFE 2 (11), 629–636.
Miller, F.G., 1962. Theory of unsteady state influx of water in linear reservoirs. J. Inst. Pet. 48 (467), 365–379.
Miller, C.C., Dyes, A.B., Hutchinson Jr., C.A., 1950. The estimation of permeability and reservoir pressure from bottom-hole pressure build-up characteristics. JPT 2 (4), 91–104.
Moghadam, S., Jeje, O., Mattar, L., 2011. Advanced gas material balance in simplified format. JCPT 50 (1), 90–98.

Mou, Z., Fan, T., 2006. Mathematical model with unsteady-state filtering flow of vertically fractured well with varying conductivity for closed circle oil reservoir. Pet. Geol. Recov. Eff. 13 (6), 66–69.

Mutalik, P.N., Joshi, S.D., 1992. Decline curve analysis predicts oil recovery from horizontal wells. Oil Gas J. 90 (36), 42–48.

Neng, Y., Yang, H., Deng, X., 2018. Structural patterns of fault broken zones in carbonate rocks and their influences on petroleum accumulation in Tazhong Paleo-uplift, Tarim Basin, NW China. Pet. Explor. Dev. 2018 (1), 40–50 + 127.

Newman, G.H., 1973. Pore-volume compressibility of consolidated, friable, and unconsolidated reservoir rocks under hydrostatic loading. JPT 25 (2), 129–134.

Nobakht, M., Clarkson, C.R., 2012a. A new analytical method for analyzing linear flow in tight shale gas reservoirs: constant flowing pressure boundary condition. SPEREE 15 (3), 370–384.

Nobakht, M., Mattar, L., Moghadam, S., et al., 2012b. Simplified forecasting of tight shale gas production in linear flow. JCPT 51 (6), 476–486.

Nobakht, M., CR, C., Kaviani, D., 2016. New type curves for analyzing horizontal well with multiple fractures in shale gas reservoirs. J. Nat. Gas Sci. Eng. 10 (1), 99–112.

Nutakki, R., Mattar, L., 1982. Pressure transient analysis of wells in very long narrow reservoirs. In: SPE11221.

Oliver, H., Viturat, D., Fjaere, O.S., et al., 2013. Dynamic Data Analysis. KAPPA.

Onur, M., Reynolds, A.C., 2002. Nonlinear regression: the information content of pressure and pressure-derivative data. SPE J. 7 (3), 243–249.

Onur, M., Serra, K.V., Reynolds, A.C., 1991. Analysis of pressure buildup data from a well in a multiwell system. SPEFE 6 (1), 101–110.

Palacio, J.C., Blasingame, T.A., 1993. Decline curve analysis using type curves analysis of gas well production data. In: SPE25909.

Papatzacos, P., 1987. Approximate partial-penetration pseudoskin for infinite-conductivity wells. SPERE 2 (2), 227–234.

Payne, D.A., 1996. Material-balance calculations in tight-gas reservoirs: the pitfalls of p/z plots and a more accurate technique. SPERE 11 (4), 260–267.

Petroleum and Natural Gas Industry Standards of China People's Republic of China, 2010. Natural Gas Reserves Calculation. SY/T6098.

Pinson Jr., A.E., 1972. Concerning the value of producing time used in average pressure determinations from pressure buildup analysis. JPT 24 (11), 1369–1370.

Podio, A.L., McCoy, J.N., Huddleston, K.L., 1987. Automatic pressure buildup data acquisition and interpretation using a microcomputer-based acoustic liquid level instrument. In: SPE16228.

Podio, A.L., McCoy, J.N., Becker, D., 1992. Integrated well performance and analysis. SPECA 4 (3), 18–23.

Prasad, R.K., 1975. Pressure transient analysis in the presence of two intersecting boundaries. JPT 27 (1), 89–96.

Pratikno, H., Rushing, J., Blasingame, T.A., 2003. Decline curve analysis using type curves fractured wells. In: SPE84287.

Proett, M.A., Chin, W.C., 1998. New exact spherical flow solution with storage and skin for early-time interpretation with applications to wireline formation and early-evaluation drillstem testing. In: SPE49140.

Pulido, H., 2006. On a well-test pressure theory of analysis for naturally fractured reservoirs, considering transient interporosity matrix, microfractures, vugs, and fractures flow. In: SPE104076.

Raghavan, R., 1993. Well Test Analysis. Prentice Hall Petroleum Engineering Series, Englewood Cliffs, NJ.

Raghavan, R., Ohaeri, C.U., 1987. Unsteady flow to a well-produced at constant pressure in a fractured reservoir. SPEFE 2 (2), 186–200.

Raghavan, R., Cady, G., Ramey Jr., H.J., 1972. Well test analysis for vertically fractured wells. JPT 24 (8), 1014–1020.
Raghavan, R., Chu, W.-C., Jones, J.R., 1999. Practical considerations in the analysis of gas-condensate well tests. SPERE 2 (3), 288–295.
Rahman, N.M.A., Mattar, L., Zaoral, K., 2006. A new method for computing pseudo-time for real gas flow using the material balance equation. JCPT 45 (10), 36–44.
Ramey Jr., H.J., 1965. Non-Darcy flow and wellbore storage effects in pressure build-up and drawdown of gas wells. JPT 17 (2), 223–233.
Ramey Jr., H.J., Agarwal, R.G., 1972. Annulus unloading rates as influenced by wellbore storage and skin effect. SPE J. 12 (5), 453–462.
Ramey Jr., H.J., Cobb, W.M., 1971. A general pressure buildup theory for a well in a closed drainage area. JPT 23 (12), 1493–1505.
Ramsay Jr., H.J., 1968. The Ability of Rate Time Decline Curves to Predict Future Production Rates (M.S. Thesis). Univ. of Tulsa, Tulsa, OK.
Saleh, A.M., Stewart, G., 1996. New approach towards understanding of near well bore behavior of perforated completions. In: SPE36866.
Schlumberger, 2002. Well Testing Interpretation. Schlumberger.
Seshadri, J., Mattar, L., 2010. Comparison of power law and modified hyperbolic decline methods. In: SPE137320.
Shen, Y., He, S., Wang, S., 2010. A new method to calculate dynamic reserves of low-permeability gas reservoir. Sci. Technol. Eng. 10 (28), 6994–6997.
Shih, M.Y., 1994. Decline Curve Analysis for Horizontal Wells (M.S. Thesis). Texas A&M University, College Station, TX.
Shiyi, Z., Fei, W., 2008. Application of deconvolution and decline-curve analysis methods for transient pressure analysis. In: SPE113323.
Slider, H.C., 1971. A simplified method of pressure buildup analysis for a stabilized well. JPT 23 (9), 1155–1160.
Spivey, J.P., Lee, W.J., 1986. The use of pseudotime: wellbore storage and the middle time region. In: SPE15229.
Spivey, J.P., Lee, W.J., 1999. Variable wellbore storage models for a dual-volume wellbore. In: SPE56615.
Spivey, J.P., Lee, W.J., 2013. Applied Well Test Interpretation. Society of Petroleum Engineers, Richardson.
Spivey, J.P., Pursell, D.A., 1998. Errors in input data, the effect on well-test interpretation results. In: SPE39773.
Spivey, J.P., Gatens, J.M., Semmelbeck, M.E., et al., 1992. Integral type curves for advanced decline curve analysis. In: SPE24301.
Spivey, J.P., Brown, K.G., Sawyer, W.K., et al., 2004. Estimating non-Darcy flow coefficient from buildup-test data with wellbore storage. SPERE 1 (4), 256–269.
Spivey, J.P., Valko, P.P., McCain, W.D., 2007. Applications of the coefficient of isothermal compressibility to various reservoir situations with new correlations for each situation. SPERE 10 (3), 43–49.
Standing, M.B., Katz, D.L., 1942. Density of natural gases. Trans. AIME 146, 140–149.
Stehfest, H., 1970. Algorithn368, numerical inversion of Laplace transforms. Commun. ACM 13 (1), 47–49.
Sun, H., 2007. Study on productivity evaluation and performance prediction method of overpressured, stress-sensitive gas reservoirs. In: SPE 108451.
Sun, H., 2011a. A balance equation of gas materials with recharge capacity and their performance prediction. Acta Pet. Sin. 32 (4), 683–686.
Sun, H., Xu, Y., 2011b. A new method for low permeability gas well performance prediction using pressure transient analysis: theory and practices. In: SPE144952.
Sun, H., 2012. Modern Well Test Analysis and Productivity Evaluation of Complex Gas Reservoirs. Petroleum Industry Press, Beijing.

Sun, H., Shi, Y., 2013a. Integrating pressure transient analysis and production analysis for dynamic reserve evaluation in Tazhong I ordovician carbonate gas field, Tarim basin. In: IPTC16469.

Sun, H., Deng, X., Han, Y., et al., 2013b. A dynamic analysis method and system for fractured vuggy carbonate gas reservoirs. China Patent: CN201310314083.7.

Sun, H., 2015a. Advanced Production Decline Analysis and Application. Petroleum Industry Press (Published by Elsevier Inc.).

Sun, H., Deng, X., Chang, B., 2015b. An analysis method and device for performance forecasting of fractured vuggy carbonate condensate gas well. China Patent: CN201510556736.1.

Sun, H., 2016. Pressure buildup analysis in multi-well systems under interferences from adjacent wells. Nat. Gas Ind. 36 (5), 62–68.

Sun, H., Ouyang, W., Zhang, M., 2018. Advanced production decline analysis of tight gas wells with variable fracture conductivity. Pet. Explor. Dev. 45 (3), 455–463.

Tan, C., 2002. Triple medium permeable-storage system and hydraulic oil-driving mechanism. Xinjiang Geol. 20 (1), 80–82.

Tao, Z., Jia, Y., Wang, H., et al., 2006. Well testing model for finite multiwall system and type curve analysis. Xinjiang Pet. Geol. 27 (1), 86–88.

Tariq, S.M., Ramey Jr., H.J., 1978. Drawdown behavior of a well with storage and skin effect communicating with layers of different radii and other characteristics. In: SPE7453.

Teng, D.T., Maloney, B.J., Mantecon, J.C., 2006. Well test by design: transient modelling to predicting behavior in extreme wells. In: SPE101872.

The People's Republic of China National Standard, 2004a. Petroleum and natural gas reserves classification. In: GBT19492.

The People's Republic of China National Standard, 2004b. Quantities and units frequently used inoil and gas exploration and development. In: SY/T 6580-2004.

The People's Republic of China National Standard, 2005. Regulation of petroleum reserves estimation. In: DZ/T 0217-2005.

Thompson, T.W., Mattar, L., 2000. Gas rate forecasting during boundary dominated flow. In: PETSOC2000-046.

Turner, R.G., Hubbard, M.G., Dukler, A.E., 1969. Analysis and prediction of minimum flowrate for the continuous removal of liquids from gas wells. JPT 21 (11), 1475–1482.

Van Everdingen, A.E., 1953. The skin effect and its influence on the productive capacity of a well. JPT 5 (6), 171–176.

Van Everdingen, A.F., Hurst, W., 1949. The application of the Laplace transformation to flow problems in reservoirs. Trans. AIME 186, 305–324.

Vega, L., Barrufet, M.A., 2005. Analysis of a non-volumetric gas-condensate reservoir using a generalized material balance equation with fluid properties from an equation of state. J. Pet. Sci. Eng. 48 (1), 105–119.

Vella, M., Veneruso, T., Lefolk, P., et al., 1992. The nuts and bolts of well testing. Oilfield Rev. 4 (2), 14–27.

Veneruso, A.F., Ehlig-Economides, C., Petitjean, L., 1991. Pressure gauge specification considerations in practical well testing. In: SPE22752.

Von Schroeter, T., Hollaender, F., Gringarten, A.C., 2004. Deconvolution of well test data as a nonlinear total least squares problem. SPE J. 9 (4), 375–390.

Waldman, N., Fair, C., Tyrrell, C., et al., 2002. Subsea well testing at the subsea tree. In: SPE77626.

Wan, Y.Z., Liu, Y.W., Liu, W.C., et al., 2016a. A numerical approach for pressure transient analysis of a vertical well with complex fractures. Acta Mech. Sin. 32 (4), 640–648.

Wan, Y.Z., Liu, Y.W., Ouyang, W.P., et al., 2016b. Numerical investigation of dual-porosity model with transient transfer function based on discrete-fracture model. Appl. Math. Mech. 37 (5), 611–626.

Wang, Z., Yao, J., Dai, W., 2007. Application of the well test interpretation method for fracture-vug reservoir in Tahe Oilfield. J. Xi'an Shiyou Univ. (Nat. Sci. Ed.) 22 (1), 72–74.
Wang, Z., Yang, H., Wang, Q., et al., 2012. Geological Theory and Exploration Technology of Giant Marine Condensate Gas Field in Tazhong Uplift. Science Press, Beijing.
Wang, Z., Yang, H., Pan, W., 2017a. et al. In: Petroleum Geology Theory and Exploration Practice of Ultra-Deep Fractured Vuggy Marine Carbonate Rocks. Petroleum Industry Press, Beijing.
Wang, Z., Lijuan, Z., Yang, H., et al., 2017b. Development Technology of Ultra-Deep Fractured Vuggy Carbonate Reservoir. Petroleum Industry Press, Beijing.
Warren, J.E., Root, P.J., 1963. The behavior of naturally fractured reservoirs. SPE J. 3 (3), 245–255.
Wei, M., Duan, Y., Fang, Q., et al., 2016. Production decline analysis method of fractured horizontal well in shale gas reservoirs based on modifying material balance. Acta Pet. Sin. 37 (4), 508–515.
Wen, Q., Zhang, S., Wang, L., et al., 2005. Influence of proppant embedment on fracture long term flow conductivity. Nat. Gas Ind. 25 (5), 65–68.
Wen, X., Deng, J., Sun, H., et al., 2012. Characteristics of well tests in the carbonate reservoirs of the eastern Tazhong1 gas field, Tarim Basin. Nat. Gas Ind. 32 (5), 36–38 +101–102.
Whittle, T.M., Lee, J., Gringarten, A.C., 2003. Will wireline formation tests replace well tests? In: SPE84086.
Winestock, A.G., Colpitts, G.P., 1965. Advances in estimating gas well deliverability. JCPT 4 (3), 111–119.
Wu, Y., Ge, J., 1983. Percolation problems in the triple media reservoir. Chin. J. Theor. Appl. Mech. 14 (1), 81–85.
Wu, Y.S., Ehlig-Economides, C.A., Qin, G., et al., 2007. A triple-continuum pressure-transient model for a naturally fractured vuggy reservoir. In: SPE110044.
Xiao, W., Li, T., Li, M., et al., 2016. Evaluation of the stress sensitivity in tight reservoirs. Pet. Explor. Dev. 43 (1), 107–114.
Xu, S., 1995. Fortran Algorithms Procedures Set, second ed. Tsinghua University Press, Beijing.
Yang, J., Liu, J., 1994. Practical Calculation Method of Gas Production. Petroleum Industry Press, Beijing.
Yang, L., Gao, C., Gao, W., 1999. Application of unbalance recovery to the gas field in central shaanganning basin. J. Xi'an Pet. Inst. 14 (2), 13–15.
Zhang, L., 1996. The concept of movable reserves and the method of determining economically recoverable reserves. Nat. Gas Explor. Dev. 19 (4), 75–76.
Zhang, D., Cai, C., Zhang, K., et al., 1991. Computational Fluid Dynamics. Sun Yat-sen University Press, Guangzhou.
Zhang, H., Lu, J., Cao, W., et al., 2017. Common formulas for well testing of gas well under different units. Well Test. 26 (6), 1–7+73.
Zhao, W., Shen, A., Pan, W., 2013. A research on carbonate karst reservoirs classification and its implication on hydrocarbon exploration: cases studies from Tarim basin. Acta Pet. Sin. 29 (9), 3213–3222.
Zhu, W., Song, H., He, D., et al., 2008. Low-velocity non-darcy gas seepage model and productivity equations of low-permeability water-bearing gas reservoirs [J]. Nat. Gas Geosci. 19 (5), 685–689.
Zhuang, H., 2013. Dynamic Well Testing in Petroleum Exploration and Development. Petroleum Industry Press (Published by Elsevier Inc.).

APPENDIX A

Unit conversions from SI to other unit systems

1. Length

\	km	m	cm	mile	ft	in.
1 × km	1	1000	10^5	0.6214	3280.84	39370.08
1 × m	0.001	1	100	6.214×10^{-4}	3.28084	39.37008
1 × cm	10^{-5}	0.01	1	6.214×10^{-6}	0.0328084	0.393701
1 × mile	1.60934	1609.34	1.60934×10^5	1	5280	63360
1 × ft	3.48×10^{-4}	0.3048	30.48	1.839×10^{-4}	1	12
1 × in.	2.54×10^{-5}	0.0254	2.54	1.5783×10^{-5}	0.08333	1

2. Area

\	m^2	cm^2	ft^2	$in.^2$
1 × m^2	1	10^4	10.7639	1550
1 × cm^2	10^{-4}	1	1.07639×10^{-3}	0.1550
1 × ft^2	0.092903	929.03	1	144
1 × $in.^2$	6.4516×10^{-4}	6.4516	6.9444×10^{-3}	1

3. Volume

\	m^3	cm^3	ft^3	bbl	L
1 × m^3	1	10^6	35.3147	6.28978	1000
1 × cm^3	10^{-6}	1	3.53147×10^{-5}	6.28978×10^{-6}	10^{-3}
1 × ft^3	0.0283168	2.83168×10^4	1	0.17811	28.3168
1 × bbl	0.158988	1.58988×10^5	5.6146	1	158.99
1 × L	10^{-3}	10^3	3.53147×10^{-2}	6.28978×10^{-3}	1

4. Pressure

\	MPa	kPa	atm	bar	kg/cm^2	psi
1 × MPa	1	1000	9.86923	10	10.1972	145.038
1 × kPa	0.001	1	9.86923×10^{-3}	0.01	0.0101972	0.145038
1 × atm	0.101325	101.325	1	1.01325	1.03323	14.6959
1 × bar	0.1	100	0.986923	1	1.01972	14.5038
1 × kg/cm^2	0.0980665	98.0665	0.967841	0.980665	1	14.2233
1 × psi	0.00689476	6.89476	0.068406	0.0689476	0.070307	1

5. Temperature

\	°C	K	°F	°R
1 × °C	t	$t+273.15$	$1.8t+32$	$1.8t+491.67$
1 × K	$T-273.15$	T	$1.8T-459.67$	$1.8T$
1 × °F	$5(f-32)/9$	$5(f+459.67)/9$	f	$f+459.67$
1 × °R	$5r/9-273.15$	$5r/9$	$r-459.67$	r

6. Oil production

\	m^3/d	cm^3/s	bbl/d
1 × m^3/d	1	$10^4/864$	6.28978
1 × cm^3/s	0.0864	1	0.543437
1 × bbl/d	0.158988	1.84014	1

7. Gas production

\	10^4 m^3/d	cm^3/s	Mscf/d	MMscf/d
1 × 10^4 m^3/d	1	$10^8/864$	353.147	0.353147
1 × cm^3/s	864×10^{-8}	1	3.05119×10^{-3}	3.05119×10^{-6}
1 × 10^3 ft^3/d	2.83168×10^{-3}	327.741	1	0.001
1 × 10^6 ft^3/d	2.83168	327741	1000	1

8. Compressibility

\	1/MPa	1/atm	$1/(kg/cm^2)$	1/psi
1/MPa	1	0.101325	0.0980665	0.00689476
1/atm	9.86923	1	0.967841	0.068406
$1/(kg/cm^2)$	10.1972	1.03323	1	0.070307
1/psi	145.038	14.6959	14.2233	1

9. Permeability

\	μm^2	cm^2	m^2	mD	D
$1 \times \mu m^2$	1	10^{-8}	10^{-12}	1.0133×10^3	1.0133
$1 \times mD$	0.98692×10^{-3}	9.8692×10^{-12}	9.8692×10^{-16}	1	10^{-3}

10. Viscosity

$1\,mPa.s = 10^{-3}\,Pa.s = 10^3\,\mu Pa.s = 1\,cP$

11. Relative density of stock tank oil, (γ_o), and °API

$$°API = \frac{141.5}{\gamma_o} - 131.5$$

$$\gamma_o = 141.5/(131.5 + °API)$$

12. Gas-oil ratio

$1\,m^3/m^3 = 5.615\,scf/STB$

$1\,scf/STB = 0.1781\,m^3/m^3$

APPENDIX B

Common formulas for gas flow

B.1 COMMON FORMULAS

Two elements must be considered for the gas well test analysis. One is the physical property of gas as the function of pressure, and the other is the skin factor relevant to the production rate or non-Darcy flow. When the pressure drawdown is small, the liquid phase diffusion equation can be used to describe the gas phase diffusion equation, and the solution is in the form of pressure. When the pressure drawdown is large, the square of pressure, pseudo-pressure, and normalized pseudo-pressure can be introduced, and the diffusion equation can be linearized; the solution is in the form of pressure square, pseudo-pressure, and normalized pseudo-pressure.

Units of measurement include Darcy units, imperial units, SI units in the petroleum industry, and various SI practical units. The units of measurement for pressure, permeability, and production rate are different, and they can be classified as 13 formats, including 9 formats derived only from SI units, as shown in Table B.1.

The transient pressure drawdown equation, transient deliverability equation, pseudo-steady deliverability equation, and pseudo-steady production equation of a gas well have four forms respectively, that is, pressure, square of pressure, pseudo-pressure, and normalized pseudo-pressure (Table B.2). Except the pseudo-steady production equation, each form of transient pressure drawdown equation, transient deliverability equation, and pseudo-steady deliverability equation include two common expressions, ln and lg. The number of coefficient combinations of common formulas on gas flow is as many as 364. They are inconvenient for field application because they can be confusing.

B.2 CONVERSION OF COEFFICIENTS IN COMMON FORMULAS ON GAS FLOW

According to the unit conversion method described by Sun et al. (2015a), the coefficients in formulas listed in Table B.2 are calculated, and the unit coefficients of physical quantities are shown in Tables B.3 and B.4. It should

Table B.1 Units of measurement of physical quantities under different unit systems

<table>
<tr><th rowspan="2">Physical quantity</th><th rowspan="2">Symbol</th><th rowspan="2">Darcy unit</th><th rowspan="2">Imperial unit</th><th colspan="9">SI practical unit</th><th rowspan="2">Industrial standard</th></tr>
<tr><th>(1)</th><th>(2)</th><th>(3)</th><th>(4)</th><th>(5)</th><th>(6)</th><th>(7)</th><th>(8)</th><th>(9)</th></tr>
<tr><td>Length</td><td>L</td><td>cm</td><td>ft</td><td colspan="9">m</td><td>m</td></tr>
<tr><td>Area</td><td>A</td><td>cm^2</td><td>ft^2</td><td colspan="9">m^2</td><td>m^2</td></tr>
<tr><td>Time</td><td>t</td><td>s</td><td>h</td><td colspan="9">h</td><td>h</td></tr>
<tr><td>Production rate</td><td>q</td><td>cm^3/s</td><td>$10^3 ft^3/d$</td><td colspan="4">m^3/d</td><td colspan="3">$10^4 m^3/d$</td><td colspan="2">$10^3 m^3/d$</td><td>$10^4 m^3/d$</td></tr>
<tr><td>Permeability</td><td>K</td><td>D</td><td>mD</td><td>μm^2</td><td>$10^{-3}\ \mu m^2$</td><td>mD</td><td>D</td><td>μm^2</td><td>$10^{-3}\ \mu m^2$</td><td>D</td><td>mD</td><td>mD</td><td>mD</td></tr>
<tr><td>Pressure</td><td>p</td><td>atm</td><td>psi</td><td colspan="8">MPa</td><td>KPa</td><td>MPa</td></tr>
<tr><td>Viscosity</td><td>μ</td><td>cP</td><td>cP</td><td colspan="9">mPa s</td><td>mPa s</td></tr>
<tr><td>Compressibility</td><td>c_g</td><td>atm^{-1}</td><td>psi^{-1}</td><td colspan="9">MPa^{-1}</td><td>MPa^{-1}</td></tr>
<tr><td>Density</td><td>ρ</td><td>g/cm^3</td><td>lbm/ft^3</td><td colspan="9">kg/m^3</td><td>kg/m^3</td></tr>
<tr><td>Temperature</td><td>T</td><td>K</td><td>°F</td><td colspan="9">K</td><td>K</td></tr>
</table>

Table B.2 Common formulas on gas flow (ln form)

Name	No.	Form	Formula
Transient pressure drawdown equation	1	Pressure	$p_{\mathrm{i}}-p_{\mathrm{wf}}=\frac{C_1}{T_{\mathrm{sc}}/p_{\mathrm{sc}}}\frac{\mu Z_{\mathrm{i}}}{p_{\mathrm{i}}}\frac{T}{Kh}\left(\ln\frac{Kt}{\phi\mu c_{\mathrm{g}}r_{\mathrm{w}}^2}+C_2+C_3S\right)q_{\mathrm{g}}$
	2	Square of pressure	$p_{\mathrm{i}}^2-p_{\mathrm{wf}}^2=\frac{C_1}{T_{\mathrm{sc}}/p_{\mathrm{sc}}}\overline{\mu Z}\frac{T}{Kh}\left(\ln\frac{Kt}{\phi\overline{\mu c}_{\mathrm{g}}r_{\mathrm{w}}^2}+C_2+C_3S\right)q_{\mathrm{g}}$
	3	Pseudo-pressure	$m(p_{\mathrm{i}})-m(p_{\mathrm{wf}})=\frac{C_1}{T_{\mathrm{sc}}/p_{\mathrm{sc}}}\frac{T}{Kh}\left(\ln\frac{Kt}{\phi\mu_{\mathrm{i}}c_{\mathrm{gi}}r_{\mathrm{w}}^2}+C_2+C_3S\right)q_{\mathrm{g}}$
	4	Normalized pseudo-pressure	$p_{p_{\mathrm{i}}}-p_{p_{\mathrm{wf}}}=\frac{C_1}{T_{\mathrm{sc}}/p_{\mathrm{sc}}}\frac{\mu_{\mathrm{i}}Z_{\mathrm{i}}}{p_{\mathrm{i}}}\frac{T}{Kh}\left(\ln\frac{Kt}{\phi\mu_{\mathrm{i}}c_{\mathrm{gi}}r_{\mathrm{w}}^2}+C_2+C_3S\right)q_{\mathrm{g}}$
Transient deliverability equation	5	Pressure	$p_{\mathrm{i}}-p_{\mathrm{wf}}=\frac{C_1}{T_{\mathrm{sc}}/p_{\mathrm{sc}}}\frac{\mu Z_{\mathrm{i}}}{p_{\mathrm{i}}}\frac{T}{Kh}\left(\ln\frac{Kt}{\phi\mu c_{\mathrm{g}}r_{\mathrm{w}}^2}+C_2+C_3S\right)q_{\mathrm{g}}+\frac{C_4}{T_{\mathrm{sc}}/p_{\mathrm{sc}}}\frac{\mu Z_{\mathrm{i}}}{p_{\mathrm{i}}}\frac{T}{Kh}Dq_{\mathrm{g}}^2$
	6	Square of pressure	$p_{\mathrm{i}}^2-p_{\mathrm{wf}}^2=\frac{C_1}{T_{\mathrm{sc}}/p_{\mathrm{sc}}}\overline{\mu Z}\frac{T}{Kh}\left(\ln\frac{Kt}{\phi\overline{\mu c}_{\mathrm{g}}r_{\mathrm{w}}^2}+C_2+C_3S\right)q_{\mathrm{g}}+\frac{C_4}{T_{\mathrm{sc}}/p_{\mathrm{sc}}}\overline{\mu Z}\frac{T}{Kh}Dq_{\mathrm{g}}^2$
	7	Pseudo-pressure	$m(p_{\mathrm{i}})-m(p_{\mathrm{wf}})=\frac{C_1}{T_{\mathrm{sc}}/p_{\mathrm{sc}}}\frac{T}{Kh}\left(\ln\frac{Kt}{\phi\mu_{\mathrm{i}}c_{\mathrm{gi}}r_{\mathrm{w}}^2}+C_2+C_3S\right)q_{\mathrm{g}}+\frac{C_4}{T_{\mathrm{sc}}/p_{\mathrm{sc}}}\frac{T}{Kh}Dq_{\mathrm{g}}^2$

Continued

Table B.2 Common formulas on gas flow (ln form)—cont'd

Name	No.	Form	Formula
	8	Normalized pseudo-pressure	$p_{p_i} - p_{p_{wf}} = \frac{C_1}{T_{sc}/p_{sc}} \frac{\mu_i Z_i}{p_i} \frac{T}{Kh} \left(\ln \frac{Kt}{\phi \mu_i c_{gi} r_w^2} + C_2 + C_3 S \right) q_g + \frac{C_4}{T_{sc}/p_{sc}} \frac{\mu_i Z_i}{p_i} \frac{T}{Kh} D q_g^2$
Pseudo-steady deliverability equation	9	Pressure	$p_i - p_{wf} = \frac{C_1}{T_{sc}/p_{sc}} \frac{\mu_i Z_i}{p_i} \frac{T}{Kh} \left(\ln \frac{r_e}{r_w} + C_2 + C_3 S \right) q_g + \frac{C_4}{T_{sc}/p_{sc}} \frac{\mu Z_i}{p_i} \frac{T}{Kh} D q_g^2$
	10	Square of pressure	$p_i^2 - p_{wf}^2 = \frac{C_1}{T_{sc}/p_{sc}} \overline{\mu Z} \frac{T}{Kh} \left(\ln \frac{r_e}{r_w} + C_2 + C_3 S \right) q_g + \frac{C_4}{T_{sc}/p_{sc}} \overline{\mu Z} \frac{T}{Kh} D q_g^2$
	11	Pseudo-pressure	$m(p_i) - m(p_{wf}) = \frac{C_1}{T_{sc}/p_{sc}} \frac{T}{Kh} \left(\ln \frac{r_e}{r_w} + C_2 + C_3 S \right) q_g + \frac{C_4}{T_{sc}/p_{sc}} \frac{T}{Kh} D q_g^2$
	12	Normalized pseudo-pressure	$p_{p_i} - p_{p_{wf}} = \frac{C_1}{T_{sc}/p_{sc}} \frac{\mu_i Z_i}{p_i} \frac{T}{Kh} \left(\ln \frac{r_e}{r_w} + C_2 + C_3 S \right) q_g + \frac{C_4}{T_{sc}/p_{sc}} \frac{\mu_i Z_i}{p_i} \frac{T}{Kh} D q_g^2$
Pseudo-steady production equation	13	Pressure	$q_g = D_1 (T_{sc}/p_{sc}) \frac{p_i}{\mu Z_i} \frac{Kh}{T} \frac{p_i - p_{wf}}{\ln \frac{r_e}{r_w} + D_2 + D_3 S}$
	14	Square of pressure	$q_g = D_1 (T_{sc}/p_{sc}) \frac{1}{\overline{\mu Z}} \frac{Kh}{T} \frac{p_i^2 - p_{wf}^2}{\ln \frac{r_e}{r_w} + D_2 + D_3 S}$
	15	Pseudo-pressure	$q_g = D_1 (T_{sc}/p_{sc}) \frac{Kh}{T} \frac{m(p_i) - m(p_{wf})}{\ln \frac{r_e}{r_w} + D_2 + D_3 S}$
	16	Normalized pseudo-pressure	$q_g = D_1 (T_{sc}/p_{sc}) \frac{\mu_i Z_i}{p_i} \frac{Kh}{T} \frac{p_{p_i} - p_{p_{wf}}}{\ln \frac{r_e}{r_w} + D_2 + D_3 S}$

Table B.3 Coefficients in common formulas on gas flow (ln form)

Transient pressure drawdown equation ($C_1 \sim C_3$) Transient deliverability equation ($C_1 \sim C_4$) (ln form)		Pressure, and normalized pseudo-pressure ($C_3 = 2$)				
		C_1	$C_1/(T_{sc}/p_{sc})$	C_2	C_4	$C_4/(T_{sc}/p_{sc})$
(a) Coefficients in transient pressure drawdown equation and transient deliverability equation-1						
Darcy unit	60°F	$1/(4\pi)$	2.7564E−04	0.8091	$1/(2\pi)$	5.5127E−04
	20°C	$1/(4\pi)$	2.7146E−04	0.8091	$1/(2\pi)$	5.4291E−04
Imperial unit		1.2575E+04	3.5561E+02	−7.4317	2.5150E+04	7.1122E+02
SI practical unit	1	9.2104E−04	3.1835E−07	2.0900	1.8421E−03	6.3670E−07
	2	9.2104E−01	3.1835E−04	−4.8177	1.8421E+00	6.3670E−04
	3	9.3324E−01	3.2257E−04	−4.8309	1.8665E+00	6.4513E−04
	4	9.3324E−04	3.2257E−07	2.0768	1.8665E−03	6.4513E−07
	5	9.2104E+00	3.1835E−03	2.0900	1.8421E+01	6.3670E−03
	6	9.2104E+03	3.1835E+00	−4.8177	1.8421E+04	6.3670E+00
	7	9.3324E+00	3.2257E−03	2.0768	1.8665E+01	6.4513E−03
	8	9.3324E+02	3.2257E−01	−4.8309	1.8665E+03	6.4513E−01
	9	9.3324E+05	3.2257E+02	−4.8309	1.8665E+06	6.4513E+02
Industrial standard (SY/T 6580–2004)		9.3324E+03	3.2257E+00	−4.8309	1.8665E+04	6.4513E+00

Transient pressure drawdown equation ($C_1 \sim C_3$) Transient deliverability equation ($C_1 \sim C_4$) (ln form)		Pressure square, and pseudo-pressure ($C_3 = 2$)				
		C_1	$C_1/(T_{sc}/p_{sc})$	C_2	C_4	$C_4/(T_{sc}/p_{sc})$
(b) Coefficients in transient pressure drawdown equation and transient deliverability equation-2						
Darcy unit	60°F	$1/(2\pi)$	5.5127E−04	0.8091	$1/\pi$	1.1025E−03
	20°C	$1/(2\pi)$	5.4291E−04	0.8091	$1/\pi$	1.0858E−03

Continued

Table B.3 Coefficients in common formulas on gas flow (ln form)—cont'd

Transient pressure drawdown equation ($C_1 \sim C_3$) Transient deliverability equation ($C_1 \sim C_4$) (ln form)		Pressure square, and pseudo-pressure ($C_3 = 2$)				
		C_1	$C_1/(T_{sc}/p_{sc})$	C_2	C_4	$C_4/(T_{sc}/p_{sc})$
Imperial unit		2.5150E+04	7.1122E+02	−7.4317	5.0300E+04	1.4224E+03
SI practical unit	1	1.8421E−03	6.3670E−07	2.0900	3.6841E−03	1.2734E−06
	2	1.8421E+00	6.3670E−04	−4.8177	3.6841E+00	1.2734E−03
	3	1.8665E+00	6.4513E−04	−4.8309	3.7330E+00	1.2903E−03
	4	1.8665E−03	6.4513E−07	2.0768	3.7330E−03	1.2903E−06
	5	1.8421E+01	6.3670E−03	2.0900	3.6841E+01	1.2734E−02
	6	1.8421E+04	6.3670E+00	−4.8177	3.6841E+04	1.2734E+01
	7	1.8665E+01	6.4513E−03	2.0768	3.7330E+01	1.2903E−02
	8	1.8665E+03	6.4513E−01	−4.8309	3.7330E+03	1.2903E+00
	9	1.8665E+06	6.4513E+02	−4.8309	3.7330E+06	1.2903E+03
Industrial standard (SY/T 6580–2004)		1.8665E+04	6.4513E+00	−4.8309	3.7330E+04	1.2903E+01

Pseudo-steady deliverability equation ($C_1 \sim C_4$) Pseudo-steady production equation ($D_1 \sim D_3$) (ln form)		Pressure, and normalized pseudo-pressure ($C_2 = D_2 = -0.75$, $C_3 = D_3 = 1$)			
		$C_1 = C_4$	$C_1/(T_{sc}/p_{sc}) = C_4/(T_{sc}/p_{sc})$	D_1	$D_1{}^*(T_{sc}/p_{sc})$
(c) Coefficients in pseudo-steady deliverability equation and pseudo-steady production equation-1					
Darcy unit	60°F	$1/(2\pi)$	5.5127E−04	2π	1.8140E+03
	20°C	$1/(2\pi)$	5.4291E−04	2π	1.8419E+03
Imperial unit		2.5150E+04	7.1122E+02	3.9762E−05	1.4060E−03
SI practical unit	1	1.8421E−03	6.3670E−07	5.4287E+02	1.5706E+06
	2	1.8421E+00	6.3670E−04	5.4287E−01	1.5706E+03

	3	1.8665E+00	6.4513E−04	5.3577E−01	1.5501E+03
	4	1.8665E−03	6.4513E−07	5.3577E+02	1.5501E+06
	5	1.8421E+01	6.3670E−03	5.4287E−02	1.5706E+02
	6	1.8421E+04	6.3670E+00	5.4287E−05	1.5706E−01
	7	1.8665E+01	6.4513E−03	5.3577E−02	1.5501E+02
	8	1.8665E+03	6.4513E−01	5.3577E−04	1.5501E+00
	9	1.8665E+06	6.4513E+02	5.3577E−07	1.5501E−03
Industrial standard (SY/T 6580–2004)		1.8665E+04	6.4513E+00	5.3577E−05	1.5501E−01

(d) *Coefficients in pseudo-steady deliverability equation and pseudo-steady production equation-2*

Pseudo-steady deliverability equation ($C_1 \sim C_4$) **Pseudo-steady production equation ($D_1 \sim D_3$) (ln form)**		**Pressure square, and pseudo-pressure ($C_2 = D_2 = -0.75$, $C_3 = D_3 = 1$)**			
		$C_1 = C_4$	**$C_1/(T_{sc}/p_{sc}) = C_4/(T_{sc}/p_{sc})$**	**D_1**	**$D_1{*}(T_{sc}/p_{sc})$**
Darcy unit	60°F	$1/\pi$	1.1025E−03	π	9.0700E+02
	20°C	$1/\pi$	1.0858E−03	π	9.2096E+02
Imperial unit		5.0300E+04	1.4224E+03	1.9881E−05	7.0301E−04
SI practical unit	1	3.6841E−03	1.2734E−06	2.7143E+02	7.8530E+05
	2	3.6841E+00	1.2734E−03	2.7143E−01	7.8530E+02
	3	3.7330E+00	1.2903E−03	2.6788E−01	7.7503E+02
	4	3.7330E−03	1.2903E−06	2.6788E+02	7.7503E+05
	5	3.6841E+01	1.2734E−02	2.7143E−02	7.8530E+01
	6	3.6841E+04	1.2734E+01	2.7143E−05	7.8530E−02
	7	3.7330E+01	1.2903E−02	2.6788E−02	7.7503E+01
	8	3.7330E+03	1.2903E+00	2.6788E−04	7.7503E−01
	9	3.7330E+06	1.2903E+03	2.6788E−07	7.7503E−04
Industrial standard (SY/T 6580–2004)		3.7330E+04	1.2903E+01	2.6788E−05	7.7503E−02

Table B.4 Coefficients in common formula on gas flow (lg form)

Transient pressure drawdown equation ($C_1 \sim C_3$) Transient deliverability equation ($C_1 \sim C_4$) (lg form)		Pressure, and normalized pseudo-pressure ($C_3 = 0.8686$) C_1	$C_1/(T_{sc}/p_{sc})$	C_2	C_4	$C_4/(T_{sc}/p_{sc})$
(a) Coefficients in transient pressure drawdown equation and transient deliverability equation-1						
Darcy unit	60°F	1.8323E−01	6.3467E−04	0.3514	$1/(2\pi)$	5.5127E−04
	20°C	1.8323E−01	6.2505E−04	0.3514	$1/(2\pi)$	5.4291E−04
Imperial unit		2.8955E+04	8.1883E+02	−3.2275	2.5150E+04	7.1122E+02
SI practical unit	1	2.1208E−03	7.3303E−07	0.9077	1.8421E−03	6.3670E−07
	2	2.1208E+00	7.3303E−04	−2.0923	1.8421E+00	6.3670E−04
	3	2.1489E+00	7.4274E−04	−2.0980	1.8665E+00	6.4513E−04
	4	2.1489E−03	7.4274E−07	0.9020	1.8665E−03	6.4513E−07
	5	2.1208E+01	7.3303E−03	0.9077	1.8421E+01	6.3670E−03
	6	2.1208E+04	7.3303E+00	−2.0923	1.8421E+04	6.3670E+00
	7	2.1489E+01	7.4274E−03	0.9020	1.8665E+01	6.4513E−03
	8	2.1489E+03	7.4274E−01	−2.0980	1.8665E+03	6.4513E−01
	9	2.1489E+06	7.4274E+02	−2.0980	1.8665E+06	6.4513E+02
Industrial standard (SY/T 6580–2004)		2.1489E+04	7.4274E+00	−2.0980	1.8665E+04	6.4513E+00

Transient pressure drawdown equation ($C_1 \sim C_3$) Transient deliverability equation ($C_1 \sim C_4$) (lg form)		Pressure square, and pseudo-pressure ($C_3 = 0.8686$) C_1	$C_1/(T_{sc}/p_{sc})$	C_2	C_4	$C_4/(T_{sc}/p_{sc})$
(b) Coefficients in transient pressure drawdown equation and transient deliverability equation-2						
Darcy unit	60°F	3.6647E−01	1.2693E−03	0.3514	$1/\pi$	1.1025E−03
	20°C	3.6647E−01	1.2501E−03	0.3514	$1/\pi$	1.0858E−03

Imperial unit		5.7910E+04	1.6377E+03	−3.2275	5.0300E+04	1.4224E+03
SI practical unit	1	4.2415E−03	1.4661E−06	0.9077	3.6841E−03	1.2734E−06
	2	4.2415E+00	1.4661E−03	−2.0923	3.6841E+00	1.2734E−03
	3	4.2977E+00	1.4855E−03	−2.0980	3.7330E+00	1.2903E−03
	4	4.2977E−03	1.4855E−06	0.9020	3.7330E−03	1.2903E−06
	5	4.2415E+01	1.4661E−02	0.9077	3.6841E+01	1.2734E−02
	6	4.2415E+04	1.4661E+01	−2.0923	3.6841E+04	1.2734E+01
	7	4.2977E+01	1.4855E−02	0.9020	3.7330E+01	1.2903E−02
	8	4.2977E+03	1.4855E+00	−2.0980	3.7330E+03	1.2903E+00
	9	4.2977E+06	1.4855E+03	−2.0980	3.7330E+06	1.2903E+03
Industrial standard (SY/T 6580–2004)		4.2977E+04	1.4855E+01	−2.0980	3.7330E+04	1.2903E+01

Pseudo-steady deliverability equation (lg form)		Pressure, and normalized pseudo-pressure ($C_2 = -0.3257$, $C_3 = 0.4343$)			
		C_1	$C_1/(T_{sc}/p_{sc})$	C_4	$C_4/(T_{sc}/p_{sc})$
(c) Coefficients in pseudo-steady deliverability equation-1					
Darcy unit	60°F	3.6647E−01	1.2693E−03	$1/(2\pi)$	5.5127E−04
	20°C	3.6647E−01	1.2501E−03	$1/(2\pi)$	5.4291E−04
Imperial unit		5.7910E+04	1.6377E+03	2.5150E+04	7.1122E+02
SI practical unit	1	4.2415E−03	1.4661E−06	1.8421E−03	6.3670E−07
	2	4.2415E+00	1.4661E−03	1.8421E+00	6.3670E−04
	3	4.2977E+00	1.4855E−03	1.8665E+00	6.4513E−04
	4	4.2977E−03	1.4855E−06	1.8665E−03	6.4513E−07
	5	4.2415E+01	1.4661E−02	1.8421E+01	6.3670E−03
	6	4.2415E+04	1.4661E+01	1.8421E+04	6.3670E+00

Continued

Table B.4 Coefficients in common formula on gas flow (lg form)—cont'd

Pseudo-steady deliverability equation (lg form)		Pressure, and normalized pseudo-pressure ($C_2 = -0.3257$, $C_3 = 0.4343$)			
		C_1	$C_1/(T_{sc}/p_{sc})$	C_4	$C_4/(T_{sc}/p_{sc})$
	7	4.2977E+01	1.4855E−02	1.8665E+01	6.4513E−03
	8	4.2977E+03	1.4855E+00	1.8665E+03	6.4513E−01
	9	4.2977E+06	1.4855E+03	1.8665E+06	6.4513E+02
Industrial standard (SY/T 6580–2004)		4.2977E+04	1.4855E+01	1.8665E+04	6.4513E+00

Pseudo-steady deliverability equation (lg form)		Pressure square, and pseudo-pressure ($C_2 = -0.3257$, $C_3 = 0.4343$)			
		C_1	$C_1/(T_{sc}/p_{sc})$	C_4	$C_4/(T_{sc}/p_{sc})$
(d) Coefficients in pseudo-steady deliverability equation-2					
Darcy unit	60°F	7.3294E−01	2.5387E−03	$1/\pi$	1.1025E−03
	20°C	7.3294E−01	2.5002E−03	$1/\pi$	1.0858E−03
Imperial unit		1.1582E+05	3.2753E+03	5.0300E+04	1.4224E+03
SI practical unit	1	8.4831E−03	2.9321E−06	3.6841E−03	1.2734E−06
	2	8.4831E+00	2.9321E−03	3.6841E+00	1.2734E−03
	3	8.5955E+00	2.9710E−03	3.7330E+00	1.2903E−03
	4	8.5955E−03	2.9710E−06	3.7330E−03	1.2903E−06
	5	8.4831E+01	2.9321E−02	3.6841E+01	1.2734E−02
	6	8.4831E+04	2.9321E+01	3.6841E+04	1.2734E+01
	7	8.5955E+01	2.9710E−02	3.7330E+01	1.2903E−02
	8	8.5955E+03	2.9710E+00	3.7330E+03	1.2903E+00
	9	8.5955E+06	2.9710E+03	3.7330E+06	1.2903E+03
Industrial standard (SY/T 6580–2004)		8.5955E+04	2.9710E+01	3.7330E+04	1.2903E+01

be noted that the standard state for measuring the gas volume flow is defined as 20°C (293.15 K) and 0.101 325 MPa in China, and 60°F (≈15.556°C, 288.706 K) and 1 atm (≈14.696 psi) in Europe and America.

B.2.1 Coefficients in the form of ln

As shown in Table B.2, common formulas on gas flow have two kinds of expressions: ln and lg. The coefficients in the form of ln under different unit combinations are shown in Table B.3.

B.2.2 Coefficients in the form of lg

In practical application, the transient pressure drawdown equation, transient deliverability equation, and pseudo-steady deliverability equation are often expressed in the form of lg, and the coefficients are shown in Table B.4.

APPENDIX C

Basis for gas-liquid two-phase flow

C.1 CONDENSATE GAS FLOW

The condensate gas reservoir is a special type of gas reservoir with both oil and gas reservoir characteristics, from which both natural gas and condensate oil (gasoline distillate to kerosene distillate and a few macromolecular hydrocarbons) can be produced. It shows the complicated phase change process of natural gas and condensate oil systems. Under the condition of certain temperatures and pressures, gas phases and liquid phases coexist, and transfer heat and mass constantly; and their volume fractions change continuously.

In view of development, dry gas fields and condensate gas fields differ the most different in the properties and phase state characteristics of formation fluids. When the formation pressure of condensate gas reservoir drops to below the dewpoint pressure, condensate oil will be separated out from the gas phase. The oil remains and is absorbed on the surface of rock particles, and begins to flow when the critical liquid saturation is exceeded, forming oil-gas two-phase flow. In different exploitation stages, 2 or 3 fluidity areas will occur in the condensate gas reservoirs (Fevang and Whitson, 1995), namely, the oil-gas two-phase flow, the oil-gas two-phase coexistence area where condensate oil does not flow and only gas phase flows, and the gas phase flow area.

C.1.1 Single-phase flow model

When the flow pressure is higher than the dewpoint pressure, the flow in the formation is a gas phase flow. When the flow pressure is lower than the dewpoint pressure, both the oil phase and gas phase occur in the near wellbore zone, but the condensate oil saturation is lower than the critical flow saturation. When the gas phase flows while the oil phase does not flow, it can be believed that the flow rate of condensate liquid in the formation is zero. Then, the gas phase flow theory can continue to be used as long as the condensate liquid volume obtained at the wellhead is converted to the corresponding condensate gas volume.

$$q_t = q_g + 542.786 q_o (1.03 - \gamma_o) \tag{C.1}$$

The expression of flowing pressure is the same as the gas phase form, so we will not repeat it here.

C.1.2 Two-phase flow model

When the formation pressure is lower than the dewpoint pressure, gas-oil two-phase flow will occur in the formation, and the flow equation satisfied is a multi-phase flow equation. The following pseudo-pressure function is used

$$\psi = \frac{\mu_{gi}}{\rho_{gi}} \int_{p_0}^{p} \left(\frac{K_{ro}\rho_o}{\mu_o} + \frac{K_{rg}\rho_g}{\mu_g} \right) dp \tag{C.2}$$

This parameter has the dimension of pressure, for the convenience of subsequent data analysis.

The gas-oil two-phase flow equation is

$$\frac{1}{r}\frac{\partial}{\partial r}\left(r \frac{\partial \psi}{\partial r} \right) = \frac{1}{3.553\eta}\frac{\partial \psi}{\partial t} \tag{C.3}$$

Where

$$\eta = \frac{\rho_o \dfrac{KK_{ro}}{\mu_o} + \rho_g \dfrac{KK_{rg}}{\mu_g}}{\phi\left(S_o \dfrac{\partial \rho_o}{\partial p} + S_g \dfrac{\partial \rho_g}{\partial p} + \rho_o \dfrac{\partial S_o}{\partial p} + \rho_g \dfrac{\partial S_g}{\partial p} \right)} \tag{C.4}$$

Initial condition:

$$\psi = \psi(r, 0) \tag{C.5}$$

Inner boundary condition at a constant rate:

$$\left(r \frac{\partial \psi}{\partial r} \right)_{r=r_w} = 1.867 \times 10^{-3} \frac{q_{tgo}\mu_{gi}}{Kh\,\rho_{gi}} \tag{C.6}$$

Infinite outer boundary condition:

$$\psi = \psi_i(r \to \infty, 0) \tag{C.7}$$

Circular closed outer boundary condition:

$$\left(\frac{\partial \psi}{\partial r} \right)_{r=r_e} = 0 \tag{C.8}$$

Outer boundary condition under constant pressure:

$$\psi = \psi_{\mathrm{i}}(r = r_{\mathrm{e}}, 0) \tag{C.9}$$

Under an infinite condition, the flowing pressure solution can be expressed as

$$\psi_{\mathrm{wf}}(t) = \psi_{\mathrm{i}} - \frac{2.1483 \times 10^{-3} q_{\mathrm{tgo}} \mu_{\mathrm{gi}}}{Kh\rho_{\mathrm{gi}}} \left(\lg \frac{\eta t}{r_{\mathrm{w}}^2} + 0.9019 + 0.8686S \right) \tag{C.10}$$

Under a circular bounded and closed condition, the average formation pressure under a pseudo-steady flow state can be expressed as

$$\overline{\psi} - \psi_{\mathrm{wf}}(t) = \frac{1.867 \times 10^{-3} q_{\mathrm{tgo}} \mu_{\mathrm{gi}}}{Kh\rho_{\mathrm{gi}}} \left(\ln \frac{r_e}{r_{\mathrm{w}}} - 0.75 + S \right) \tag{C.11}$$

The calculation of the pseudo-pressure function should give consideration to the fact that condensate oil may separate out if the formation pressure is lower than the dewpoint pressure when fluids flow in a condensate gas reservoir, and it only flows when its saturation is higher than the critical flow saturation, so the pseudo-pressure function consists of the following three parts:

$$\begin{aligned}\psi(p) &= \frac{\mu_{\mathrm{gi}}}{\rho_{\mathrm{gi}}} \int_{p_0}^{p} \left(\frac{K_{\mathrm{ro}}\rho_{\mathrm{o}}}{\mu_{\mathrm{o}}} + \frac{K_{\mathrm{rg}}\rho_{\mathrm{g}}}{\mu_{\mathrm{g}}} \right) dp \\ &= \frac{\mu_{\mathrm{gi}}}{\rho_{\mathrm{gi}}} \left[\begin{array}{l} \displaystyle\int_{p_0}^{p_{\mathrm{cr}}} \left(\frac{K_{\mathrm{ro}}\rho_{\mathrm{o}}}{\mu_{\mathrm{o}}} + \frac{K_{\mathrm{rg}}\rho_{\mathrm{g}}}{\mu_{\mathrm{g}}} \right) dp + \int_{p_{\mathrm{cr}}}^{p_{\mathrm{dew}}} \left(\frac{K_{\mathrm{ro}}\rho_{\mathrm{o}}}{\mu_{\mathrm{o}}} + \frac{K_{\mathrm{rg}}\rho_{\mathrm{g}}}{\mu_{\mathrm{g}}} \right) dp \\ \displaystyle + \int_{p_{\mathrm{dew}}}^{p} \left(\frac{K_{\mathrm{ro}}\rho_{\mathrm{o}}}{\mu_{\mathrm{o}}} + \frac{K_{\mathrm{rg}}\rho_{\mathrm{g}}}{\mu_{\mathrm{g}}} \right) dp \end{array} \right]\end{aligned} \tag{C.12}$$

The relationship between relative permeability and pressure in Eq. (C.10) can be obtained using the steady state theory, that is:

$$\frac{K_{\mathrm{ro}}}{K_{\mathrm{rg}}} = \frac{\rho_{\mathrm{o}} L \mu_{\mathrm{o}}}{\rho_{\mathrm{g}} V \mu_{\mathrm{g}}} \tag{C.13}$$

Where, $L = \frac{\rho_{\mathrm{o}} S_{\mathrm{o}}}{\rho_{\mathrm{o}} S_{\mathrm{o}} + \rho_{\mathrm{g}} S_{\mathrm{g}}}$ indicates the molar fraction of condensate oil, dimensionless; $V = \frac{\rho_{\mathrm{g}} S_{\mathrm{g}}}{\rho_{\mathrm{o}} S_{\mathrm{o}} + \rho_{\mathrm{g}} S_{\mathrm{g}}}$ indicates the molar fraction of condensate gas, dimensionless; ρ_{o}, ρ_{g}, μ_{o}, μ_{g}, L, V change with the pressure and temperature, and is obtained through the phase state flash calculation.

C.2 GAS-WATER TWO-PHASE FLOW

Gas flow in most of the gas wells contains a little condensate oil and water. So, the flow in the wellbore is a gas-liquid two-phase flow. As compared with oil wells, the gas-liquid ratio (GLR) in gas wells is very high (greater than 2000 m^3/m^3) and the flow pattern is in a spray flow, amid a multi-phase flow. To simplify the calculation, the single-phase flow model can be used. When the GLR is low, the two-phase flow model is needed.

C.2.1 Single-phase flow model

The problem that natural gas contains condensate oil and water is regarded as the two-phase problem between complex hydrocarbon gas and water. When the flow in the wellbore is in a spray flow, the flow in the formation can still be regarded as a single-phase flow. Then, the gas phase flow theory can still be followed, as long as the condensate liquid volume and water volume obtained at the wellhead are converted to the corresponding gas volume. First, calculating the hydrocarbon flow rate, we have

$$q_{go} = q_g + 542.786(1.03 - \gamma_o)q_o \tag{C.14}$$

Then, water production is corrected. In case of spray flow, the volumetric flow rates of gas and water are the same; in other words, the relative velocity is zero. Then, the converted natural gas production is

$$q_{tgow} = q_{gO} f_w \tag{C.15}$$

$$f_w = 1 + \frac{W_w}{W_g} \tag{C.16}$$

The expression of flow pressure is the same as the single-phase gas form, so we will not repeat it here.

C.2.2 Two-phase flow model

The gas-water two-phase pseudo-pressure is defined as

$$\psi = \frac{\mu_{gi}}{\rho_{gi}} \int_{p_0}^{p} \left(\frac{K_{rw}\rho_w}{\mu_w} + \frac{K_{rg}\rho_g}{\mu_g} \right) dp \tag{C.17}$$

The control equation is

$$\frac{1}{r}\frac{\partial}{\partial r}\left(r\frac{\partial \psi}{\partial r} \right) = \frac{1}{3.553\eta}\frac{\partial \psi}{\partial t} \tag{C.18}$$

Where

$$\eta = \frac{\rho_w \frac{KK_{rw}}{\mu_w} + \rho_g \frac{KK_{rg}}{\mu_g}}{\phi\left(S_w \frac{\partial \rho_w}{\partial p} + S_g \frac{\partial \rho_g}{\partial p} + \rho_w \frac{\partial S_w}{\partial p} + \rho_g \frac{\partial S_g}{\partial p}\right)} \tag{C.19}$$

Initial condition

$$\psi = \psi(r, 0) \tag{C.20}$$

Inner boundary condition at a constant rate

$$\left(r \frac{\partial \psi}{\partial r}\right)_{r=r_w} = 1.867 \times 10^{-3} \frac{q_{tgw}\,\mu_{gi}}{Kh\,\rho_{gi}} \tag{C.21}$$

Infinite outer boundary condition

$$\psi = \psi_i(r \to \infty, 0) \tag{C.22}$$

Circular closed outer boundary condition

$$\left(\frac{\partial \psi}{\partial r}\right)_{r=r_e} = 0 \tag{C.23}$$

Outer boundary condition under constant pressure

$$\psi = \psi_i(r = r_e, 0) \tag{C.24}$$

Under an infinite condition, the flowing pressure solution can be expressed as

$$\psi_{wf}(t) = \psi_i - \frac{2.1483 \times 10^{-3} q_{tgw} \mu_{gi}}{Kh \rho_{gi}} \left(\lg \frac{\eta t}{r_w^2} + 0.9019 + 0.8686 S \right) \tag{C.25}$$

Under a circular bounded and closed condition, the average formation pressure under a pseudo-steady flow state can be expressed as

$$\overline{\psi} - \psi_{wf}(t) = \frac{1.867 \times 10^{-3} q_{tgw} \mu_{gi}}{Kh \rho_{gi}} \left(\ln \frac{r_e}{r_w} - 0.75 + S \right) \tag{C.26}$$

INDEX

Note: Page numbers followed by *f* indicate figures and *t* indicate tables.

M

N

O

P

CPI Antony Rowe
Eastbourne, UK
June 18, 2019